Error-Correction Coding for Digital Communications

Applications of Communications Theory
Series Editor: R. W. Lucky, *Bell Laboratories*

INTRODUCTION TO COMMUNICATION SCIENCE AND SYSTEMS
John R. Pierce and Edward C. Posner

OPTICAL FIBER TRANSMISSION SYSTEMS
Stewart D. Personick

TELECOMMUNICATIONS SWITCHING
J. Gordon Pearce

ERROR-CORRECTION CODING FOR DIGITAL COMMUNICATIONS
George C. Clark, Jr., and J. Bibb Cain

A Continuation Order Plan is available for this series. A continuation order will bring delivery of each new volume immediately upon publication. Volumes are billed only upon actual shipment. For further information please contact the publisher.

Error-Correction Coding for Digital Communications

George C. Clark, Jr.
and
J. Bibb Cain

Harris Corporation
Melbourne, Florida

PLENUM PRESS • NEW YORK AND LONDON

Library of Congress Cataloging in Publication Data

Clark, George C. (George Cyril), 1938-
    Error-correction coding for digital communications.

    Bibliography: p.
    Includes index.
    1. Data transmission systems. 2. Error-correcting codes (Information theory). I.
Cain, J. Bibb. II. Title.
TK5102.5.C52   621.38′0413   81-1630
ISBN 0-306-40615-2   AACR2

First Printing—June 1981
Second Printing—August 1982
Third Printing—March 1988

©1981 Plenum Press, New York
A Division of Plenum Publishing Corporation
233 Spring Street, New York, N.Y. 10013

Printed in the United States of America

Preface

Error-correction coding is being used on an almost routine basis in most new communication systems. Not only is coding equipment being used to increase the energy efficiency of communication links, but coding ideas are also providing innovative solutions to many related communication problems. Among these are the elimination of intersymbol interference caused by filtering and multipath and the improved demodulation of certain frequency modulated signals by taking advantage of the "natural" coding provided by a continuous phase. Although several books and numerous articles have been written on coding theory, there are still noticeable deficiencies. First, the practical aspects of translating a specific decoding algorithm into actual hardware have been largely ignored. The information that is available is sketchy and is widely dispersed. Second, the information required to evaluate a particular technique under situations that are encountered in practice is available for the most part only in private company reports.

This book is aimed at correcting both of these problems. It is written for the design engineer who must build the coding and decoding equipment and for the communication system engineer who must incorporate this equipment into a system. It is also suitable as a senior-level or first-year graduate text for an introductory one-semester course in coding theory.

The book uses a minimum of mathematics and entirely avoids the classical theorem/proof approach that is often seen in coding texts. Whenever possible heuristic arguments are advanced and concepts are developed by drawing analogies. We have intentionally constrained the level of mathematical rigor used in presenting error-correction concepts. Notwithstanding, coding is an extremely mathematical subject, and it would be an impossible task to attempt to present the material with no mathematics at all. The assumption is made that the practicing engineer is mostly interested in using the mathematics and not in constructing rigorous proofs. Thus, the goal is to develop a manipulative and intuitive understanding through numerous examples. Although there are many perils in this approach, a capable engineer can usually verify that a particular procedure is correct by trying it on a few test cases. The serious student may later wish to round out his or her understanding by pursuing the subject to its very roots. In addition to the large number of papers in the open literature, the excellent texts by Peterson and Weldon,[1] Berlekamp,[2] and Gallager,[3] are highly recommended for this purpose.

Not all aspects of coding theory are treated in equal depth. Some topics, such as coding bounds, are treated very lightly while some others, such as certain classes of rarely used codes, are omitted entirely. Numerous classes of codes of practical interest are discussed in terms of methods of construction and code properties. However, there exist additional interesting properties of these codes as well as other code classes which we have chosen to omit for reasons of brevity. Instead, we have chosen to emphasize the formulation and implementation of decoding algorithms. The viewpoint, the selection of topics, and the degree of emphasis largely reflects the authors' personal preference and is based upon 15 years of practical experience in designing and evaluating coding equipment. However, we feel that the approach we have taken will also provide the reader with the insight and motivation to undertake some of the more difficult topics in other texts.

In this book the application of error-correction coding is presented in the context of communication system design. Again, this reflects the authors' personal preferences. These techniques may also be used in other applications such as in computer memory systems. We have attempted to develop the tools necessary to enable the reader to evaluate such applications and to select an appropriate coding/decoding technique.

An introduction to the use of error-correction coding in digital communication systems is presented in Chapter 1. Following this, the next four chapters are devoted to block codes and the associated decoding techniques. Fundamental concepts of linear codes (parity check matrix, syndrome, shift-register encoders, etc.) are developed in Chapter 2, and several useful classes of codes are presented. The simple hard decision decoding techniques for block codes are discussed in Chapter 3. These include Meggitt decoders, information-set decoders, and threshold decoders. Special attention is given to those features which result in minimum implementation complexity. Chapter 4 presents a collection of techniques for decoding block codes utilizing demodulator "soft" decisions. These include techniques developed by Massey, Hartmann and Rudolph, Chase, and Weldon. Various hardware realizations are presented, and performance/complexity considerations are discussed. Algebraic techniques for multiple-error correction are developed in Chapter 5. The theory is developed using a transform approach. Numerous examples are used, and various hardware configurations are discussed. Convolutional coding techniques are presented in the next two chapters. Basic structural properties of convolutional codes and the Viterbi algorithm are developed in Chapter 6. Numerous performance curves are presented, and special attention is given to performance/complexity tradeoffs. A similar treatment is provided in Chapter 7 for the table look-up, threshold, and sequential decoding techniques. Finally, in Chapter 8 we discuss a number of important system applications including concatenated codes, coding for the white Gaussian noise channel, interleaver structures, burst-noise channels, coding for spread spectrum systems, and coding for bandwidth-constrained channels. By including an extensive treatment of applications, we believe this book will be extremely useful to the communication system engineer. Numerous lists of good codes are provided in Appendices A and B.

There are a number of acknowledgments which we should make. A book of this nature relies heavily on the pioneering work of the many researchers in this field. We have tried to adequately acknowledge this work in the references cited throughout the text. The authors are grateful for having had the opportunity to work for the Harris Corporation, Government Systems Group where they have been able to develop error-correction hardware and to apply coding techniques in a variety of communication systems. We would also like to thank C. J. Palermo and R. E. Totty for their encouragement during the writing of this book. Finally, we would like to acknowledge the technical contributions provided by our colleagues J. M. Geist, R. C. Davis, R. W. Boyd, and G. H. Thaker, the reviews provided by D. B. Bradley, A. M. Bush, and L. D. Rudolph, and the long hours of assistance in preparation of the manuscript provided by Barbara Snow and Sandra Wilson. Special thanks must also go to our wives Ann and Sissie for their encouragement and support during the past year.

Harris Corporation   George C. Clark, Jr.
                                          J. Bibb Cain

Contents

1. Fundamental Concepts of Coding 1

  1.1. Basic Principles 2
  1.2. Practical Constraints 8
    1.2.1. Data Source 9
    1.2.2. Encoder 9
    1.2.3. Modulator 10
    1.2.4. Waveform Channel 11
    1.2.5. Demodulator 12
    1.2.6. Decoder 14
    1.2.7. Discrete Data Channel 15
  1.3. Performance Computations 18
    1.3.1. Exact Computation of Error Rate 18
    1.3.2. Union Bound Computation of Error Rate 23
    1.3.3. Performance of Error Detection Systems 24
    1.3.4. Soft Decision Decoding 26
    1.3.5. Quantization Effects 30
    1.3.6. Effect of Noncoherent Combining 34
    1.3.7. Coding Gain 35
  1.4. Coding Bounds 37
    1.4.1. Bounds on Minimum Distance 38
    1.4.2. Random Coding Bound 41
  1.5. Remarks 46
  Problems 47

2. Group Codes 49

  2.1. Generalized Parity Check Codes 50
    2.1.1. Parity Check Matrix 52
    2.1.2. Minimum Distance Considerations 53
    2.1.3. Hamming Codes 54

   2.1.4. Generator Matrix 56
   2.1.5. Dual Codes 57
   2.1.6. The Syndrome 58
   2.1.7. The Penny-Weighing Problem 59
   2.1.8. A Simple Iterated Code 61
  2.2. Polynomial Codes 62
   2.2.1. Finite Field Arithmetic 63
   2.2.2. Generation of Polynomial Codes 68
   2.2.3. Cyclic Codes 72
   2.2.4. Alternate Encoder for Cyclic Codes 73
   2.2.5. Additional Properties of Polynomials and Galois Field
      Elements 75
   2.2.6. Codes Specified by Roots 78
   2.2.7. Syndrome Polynomials 79
   2.2.8. Modifications of Codes 84
  2.3. Important Classes of Group Codes 85
   2.3.1. BCH Codes 86
   2.3.2. Golay Codes 88
   2.3.3. Maximal-Length Codes 89
   2.3.4. Reed–Muller Codes 91
   2.3.5. Quadratic Residue Codes 92
   2.3.6. Remarks 94
  Problems 94

3. Simple Nonalgebraic Decoding Techniques for Group Codes 97

  3.1. Meggitt Decoders 97
  3.2. Information Set Decoding 102
   3.2.1. General Decoding Algorithm 102
   3.2.2. Information Set Decoding Using the **H** Matrix 107
   3.2.3. Predetermined Covering Sets 109
   3.2.4. Random Search Techniques 119
  3.3. Threshold Decoding 131
  3.4. Remarks 137
  Problems 138

4. Soft Decision Decoding of Block Codes 141

  4.1. APP Threshold Decoding 143
   4.1.1. Derivation of the APP Decoding Rule 144
   4.1.2. Computation of the Weighting Terms 146
   4.1.3. Approximate Computation of w_i 148
   4.1.4. Implementation of the APP Decoder 150
  4.2. Optimum Symbol-by-Symbol Decoding 154
   4.2.1. Derivation of the Hartmann–Rudolph Algorithm 155
   4.2.2. An Alternate Form of the HR Algorithm 156
   4.2.3. A Design Example 158
   4.2.4. Greenberger Approximation to the HR Algorithm 161

4.3.	The Weldon Algorithm	162
4.4.	The Chase Algorithm	167
	4.4.1. Standard Chase Algorithms	167
	4.4.2. Variations on the Chase Algorithms	169
4.5.	Information-Set-Decoding Algorithms	172
	4.5.1. Information-Set-Only Algorithms	172
	4.5.2. Omura-Type Algorithms	175
	4.5.3. Partial Syndrome Decoding	177
	4.5.4. Some Performance Predictions	178
	Problems	179

5. Algebraic Techniques for Multiple Error Correction   181

5.1.	Finite-Field Transforms	183
5.2.	BCH Codes	185
5.3.	BCH Decoding Techniques	188
	5.3.1. Frequency Domain Decoding	188
	5.3.2. Time Domain Decoding	191
5.4.	Solution of the Key Equation	195
	5.4.1. Euclid's Algorithm	195
	5.4.2. Berlekamp's Algorithm	201
5.5.	Implementation Considerations	209
5.6.	Correction of Errors and Erasures	214
5.7.	Performance Results	219
5.8.	Remarks	222
	Problems	224

6. Convolutional Code Structure and Viterbi Decoding   227

6.1.	Binary Rate-$1/2$ Convolutional Codes	228
6.2.	Viterbi Decoding Algorithm	231
6.3.	Rate-m/n Convolutional Codes	235
6.4.	Finite-State Machine Description and Distance Properties of Convolutional Codes	238
6.5.	Performance of Convolutional Codes with Viterbi Decoding	242
	6.5.1. Union Bounds	243
	6.5.2. Optimum Codes	246
	6.5.3. Performance with Coherent PSK	247
	6.5.4. Performance with Orthogonal Signaling and Noncoherent Detection	250
6.6.	Implementation Considerations	253
	6.6.1. Synchronizer	255
	6.6.2. Branch Metric Computer	256
	6.6.3. Path Metric Storage and Updating	258
	6.6.4. Storage and Updating of Hypothesized Information Sequences	261
	6.6.5. Output Device	263
	6.6.6. Demodulator Quantization and AGC	264
6.7.	Remarks	265
	Problems	266

7. Other Convolutional Decoding Techniques 267

    7.1. Syndrome Decoding Techniques 268
        7.1.1. Basic Concepts 268
        7.1.2. Table Look-up Decoding 278
        7.1.3. Threshold Decoding 283
    7.2. Sequential Decoding Techniques 297
        7.2.1. Sequential Decoding via the Fano Algorithm 299
        7.2.2. Selection of Sequential Decoding Metric 302
        7.2.3. Code Selection 304
        7.2.4. Sequential Decoder Computational Problem 304
        7.2.5. Performance of Sequential Decoders 313
        7.2.6. Implementation Considerations 318
        7.2.7. Stack Sequential Decoding 326
    Problems 328

8. System Applications 331

    8.1. Concatenated Codes 331
        8.1.1. Concatenated Coding Concepts 332
        8.1.2. Reed–Solomon/Orthogonal Code Systems 333
        8.1.3. Reed–Solomon/Short-Block Code Systems 337
        8.1.4. Reed–Solomon/Convolutional Code Systems 339
    8.2. Coding for the White Gaussian Noise Channel 341
    8.3. Interleaver Structures for Coded Systems 345
        8.3.1. Periodic Interleavers 346
        8.3.2. Pseudorandom Interleavers 349
    8.4. Coding for Burst Noise Channels 352
        8.4.1. Burst Noise Processes 353
        8.4.2. Convolutional Code Performance with Random Erasures .. 354
        8.4.3. Convolutional Code Performance with Periodic Burst
            Erasures 358
        8.4.4. Performance Degradation Due to Random Erasure Burst .. 363
        8.4.5. Interleaving Implications 364
    8.5. Coding for Spread Spectrum Systems 366
        8.5.1. Direct Sequence PN Spread Systems 367
        8.5.2. Frequency Hop Systems 375
    8.6. Coding for Bandwidth-Constrained Channels 381
    Problems 390

Appendix A. Code Generators for BCH Codes 393
Appendix B. Code Generators for Convolutional Codes 399

    B.1. Viterbi Decoding 399
    B.2. Table Look-up Decoding 400
    B.3. Threshold Decoding 400
    B.4. Sequential Decoding 401

References 409
Index 415

Error-Correction Coding for Digital Communications

Fundamental Concepts
of Coding

When digital data are transmitted over a noisy channel, there is always a chance that the received data will contain errors. The user generally establishes an error rate above which the received data are not usable. If the received data will not meet the error rate requirement, error-correction coding can often be used to reduce errors to a level at which they can be tolerated. In recent years the use of error-correction coding for solving this type of problem has become widespread.

The utility of coding was demonstrated by the work of Shannon.[4] In 1948 he proved that if the data source rate is less than a quantity called the channel capacity, communication over a noisy channel with an error probability as small as desired is possible with proper encoding and decoding. Essentially, Shannon's work states that signal power, channel noise, and available bandwidth set a limit only on communication rate and not on accuracy.

It soon became clear that the *real* limit on communication rate was set not by channel capacity but by the *cost* of implementation of coding schemes. Cost constraints force communication at rates substantially below capacity. In recent years much research has been directed toward finding efficient and practical coding schemes for various types of noisy channels. Most of the progress toward finding practical schemes has come in the last fifteen years, and it is now clear that coding can provide significant performance improvements in many applications. There have been a number of applications where coding equipment has been built and used quite successfully. The increasing practicality of coding is due to new developments within the field of error-correcting codes and the dramatic reductions in cost and size of solid state electronic devices.

1.1. Basic Principles

Error-correction coding is essentially a signal processing technique that is used to improve the reliability of communication on digital channels. Although individual coding schemes take on many different forms and have their roots in diverse mathematical disciplines, they all have two common ingredients. One is the use of *redundancy*. Coded digital messages always contain extra or redundant symbols. These symbols are used to accentuate the uniqueness of each message. They are always chosen so as to make it very unlikely that the channel disturbance will corrupt enough of the symbols in a message to destroy its uniqueness. The second ingredient is *noise averaging*. This averaging effect is obtained by making the redundant symbols depend on a span of several information symbols. Some valuable insight into the coding process can be obtained by examining each of these ingredients separately.

First, consider a binary communication channel that is contaminated by an unwanted disturbance such that errors occur independently on each symbol and with an average error rate of $P_e = 0.01$. If one were to examine a block of 10 symbols from this channel, it would be extremely difficult to identify specific erroneous symbols. On the other hand, one could make the statement that the block contains three or fewer errors and only be wrong twice out of every million blocks. Furthermore, the ability to make statements of this type improves as the block size increases. Not only does the fraction of symbols that are in error in a block approach the average channel error rate, but more importantly the fraction of blocks that contain a number of errors that differ substantially from this average also becomes very small. One may get a feeling for the extent to which this statement is valid by making some simple calculations. Suppose for the same channel we compute the probability that the fraction of symbols in error exceeds a value ρ and plot this function for several different block lengths. The result is shown in Fig. 1-1. Observe that the "curves" get progressively steeper as the block length increases, and that they approach a "step" function which occurs at an abscissa of 0.01.

The curves in Fig. 1-1 suggest that if we are willing to process symbols in blocks rather than one at a time, it might be possible to reduce the overall error rate. What is required is the existence of a "coding" scheme that can tolerate some fraction of the symbols in a block being in error without destroying the "uniqueness" of the message that the block conveys, thereby causing a block error. Figure 1-1 indicates explicitly, for several different block lengths, exactly what fraction of errors must be corrected in order to produce a given block error rate. It also indicates that for a fixed block error rate the fraction of errors that must be corrected decreases with in-

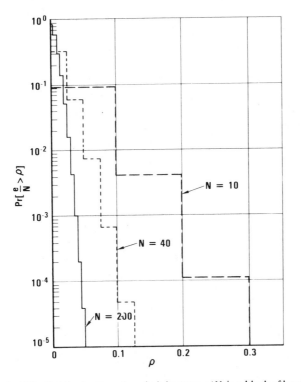

Figure 1-1. Probability that the fraction of symbols in error, e/N, in a block of length N exceeds ρ for $P_e = 0.01$.

creasing block length. These results indicate the potential for performance improvement that is obtained through noise averaging and that this potential increases with block length. Thus, longer block codes should be more "effective" than short ones.

Having established the desirability of being able to correct symbol errors, the next logical question is: how can this be accomplished? The key is redundancy. Some reflection on the part of the reader will reveal that if one is going to attempt to correct errors in a message represented by sequence of n binary symbols, then it is absolutely essential not to allow the use of all 2^n possible sequences as being legitimate messages. If, in fact, every possible received sequence of n symbols were a legitimate message, then one would have no possible basis for determining if one sequence were any more valid than any other sequence. Carrying this reasoning a little further, it becomes clear that if one wishes to correct all patterns of t or fewer errors, it is both necessary and sufficient for every legitimate message sequence to differ from every other legitimate message sequence in at least $2t + 1$ positions. For example, if one wished to correct all single and double symbol errors, it is necessary that all pairs of message sequences differ in at least five symbols. Any received sequence which contains two errors and, therefore, differs from the correct sequence in exactly two places will always differ from all other message sequences in at least three places. We refer to the number of positions in which any two sequences differ from each other as the *Hamming distance*, d, between the two sequences. The smallest value of d for all pairs of code sequences is called the *minimum distance* of the code and is designated d_{min}. Since d_{min} must always be one more than twice the number of errors that are to be corrected, one may write

$$t = \left\lfloor \frac{d_{min} - 1}{2} \right\rfloor \tag{1-1}$$

where $\lfloor \ \rfloor$ is used to designate the integer part. The parameter, t, indicates that all combinations of t or fewer errors in any received sequence can be corrected. [There are some channel models for which t larger than indicated in (1-1) is possible, as demonstrated in Problem 1-8.]

Example. Consider a code which consists of the four code words 0 0 0 0 0, 0 0 1 1 1, 1 1 1 0 0, and 1 1 0 1 1. Each code word can be used to represent one of four possible messages. Note that the code includes only a small fraction of the 32 possible sequences of length five, and as a result, one can select code words which all differ from each other in at least three positions. Thus, this code has a minimum distance of 3 and can correct a single error in any position. In order to carry out the correction process for this code, it is necessary to associate each of the 28 nonallowable sequences with the "closest" of the four allowable sequences. This process implies the creation of a "decoding" table, which is constructed by first listing under each code word all of the possible received sequences which differ in only one position. The result is the portion of Table 1-1 that is enclosed in the dashed lines. Note that after constructing this portion, there are eight sequences left over. Each of these eight differ in at least two positions from each code word. Unlike the other sequences, however, there does not exist a unique way to assign these in the table. The sequence 1 0 0 0 1, for example, could have been placed under either the fourth column or the first column. The table is used in the decoding process by finding the column of the table which contains the received sequence and selecting the code word at the head of the column as the decoder output.

There is a very good reason the table is constructed in this manner. The probability of any particular pattern of i errors occurring is $P_e^i(1 - P_e)^{5-i}$. Then for $P_e < 1/2$ we note that

$$(1 - P_e)^5 > P_e(1 - P_e)^4 > P_e^2(1 - P_e)^3 > \cdots$$

Table 1-1. Decoding Table for a Four-Word Code

00000	11100	00111	11011
10000	01100	10111	01011
01000	10100	01111	10011
00100	11000	00011	11111
00010	11110	00101	11001
00001	11101	00110	11010
10001	01101	10110	01010
10010	01110	10101	01001

Thus, a particular single-error pattern is more likely than a particular double-error pattern, etc. This means that a decoder which decodes a particular received sequence into the code word that is nearest in Hamming distance actually selects the most likely transmitted word (assuming all transmitted words are equally likely). A decoder implementing this decoding rule is a *maximum-likelihood decoder*, and it provides minimum probability of sequence error under these conditions. In this sense it is an optimum decoder. This is a very important concept since maximum-likelihood decoders are used frequently for short codes. In addition, maximum-likelihood decoder performance provides a benchmark with which the performance of nonoptimum decoding techniques can be compared. If one actually decodes using a decoding table, then the entries in the table can be assigned using the above procedure to implement maximum-likelihood decoding. Unfortunately, the size of the table grows exponentially with the code block length, and thus the direct use of a decoding table is impractical for long codes. As a conceptual device, however, the decoding table is quite useful in illuminating some of the important properties of block codes.

Within the decoding table the set of legitimate code words is a subset (the first row of the decoding table) of the set of all 2^n n-symbol sequences (or n-tuples). Additionally, the process of creating a decoding table divides the set of n-tuples into distinct subsets (the columns of the decoding table). For a t-error-correcting code, the number of n-tuples, N_e, in each subset obeys the inequality

$$N_e \geq 1 + n + \binom{n}{2} + \cdots + \binom{n}{t} \qquad (1\text{-}2)$$

where $\binom{n}{i}$ is the ith binomial coefficient $n!/[i!(n-i)!]$.

This inequality follows immediately by noting that there are exactly n patterns that differ from the correct one in one position, $\binom{n}{2}$ patterns that differ in two positions, etc. As in the simple example given previously, there will almost always be some patterns that are left over after assigning all those that differ in t or fewer places (thus accounting for the inequality).

At this point we are in a position to relate the amount of redundancy in a code to the number of errors that are correctable. First observe that there are 2^n possible sequences. Each column of the decoding table contains N_e of these sequences so that the number of code words, N_c, must obey the inequality

$$N_c \leq 2^n \left/ \left[1 + n + \binom{n}{2} + \cdots + \binom{n}{t} \right] \right. \qquad (1\text{-}3)$$

This is called a *Hamming bound* or "sphere-packing" bound. The equality in this bound can be achieved only for so-called *perfect codes*. These are codes which can correct all patterns of t or fewer errors and no others. There are only a small number of perfect codes which have been found and consequently the equality in (1-3) is almost never achieved.

At the encoder we envision a process by which a k-symbol information sequence is mapped into an n-symbol code sequence. Although the terminology is usually restricted to the so-called linear codes (to be discussed), we shall refer to any such mapping as an (n, k) code. Since the k-symbol sequence can take on 2^k distinct values, inequality (1-3) can be written

$$2^k \leq 2^n \left/ \left[1 + n + \binom{n}{2} + \cdots + \binom{n}{t} \right] \right. \qquad (1\text{-}4)$$

A measure of the efficiency implied by a particular code choice is given by the ratio

$$R = k/n \qquad (1\text{-}5)$$

where R is defined as the *code rate*. The fraction of transmitted symbols that are redundant is $1 - R$.

The mapping implied by the encoder can be described by a look-up table. For example, the four-word code discussed previously is described in Table 1-2. The portion of the code sequence contained between the dashed lines is identical to the input sequence. Thus, each code sequence is easily and uniquely related to the input. Not all block codes exhibit this property. Those which do are referred to as *systematic* codes. For systematic codes, the concept of redundant digits becomes very clear and in Table 1-2 consists of the digits in positions 1, 4, and 5. Conversely, codes which do not exhibit this property are called *nonsystematic* codes.

Table 1-2.
Encoding Look-up Table

Input sequence	Code sequence
0 0	0 ¦ 0 0 ¦ 0 0
0 1	0 ¦ 0 1 ¦ 1 1
1 0	1 ¦ 1 0 ¦ 1 1
1 1	1 ¦ 1 1 ¦ 0 0

Many good constructive coding procedures have been found which permit the correction of multiple errors and allow one to achieve rather remarkable improvements in the symbol error rate. These codes are easy to generate and with the availability of modern semiconductor devices are relatively straightforward to decode. As an example, there is a block code of length 40 that contains 50 % redundant symbols and is capable of correcting up to four randomly occurring errors. An examination of Fig. 1-1 reveals that if $P_e = 0.01$ this code provides an overall block error rate of less than 10^{-4}. If this were not good enough, the engineer would have the option of increasing the amount of redundancy and correcting more errors or of going to a somewhat longer block length and taking advantage of more averaging. In either case, there are associated costs that must be evaluated. Both options, however, are available and may represent practical alternatives.

Before proceeding, there is an interesting digression that is of little practical importance but which frustrated researchers in the coding area for many years. The form of the curves in Fig. 1-1 suggests that if one had a scheme for correcting a fixed fraction, t/n, of erroneous symbols per block (in this case t/n is slightly greater than 0.01), then the error rate could be made arbitrarily small by simply choosing the block to be long enough. Unfortunately, this turns out to be a very difficult task. Most constructive procedures encounter the problem that they can only maintain a constant ratio t/n at the expense of an increasing percentage of redundant symbols (or equivalently, $R \to 0$ as $n \to \infty$). Thus, a loss of efficiency occurs because the relative number of useful messages that these schemes convey becomes vanishingly small for long block lengths. A partial solution to this problem was given by Justesen[5] in 1972. Justesen has shown that it is possible to construct a class of codes which is asymptotically good (in the sense described above) and for which a decoding procedure can be specified. To the authors' knowledge, however, these codes have not been used in any real communication systems.

1.2. Practical Constraints

In the previous section we introduced the concept of redundancy and how it is used in conjunction with averaging to improve the reliability of a digital communication system. We also defined a property of codes called the minimum distance, showed how this parameter is related to the number of errors that can be corrected, and how the number of errors to be corrected determines the percentage of redundant symbols that are required for a specific code length. In developing these concepts we intentionally used a very simple channel model that ignores many of the practical constraints that are present in the real world. In this section a general model for a digital communication system is presented. This model shows the relationship of the coder and decoder to the rest of the system and describes the constraints that the system imposes on the coding process. Two important notions will be developed. The first is the fact that there is a cost associated with the addition of redundancy and hence one would like to minimize the percentage of redundant symbols that are required to achieve a given minimum distance. This leads to a desire to find optimal or near optimal codes and decoding techniques. The second notion is that there is often additional information available to the decoder which indicates the reliability of individual symbol decisions. This reliability information can often be used either to simplify the decoding process or to improve its performance.

A block diagram which describes the digital communication process is shown in Fig. 1-2. This model is sufficiently general to describe most of the situations that will be discussed in this text. The function of each block will be discussed in succeeding paragraphs.

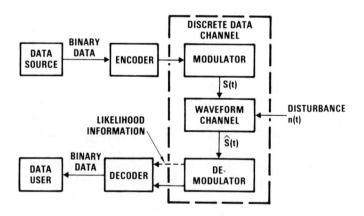

Figure 1-2. General model of digital communication process.

1.2.1. Data Source

The data source generates data in the form of binary symbols. The normal assumption will be that these data have been processed such that the individual "ones" and "zeros" occur independently and with equal probability. However, in some real-world situations certain data patterns may be more prevalent than others, and this possibility can sometimes invalidate conclusions that are based on completely random data. Hence, the reader should exercise caution when designing an actual system.

1.2.2. Encoder

Two kinds of codes are discussed in this text. These are *block codes* and *tree codes*. The distinguishing difference between the encoders for these two codes is the presence or absence of memory. Conceptually, the encoder for a block code is a *memoryless* device which maps a k-symbol input sequence into an n-symbol output sequence. The term "memoryless" indicates that each n-symbol block depends only upon a specific k-symbol block and on no others. It does not mean that the encoder does not contain memory elements. The usual distinguishing parameters for a block code are the values of n, k, $R = k/n$, and d_{min}. Practical values for k range from 3 to several hundred and for R from $1/4$ to $7/8$. Values outside this range are possible but often present certain practical difficulties. The input and output sequences are usually in the form of binary symbols but occasionally may be symbols from some higher-order alphabet. The encoder for a tree code is a device with *memory* that accepts binary symbols in sets of m and outputs binary symbols in sets of n. Each set of n output symbols is determined by the current input set and a span of v of the preceding input symbols. The memory span of the encoder is, therefore, $v + m$ input symbols. The parameter $v + m$ is often referred to as the constraint length of the code and designated by the symbol $k = v + m$ (not to be confused with the k in a block code). There is little consistency as to usage in the coding literature. Some authors refer to the constraint length as k while others refer to it as v. In this text we represent the constraint length by v since it is less confusing for codes with $m > 1$. The parameter $k = v + m$ is used sparingly. Tree codes are also characterized by a rate $R = m/n$ and a minimum free distance, d_{free}. The precise definition for d_{free} is slightly more involved than that of d_{min} for a block code, but essentially conveys the same information. Typical parameters for tree codes are m and n in the range of 1 to 8, R in the range of $1/4$ to $7/8$, and v in the range of 2 to 60.

Another manner of classifying codes is as *linear* or *nonlinear*. Linear codes form a linear vector space and have the very important property that two code words can be added using a suitable definition for addition to produce a third code word. In the usual case of a binary code, this operation is symbol-by-symbol modulo-2 addition of the two code words (i.e., $1 + 1 = 0, 1 + 0 = 1, 0 + 0 = 0$). As we will discover subsequently, this property has two far-reaching effects. The first is that it vastly simplifies the encoding and decoding problem by allowing one to express any code word as the "linear" combination of a small set of reference code words known as *basis vectors*. The second effect is that it also significantly simplifies the problem of computing performance by making the distance between any two code words equivalent to the distance between the all-zero code word and some other code word. Thus, when one is computing the performance of the code, it is only necessary to consider the effect of transmitting the all-zero code word. This performance computation is further facilitated by noting that the Hamming distance between a given code word and the all-zero code word is equivalent to the number of nonzero elements in the word. This number is referred to as the *Hamming weight* of the code word, and a list which contains the number of code words of each weight can be used to estimate code performance via the union bound of Section 1.3.2. Such a list is referred to as the code *weight structure*.

Almost all of the coding schemes used in practical applications are linear codes. Linear block codes are usually referred to as *group codes* since the code words form a mathematical structure called a group. Linear tree codes are commonly called *convolutional codes* since the encoding operation can be thought of as the discrete-time convolution of the input sequence with the impulse response of the encoder. Throughout this text we will deal almost exclusively with group codes and convolutional codes.

Finally, codes may be classified as either random-error-correcting codes or burst-error-correcting codes. We will deal almost exclusively with codes designed for the correction of random or independent errors. Codes have been designed for the correction of burst errors with well-defined burst characteristics. However, it is often desirable in a burst error environment to use a random-error-correcting code with an interleaver/deinterleaver. This approach involves scrambling the coded sequence prior to the transmission followed by unscrambling the sequence after reception in order to randomize the burst error process. We will discuss this approach in succeeding chapters.

1.2.3. Modulator

The modulator generates a set of finite time duration waveforms and provides a mapping between the encoder output and this set of waveforms. For binary modulation schemes, each output symbol from the encoder is used to select one of the two possible waveforms. For M-ary modulation schemes, the output of the encoder is segmented into sets of j symbols (where $M = 2^j$) and each set is used to select one of the M waveforms. In this case several different mappings are possible and the particular choice depends upon the objectives one is attempting to achieve. Two common situations are where the j-symbol group corresponds to an M-ary symbol in a block code or where the j-symbol group corresponds to the n output symbols in a tree code.

There are several specific cases of interest. In systems where coherent demodulation is possible (i.e., a carrier reference can be obtained), the binary modulation scheme *phase-shift keying* (PSK) is often used. With this technique a transmitted 1 is represented by the waveform $s_1(t) = A(t) \times \cos \omega_0 t$ while a transmitted 0 is represented by the negative or antipodal signal $s_0(t) = -s_1(t) = A(t) \cos(\omega_0 t + \pi)$. In noncoherent systems, the determination of the sign of the transmitted waveform is not possible so that typically a pair of orthogonal[†] signals $s_0(t)$ and $s_1(t)$ are used to represent transmitted 0 and 1, respectively. The case of M-ary signaling in noncoherent systems is a simple extension in that a set of M mutually orthogonal waveforms $s_0(t), s_1(t), \ldots, s_{M-1}(t)$ are used.

1.2.4. Waveform Channel

The waveform channel consists of all the hardware and physical media that the waveform passes through in going from the output of the modulator to the input of the demodulator. It does not have to be an actual real time communication system but might represent a data storage or recording system. In the usual case, the output, $\hat{s}(t)$, of the channel is a scaled replica of the input, $s(t)$, to which some random disturbance, $n(t)$, has been added. Much more general situations are possible. Distortion may be present due to heavy filtering or multiple signal paths. The disturbance may cause signal suppression which could in turn cause the amplitude of the received signal to vary or the channel itself may be time varying. The disturbance $n(t)$ may be simple receiver noise which can be modeled as an additive Gaussian process or it may be urban noise of various kinds or it may be intentional jamming by an unfriendly party. Although much of the existing work in coding has been tailored to the simple case of additive Gaussian noise, it is often in the more complicated situations where coding concepts prove to be most effective. Several examples of this are given in Chapter 8.

[†] To say that $s_0(t)$ and $s_1(t)$ are orthogonal means that $\int_0^T s_0(t) s_1(t) \, dt = 0$, where T is the symbol duration.

1.2.5. Demodulator/Detector

The demodulator is a device which estimates which of the possible symbols was transmitted based upon an observation of the received signal $\hat{s}(t)$. The probability that this estimate is correct depends upon the ratio of signal power to noise power in the data bandwidth, the amount of signal distortion due to filtering and nonlinear effects, and the detection scheme that is being used. In a coded system the demodulator sometimes performs a second function. That function is to supply information to the decoder as to the reliability of each individual symbol decision. This information may be obtained in a number of different ways, and the approach utilized depends strongly on the nature of the disturbance, $n(t)$. One possibility occurs when there is a known interference such as a radar signal whose presence can be determined independently. The reliability information would then be a single symbol that indicates whether the radar is on or off. Another possibility occurs when the detector is a sampled matched filter, and $n(t)$ is additive Gaussian noise. In this case, the magnitude of the sampled voltage relative to the decision threshold is a strong indication of the reliability of the decision. We are interested in several specific demodulator types. However, we are not going to derive the optimum receiver structures but simply give the characteristics of these demodulators that are important in relation to the coding problem. The reader is referred to a good book on communication theory for more detail (e.g., Wozencraft and Jacobs[6]).

1.2.5.1. Coherent Systems. As mentioned previously, we will assume the use of antipodal transmitted signals $\pm s(t)$. The fact that a carrier reference is available will allow the sign of the transmitted waveform to be determined. The channel noise $n(t)$ is assumed to be white Gaussian noise with double-sided spectral density $N_0/2$ W/Hz. The optimum receiver is a filter matched to $s(t)$ which is sampled each symbol time to determine its polarity [i.e., $h(t) = Ks(T - t)$, where T is the symbol duration]. The voltage at the matched filter output at the sample time is a Gaussian random variable with mean $\pm E_s^{1/2}$ (depending on whether a 1 or 0 was transmitted) and variance $\sigma^2 = N_0/2$. [It is assumed that $\int_{-\infty}^{\infty} h^2(t)\,dt = 1$ and the received signal energy per symbol time is E_s.] Thus, the probability density function of this voltage ρ can be written as

$$p(\rho\,|\,1) = \frac{\exp\left[-(\rho - E_s^{1/2})^2/N_0\right]}{(\pi N_0)^{1/2}} \qquad (1\text{-}6)$$

and

$$p(\rho\,|\,0) = \frac{\exp\left[-(\rho + E_s^{1/2})^2/N_0\right]}{(\pi N_0)^{1/2}} \qquad (1\text{-}7)$$

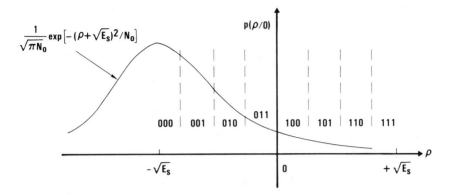

Figure 1-3. Probability density function of the matched filter output for a coherent PSK system.

for transmitted 1 and 0, respectively. The second probability density function is shown in Fig. 1-3. Assuming equally likely transmitted symbols, the optimum decision threshold is at zero. Then the demodulator output is a 0 if the voltage at the matched filter output is negative. Otherwise, the output is a 1. Thus, given that a 0 was transmitted, the probability of error is the probability that $\rho > 0$. It can easily be shown that this is given by

$$P_e = Q[(2E_s/N_0)^{1/2}] \tag{1-8}$$

with $Q(\alpha)$ defined as

$$Q(\alpha) = \int_{\alpha}^{\infty} \frac{1}{(2\pi)^{1/2}} e^{-\beta^2/2} \, d\beta \tag{1-9}$$

The same result is obtained for a transmitted 1 so the demodulator output error rate is given by (1-8).

1.2.5.2. Noncoherent Systems. We assume that the noncoherent demodulator used for binary orthogonal signaling utilizes two matched filters [matched to $s_0(t)$ and $s_1(t)$] followed by square-law envelope detectors. When these filters are sampled at the symbol time, the probability density function for the detector with signal present is

$$p_{s+n}(\rho) = \rho I_0(a\rho) \exp[-(\rho^2 + a^2)/2], \qquad \rho \geq 0 \tag{1-10}$$

where $I_0(\cdot)$ is the zero-order modified Bessel function of the first kind and $a = (2E_s/N_0)^{1/2}$.[6] For the case of signal absent, the probability density function is

$$p_n(\rho) = \rho \exp(-\rho^2/2), \qquad \rho \geq 0 \tag{1-11}$$

[Since $s_0(t)$ and $s_1(t)$ are orthogonal, one envelope detector output will always have signal present and be described by (1-10) at the proper sampling time while the other will have signal absent and be described by (1-11).]

A symbol error is made by the demodulator if the sampled voltage of the envelope detector with signal absent exceeds that of the envelope detector with signal present. It can be shown that the probability of this event is

$$P_e = \tfrac{1}{2} \exp(-E_s/2N_0) \tag{1-12}$$

Performance of a noncoherent system with binary orthogonal signaling as given by (1-12) is substantially inferior to that given by (1-8) for a coherent system employing binary antipodal signals.

The efficiency of noncoherent systems may be improved in several ways. In some systems differentially coherent PSK (DPSK) can be used, i.e., the data are transmitted as either no phase change or a change in carrier phase of π. Symbol decisions are made on the basis of whether the new received phase differs from the preceding received phase in magnitude by more or less than $\pi/2$. DPSK is statistically equivalent to binary orthogonal signaling but with 3 dB more efficiency (because the message-bearing waveform is two symbols in length). Thus, the DPSK error rate is

$$P_e = \tfrac{1}{2} \exp(-E_s/N_0) \tag{1-13}$$

Another technique that is more efficient than binary orthogonal signaling is the use of M-ary orthogonal signaling. The demodulator generalizes in a natural way from the binary demodulator. In this case, at each sample time there will be one envelope detector with signal present whose output voltage is described by the probability density function (1-10), and $M - 1$ envelope detectors with no signal present whose output voltage is described by the probability density function (1-11). It can be shown that the M-ary symbol error rate at the demodulator output is given by

$$P_s = \frac{\exp(-E_s/N_0)}{M} \sum_{j=2}^{M} (-1)^j \binom{M}{j} \exp(E_s/jN_0) \tag{1-14}$$

1.2.6. Decoder

The decoder is the device which inverts the operation of the encoder. Because the sequence of symbols that are generated by the demodulator may contain errors, the decoder must perform a significantly more complex mapping than the encoder. Although this could in principle be performed using optimal techniques such as a look-up table similar to Table 1-1, this approach rapidly becomes impractical as the size of the code grows. In practice, one must often resort to a nonoptimal technique. To make the decoding operation feasible, one must devise computational procedures for realizing this process. The bulk of this text is devoted to a detailed exposition of the most practical techniques that have been found. The decoding algorithms can be categorized according to whether the techniques apply to block codes or tree codes. The two major classes of block decoding algorithms to be discussed are

1. procedures based on elementary code structures which include information set decoders, Meggitt decoders, and threshold decoders; and

2. algebraic procedures which are based on the detailed algebraic structure of certain codes and basically involve solving sets of algebraic equations.

The four major classes of tree decoding algorithms are

1. the Viterbi algorithm (maximum-likelihood decoding);
2. sequential decoding algorithms;
3. threshold decoding; and
4. table look-up decoding.

The use of reliability information or so-called "soft decisions" is applicable to both block codes and tree codes. Most of the algorithms for decoding tree codes make use of this information in a straightforward manner. The use of "soft decisions" in block codes is somewhat more involved and generally requires significant changes in the algorithms.

1.2.7. Discrete Data Channel

From the viewpoint of the encoder and decoder, the segment of Fig. 1-2 enclosed in dashed lines is the most important. This channel is characterized by a set of input symbols, output symbols, and transition probabilities. In the simplest case, the transition probabilities are time invariant and independent from symbol to symbol. This is the so-called discrete memoryless channel (DMC). The most commonly encountered case of the DMC is the so-called binary symmetric channel (BSC). One method of realizing this channel is to use a coherent system with antipodal waveforms. In this case the probability of error is given by (1-8). Likewise, one could use binary orthogonal waveforms in a noncoherent system and achieve a probability of error given by (1-12). This channel is often represented diagramatically as in Fig. 1-4.

Although practical demodulators rarely achieve the performance indicated by (1-8) it is possible to build a real demodulator using binary phase shift keying such that the performance curve (probability of error versus E_s/N_0) is nearly parallel to the ideal curve defined by (1-8) and to make the errors independent or nearly independent from symbol to symbol. For

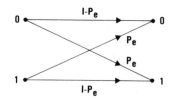

Figure 1-4. Binary symmetric channel.

this reason the channel whose performance is defined by (1-8) is often used as a standard by which various coding schemes are measured and compared.

This signaling scheme can also be used to provide an example of a typical "soft decision" data channel. The probability density function for the sampled voltage on the output of the matched filter when a data 0 is transmitted is shown in Fig. 1-3. If a simple analog-to-digital converter with slicing levels, as shown in Fig. 1-3, is used, then the binary symmetric channel is converted to a binary input, 8-ary output channel. The matched filter output has simply been quantized further. This channel is also symmetric with the transition probabilities being computed as the area under the density function between the appropriate slicing levels. This binary input, 8-ary output channel is often shown diagramatically as in Fig. 1-5. The eight-level quantization scheme is also the one most frequently used in the soft decision decoding systems presently being built. For reasons which will subsequently become clear, this is a near ideal choice and approaches very closely the performance that can be achieved using an infinite number of quantization levels.

It is obvious that one can make similar discrete data channel models for other signaling and detection schemes, and this will be done later in the

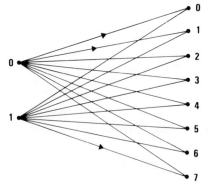

Figure 1-5. Binary input, 8-ary output channel produced by an eight-level quantizer on a Gaussian channel.

text when it becomes appropriate. Although the probability of error in these other models may not be representable using a Q function of $(2E_s/N_0)^{1/2}$, it will typically exhibit a monotonic dependence on E_s/N_0. That is, the probability of error will almost always vary inversely with signal-to-noise ratio. This leads us to an important observation, namely, that the use of redundancy is not free. If redundant symbols are included in a digital transmission and simultaneously the rate of information transmission and the power remain constant, then the energy associated with each transmitted symbol must be reduced and, consequently, the error rate must increase. For example, if a 12-bit message is to be transmitted at the rate of one per second using 1 W of transmitter power, then each transmitted binary symbol uses $1/12$ of a joule. However, if this same message were encoded using a length 24 block code (50% redundant symbols) and transmitted at the same message rate, then each transmitted binary symbol uses $1/24$ of a joule. It is obvious that the probability of symbol error for the coded case will be worse than for the uncoded case. If the coding scheme is to produce a net gain in performance, it must improve the error rate enough to make up for the loss that was incurred by introducing redundancy in the first place. Since the improvement is obviously a strong function of the number of errors that are to be corrected, it is important to correct as many errors as possible for a fixed block length and fixed code rate.

In some types of systems one might add redundancy while keeping the power and symbol rate constant. Note that in this case the symbol error rate for both the coded and the uncoded case are equal, and the addition of redundancy forces a reduction in information rate to a factor R times the symbol rate. Obviously, it is desirable to add as little redundancy as necessary to achieve the desired error rate in order to maximize the rate of information transmission.

So far we have only discussed examples in which the symbol errors are independent from one symbol to the next. Often in practical structures this will not be true and closely spaced or "burst" errors may be more common than widely spaced errors. This could be caused by a periodic noise source such as a nearby radar or a rotating machine or by fading in the communication link. One can extend the modeling ideas discussed thus far by making the transition probability depend upon the previous transitions or by making them functions of time. Typical solutions for situations of this type have included either devising codes which exhibit good word separation properties based upon burst lengths rather than on the number of random errors, or interleaving several code words from a random-error-correcting code. Unfortunately, except in a few isolated situations, it is almost impossible to predict the performance of such schemes analytically. One must often resort to simulation techniques, and thus the models lose some of their utility.

1.3. Performance Computations

The usual measure of performance of a coded system is the average error rate that is achieved at a specified signal-to-noise ratio. The large variety of codes, modulation schemes, and decoding techniques that are available make it virtually impossible to catalog all potentially useful situations in advance. Thus, the design engineer is frequently called upon to compute the performance of a particular coding scheme under a set of conditions that is different from those encountered on previous occasions. There are basically two methods by which the performance may be determined. One is an exact procedure which involves a detailed enumeration of all possible error situations. Thus, it produces a result which is valid at all signal-to-noise ratios. In the great majority of practical applications, the designer is only interested in the high signal-to-noise ratio performance. That is, he is interested in those situations where the probability of error is less than 10^{-3} or 10^{-4}. In this case an approximate procedure is available which makes use of union bound techniques. Both of these procedures are developed in this section using binary block codes. The extension to convolutional codes will be discussed in Chapter 6 after some of the important structural properties of these codes have been developed.

The exact procedure is first developed for an arbitrary block code operating on a general binary channel. The resulting formulas are then simplified in two distinct steps. The first simplification occurs if the block code is restricted to be a group code. The second simplification occurs when the group code is used to correct all patterns of t or fewer errors. The union bound technique is illustrated only for the case of a group code. The reader should have no difficulty in applying the union bound technique to other cases. Finally, the union bound approach is used to discuss the effects of demodulator quantization and the difference in performance between hard and soft decisions.

1.3.1. Exact Computation of Error Rate

There are two quantities of interest regarding the error performance of block codes. These are the sequence error rate and the equivalent output symbol or bit error rate. In computing these quantities it is useful to keep in mind the decoding table concept discussed in Section 1.1 and shown in Table 1-1. A generalized version of this table is shown in Fig. 1-6. Recall that this table is constructed by listing each of the code words in the first row and under each code word listing all of the possible received patterns that would be decoded as that particular code word. Note that these received patterns could equally well consist of either hard decisions or soft decisions. In order to keep the discussion as simple as possible, we will

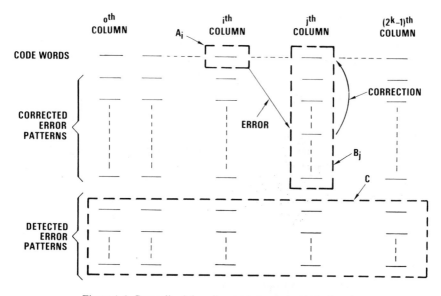

Figure 1-6. Generalized decoding table for an (n, k) block code.

assume that the entries are hard decisions. All of the arguments to be presented are easily extended to the soft decision case. There is one additional feature shown in this table that was not discussed previously. The decoder has the option of not attempting to decode all of the possible received error patterns. For instance, in the decoding table illustrated by Table 1-1, the decoder could be constructed to decode only single symbol errors and to flag any other received pattern as a *detected but uncorrected error*. Decoding algorithms which use this strategy are said to be incomplete while decoding algorithms which decode all possible received patterns are said to be complete. If the incomplete algorithm is such that it corrects all patterns of t or fewer errors but none of higher weight, it is said to be a *bounded-distance decoder*.

Referring to Fig. 1-6, the decoder will make an error if, and only if, a code word at the head of a column i is transmitted, and a sequence that is not located in the correction portion of column i is received. Conversely, it will decode correctly if, and only if, the received sequence falls within the correction portion of the column corresponding to the transmitted word. This latter statement provides the simplest basis for computing sequence error probability. We define the following events:

event A_i: The ith code word is selected for transmission

event B_j: The received pattern falls in the correction portion of column j (upper portion of column j)

event C:　The received pattern falls in the detection portion of Fig. 1-6.

The sequence error probability P_s may be found by subtracting the probability of correct decoding from 1, i.e.,

$$P_s = 1 - \sum_{i=0}^{2^k-1} \Pr(A_i)\Pr(B_i|A_i) \tag{1-15}$$

The sequence error can also be broken down into two components. One is the probability of undetected sequence error, and the other is the probability of detected sequence error. Thus,

$$P_s = P_{s_{ud}} + P_{s_d} \tag{1-16}$$

where

$$P_{s_d} = \sum_{i=0}^{2^k-1} \Pr(A_i)\Pr(C|A_i) \tag{1-17}$$

$P_{s_{ud}}$ can obviously be computed using (1-15), (1-16), and (1-17).

The expression for average bit error rate P_b is somewhat more complicated. In this case we must enumerate the actual error events and weight the probability of each event by the number of symbol errors that occur. Thus,

$$P_b = \sum_{i=0}^{2^k-1}\sum_{\substack{j=0\\(j\neq i)}}^{2^k-1} \frac{\delta_{ij}}{k}\Pr(A_i)\Pr(B_j|A_i) + \sum_{i=0}^{2^k-1}\frac{\varepsilon_i}{k}\Pr(A_i)\Pr(C|A_i) \tag{1-18}$$

In this expression, δ_{ij} is the number of information symbol errors between the ith code word and the jth code word, and ε_i is the average number of errors associated with transmitting the ith code word, receiving an n-tuple in the lower portion of Fig. 1-6, and interpreting that n-tuple as some information sequence. The manner in which this interpretation is carried out determines the value of ε_i. For instance, one strategy is to simply *delete* the message and deliver an appropriate indication to the user. In this case ε_i would be zero; however, there would be a nonzero *probability of deletion*. Alternately, the decoder might not attempt any correction. For the case of systematic codes, the decoder would simply output the received information symbols. This last strategy is equivalent to not performing error detection.

Equations (1-15) through (1-18) are probably impossible to evaluate except for very small values of k. Fortunately, when the block code is also a group code, they can be simplified significantly. In order to carry out this simplification process, we will have to use properties of group codes that have not yet been developed. Hopefully, the reader will be willing to proceed on faith until the necessary justification is supplied in Chapter 2.

There are three properties of group codes that will be required. They are:

1. The modulo-2 sum of any two code words is a code word.
2. Group codes always contain the all-zero word.
3. If a fixed code word is added modulo-2 to a list of all of the code words, the list will be regenerated in a different order.

Actually, properties 2 and 3 are implied by 1. In constructing the decoding table for a group code it is customary to let the all-zero code word head the zeroth column. The other entries in this column are then the complete set of correctable error patterns. Each remaining column in the table is the sum of each of the correctable error patterns in column zero and the code word which heads the column.

Let $\mathbf{X}_i$, $\mathbf{X}_j$, and $\mathbf{X}_\xi$ designate the code words at the head of columns i, j, and ξ, where $\mathbf{X}_\xi = \mathbf{X}_i + \mathbf{X}_j$ and $+$ indicates symbol-by-symbol modulo-2 addition.

An arbitrary entry in column j is $\mathbf{e} + \mathbf{X}_j$, where $\mathbf{e}$ is the correctable error pattern. The probability of receiving $\mathbf{e} + \mathbf{X}_j$ given that $\mathbf{X}_i$ was transmitted is a function of the difference pattern between $\mathbf{X}_i$ and $\mathbf{e} + \mathbf{X}_j$. Since $\mathbf{X}_i + \mathbf{X}_j = \mathbf{X}_\xi$, this difference pattern is the same as the difference pattern between the all-zero word and $\mathbf{X}_\xi + \mathbf{e}$. Consequently, for each value of i and j there is a ξ such that

$$\Pr(B_\xi | A_0) = \Pr(B_j | A_i) \qquad (1\text{-}19)$$

Furthermore, if one holds i fixed and lets j range through all possible values, then ξ will also range through all possible values. Using similar arguments, it is easy to show

$$\Pr(B_0 | A_0) = \Pr(B_i | A_i)$$

$$\delta_{0\xi} = \delta_{ij}$$

and

$$\varepsilon_0 = \varepsilon_i$$

Finally, noting that

$$\sum_{i=0}^{2^k - 1} \Pr(A_i) = 1$$

it follows immediately that

$$P_s = 1 - \Pr(B_0 | A_0) \qquad (1\text{-}20)$$

$$P_{s_d} = \Pr(C | A_0)$$

and

$$P_b = \sum_{\varsigma=1}^{2^k-1} \frac{\delta_{0\varsigma}}{k} \Pr(B_\varsigma|A_0) + \frac{\varepsilon_0}{k} \Pr(C|A_0) \tag{1-21}$$

One final simplification of (1-21) is possible when the decoder is correcting all patterns of t or fewer errors. In this case we can divide the code words into sets according to the weight of the code word. The summation in (1-21) can then be rearranged so that one sums over all words of weight 1, then all words of weight 2, etc. Define the following:

event B'_j = received pattern falls in the correction portion of a code word of weight j

n_j = number of code words of weight j (the previously defined code weight structure)

δ'_{0j} = average number of nonzero information symbols associated with a code word of weight j

Equation (1-20) may now be written as

$$P_s = 1 - \sum_{i=0}^{t} \binom{n}{i} p^i q^{n-i} \tag{1-22}$$

where p is the symbol error probability and $q = 1 - p$. In addition, the probability of undetected sequence error may be written as

$$P_{s_{ud}} = \sum_{j=1}^{n} n_j \Pr(B'_j|A_0) \tag{1-23}$$

The term $\Pr(B'_j|A_0)$ is just the probability that the received pattern falls within t symbols of the weight-j code word. By enumerating all possibilities, this is easily computed as (see problems)

$$\Pr(B'_j|A_0) = \sum_{v=0}^{t} \sum_{r=0}^{v} \binom{j}{v-r} \binom{n-j}{r} p^{j-v+2r} q^{n-j+v-2r} \tag{1-24}$$

From these expressions the probability of detected sequence error is found to be

$$P_{s_d} = \Pr(C|A_0)$$
$$= P_s - P_{s_{ud}} \tag{1-25}$$

Finally, the bit error probability is found by rewriting (1-21) as

$$P_b = \sum_{j=1}^{n} \frac{\delta'_{0j}}{k} n_j \Pr(B'_j|A_0) + \frac{\varepsilon_0}{k} \Pr(C|A_0) \tag{1-26}$$

The term δ'_{0j}/k can be very closely approximated by j/n. This approximation becomes exact for a very important subclass of group codes called cyclic codes.

At high signal-to-noise ratios one can write a very good approximation to the average bit error probability as follows. For a distance $d = 2t + 1$ code that is used to correct random errors, the most prevalent undetected error will be the case in which the error pattern contains exactly $t + 1$ errors. In this case the algorithm will erroneously assume the received sequence contains t errors and will change these t so that the sequence now contains exactly $d = 2t + 1$ errors. Since these errors can be anywhere in the n-symbol sequence the probability of bit error given that a sequence error occurred is d/n and P_b may be written approximately

$$P_b \approx \frac{d}{n} P_s \tag{1-27}$$

This expression is also seen to be the leading term in (1-26) if we interpret P_s as $n_d \Pr(B'_d | A_0)$.

1.3.2. Union Bound Computation of Error Rate

This technique can be used on any block or convolutional code with maximum-likelihood decoding, but it will be illustrated only for the case of a group code used to correct random errors. It is based on the following idea. If an event can be expressed as the union of several subevents, then the probability of that event occurring is always less than or equal to the sum of the probabilities of all subevents. This sum is obviously an over-bound since it counts the contribution due to overlapping events more than once. In the case of a group code, the probability of error can be computed by considering the effect of transmitting the all-zero code word. An error will be made if the received sequence is closer to one of the other code words than it is to the all-zero word. Thus, the probability of error can be over-bounded as the sum of the probabilities of each of these individual error events. Using the same notation as previously defined we have

event B''_j = the distance between the received sequence and code word of weight j is smaller than the distance between the received sequence and the all-zero code word.

Using this approach the sequence error probability for a maximum-likelihood decoder is upper bounded by

$$P_s \leq \sum_{j=1}^{n} n_j \Pr(B''_j | A_0) \tag{1-28}$$

In a similar fashion the average bit error probability is upper bounded by

$$P_b \leq \sum_{j=1}^{n} \frac{\delta'_{0j}}{k} n_j \Pr(B''_j | A_0) \tag{1-29}$$

This equation bears a very strong resemblance to (1-26). The essential difference is that $\Pr(B'_j|A_0)$ is computed by summing over all patterns that are t errors or less from the given code word and $\Pr(B''_j|A_0)$ is computed by summing over all patterns that differ in $j/2$ or fewer of the j nonzero digits of the given code word. That is,

$$\Pr(B''_j|A_0) = \begin{cases} \displaystyle\sum_{i=(j+1)/2}^{j} \binom{j}{i} p^i q^{j-i}, & j \text{ odd} \\ \displaystyle\frac{1}{2}\binom{j}{j/2} p^{j/2} q^{j/2} + \sum_{i=j/2+1}^{j} \binom{j}{i} p^i q^{j-i}, & j \text{ even} \end{cases} \tag{1-30}$$

At high signal-to-noise ratios the contribution due to error terms of weight greater than $t+1$ becomes vanishingly small and thus (1-29) and (1-26) produce identical results. The principal advantage to the union bound is that it is often much simpler to compute $\Pr(B''_j|A_0)$ than to compute $\Pr(B'_j|A_0)$, particularly for the case of soft decisions. Since maximum-likelihood decoding provides optimum performance by achieving minimum P_s for any code, union bound performance calculations often provide a performance goal with which to measure the utility of nonoptimal decoding techniques. The only difficulty is that knowledge of the code weight structure is necessary. While the weight structure is known for many codes of interest, it is usually quite difficult to compute the weight structure of an arbitrary code.

1.3.3. Performance of Error Detection Systems

A degenerate case of these performance calculations provides the performance of a pure error detection system. In such a situation no error correction is attempted, and the table in Fig. 1-6 consists of code words in the first row and all other patterns are in the detection portion. This technique has been used for years in computer systems and other applications where one may request a retransmission if any channel errors are detected. The sequence error rate given by (1-20) becomes

$$P_s = 1 - (1-p)^n \tag{1-31}$$

and the probability of undetected sequence error as given by (1-23) becomes

$$P_{s_{ud}} = \sum_{j=1}^{n} n_j p^j (1-p)^{n-j} \tag{1-32}$$

Thus, the probability of detected sequence error (sometimes called the probability of deletion) is given by

$$P_{s_d} = 1 - \sum_{j=0}^{n} n_j p^j (1-p)^{n-j} \tag{1-33}$$

The only difficulty with the calculations of (1-32) and (1-33) is that knowledge of the code weight structure is required.

An interesting asymptotic result is the error detection performance of an (n, k) code as $p \to 1/2$. In this case all received words in Fig. 1-6 are equally likely with probability 2^{-n}. Thus the probability of undetected sequence error is

$$P_{s_{ud}} = (2^k - 1) 2^{-n}$$
$$\approx 2^{-(n-k)} \tag{1-34}$$

for $n - k$ large. In addition, we see that

$$P_{s_d} = 1 - 2^{-(n-k)} \tag{1-35}$$

This means that while the throughput is very small, the undetected sequence error rate can be made as small as desired by sufficiently increasing the number of redundant symbols in each code word. For codes for which either the weight structure or noise statistics are not known, (1-34) is often used as a worst-case performance estimate. With well-designed codes the peak value of $P_{s_{ud}}$ (as a function of p for $p < 1/2$) is reasonably close to that predicted by (1-34). This is not true for all codes, however, and one should be careful in using this result.

Implementation of pure error detection systems is trivial since one needs only to test the received word to determine if it is a code word. Any of the cyclic codes to be described in Chapter 2 are suited for this purpose. We will not discuss this technique further. In this book we are interested in the more difficult task of designing techniques for forward error correction (though as Fig. 1-6 indicates, some error detection may also be employed).

Example. The code and decoding table of Table 1-1 is to be used over a BSC with error rate p, and a complete algorithm is used in decoding. Since the code is a group code, the sequence error rate may be computed exactly from (1-20) as

$$P_s = 1 - q^5 - 5pq^4 - 2p^2q^3$$
$$= 8p^2q^3 + 10p^3q^2 + 5p^4q + p^5,$$

where the second line follows from

$$\sum_{i=0}^{n} \binom{n}{i} p^i q^{n-i} = 1$$

A union bound on sequence error rate is also easily computed using (1-28) and (1-30). The code weight structure needed in (1-28) is obtained by inspection of the list of code words as $n_3 = 2$, $n_4 = 1$, and $n_1 = n_2 = n_5 = 0$. Then the union bound is

$$P_s \leq 2 \sum_{i=2}^{3} \binom{3}{i} p^i q^{3-i} + \frac{1}{2} \binom{4}{2} p^2 q^2 + \sum_{i=3}^{4} \binom{4}{i} p^i q^{4-i}$$

$$= 3p^2(2q + q^2) + 2p^3(1 + 2q) + p^4$$

Note that at high signal-to-noise ratios ($p \to 0$ and $q \to 1$) the exact expression for P_s is closely approximated by $8p^2$ while the union bound is approximately $9p^2$. In addition to providing a close performance estimate as $p \to 0$ (and hence as $P_s \to 0$), union bound estimates are also quite useful at more interesting values (e.g., at $P_s = 10^{-5}$). Now suppose that the same code is used for error detection only. In this case the probability of undetected sequence error can be computed from (1-32) as

$$P_{s_{ud}} = 2p^3 q^2 + p^4 q$$

Obviously, the undetected error rate for error detection is substantially better than that for complete decoding. Note that as $p \to 0$ the undetected error rate is approximated by $8p^2$ for complete decoding compared with $2p^3$ for error detection.

1.3.4. Soft Decision Decoding

In a coded system one would normally wish to structure the decoder such that for a given code the probability of error is minimized. In the hard decision case we found that an optimum procedure was to decode by picking the code word which differs from the received sequence in the smallest number of positions. That is, we choose the code word which minimizes the "distance" between the code word and the received sequence. This is the maximum-likelihood decoder, and it generalizes easily to the soft decision case. The trick is to define a suitable distance function to utilize soft decisions.

It is instructive to consider the decoding problem from a slightly different viewpoint. Let us assume that we are using binary antipodal signaling on an unquantized Gaussian channel, and that an (n, k) block code is employed. The demodulator presents the decoder with a sequence of n numbers which represent the actual sample voltages from the matched filter. Let ρ be the actual received sequence and let $S_l(l = 0, \ldots, 2^k - 1)$ be the lth potential transmitted code word as it would be received in the absence of noise. An optimum decoder would choose the S_l which maximizes $\Pr(S_l | \rho)$, where $\Pr(S_l | \rho)$ is the conditional probability that S_l is the actual transmitted sequence given that ρ has been received. Invoking Bayes' rule, we write

$$\Pr(\mathbf{S}_l \mid \boldsymbol{\rho}) = \frac{p(\mathbf{S}_l, \boldsymbol{\rho})}{p(\boldsymbol{\rho})} = p(\boldsymbol{\rho} \mid \mathbf{S}_l) \frac{\Pr(\mathbf{S}_l)}{p(\boldsymbol{\rho})}$$

If we assume that all messages are equally likely, then maximizing $\Pr(\mathbf{S}_l \mid \boldsymbol{\rho})$ is equivalent to maximizing $p(\boldsymbol{\rho} \mid \mathbf{S}_l)$. Since the noise affecting each symbol is independent and Gaussian with zero mean and variance $\sigma^2 = N_0/2$ [assuming $\int_{-\infty}^{\infty} h^2(t)\, dt = 1$], we may write $p(\boldsymbol{\rho} \mid \mathbf{S}_l)$ as the product of n Gaussian density functions as follows:

$$p(\boldsymbol{\rho} \mid \mathbf{S}_l) = \prod_{i=1}^{n} \left[\frac{1}{(\pi N_0)^{1/2}} \right] \exp\left[-(\rho_i - S_{li})^2 / N_0 \right]$$

$$= \left[\frac{1}{(\pi N_0)^{1/2}} \right]^n \exp\left[-\sum_{i=1}^{n} (\rho_i - S_{li})^2 / N_0 \right]$$

where S_{li} are the components of $\mathbf{S}_l$ and are $\pm(E_s^{1/2})$. This expression is maximized when

$$d^2 = \sum_{i=1}^{n} (\rho_i - S_{li})^2 \tag{1-36}$$

is minimized. This summation is simply the Euclidian distance between the hypothesized sequence and the received signal. Thus the concept of distance that we have discussed previously generalizes to the unquantized channel. One could in principle, at least, describe the decoder structure by a continuous version of the decoding table of Fig. 1-6 (i.e., the columns would become regions in an n-dimensional space) using the d given by (1-36). The utility of this formulation is that the distance function is a simple addition of numbers rather than a product of exponential functions. This has certain engineering value, as will become evident when we examine the details of various decoder implementations.

In practical communication systems, we rarely have the luxury of being able to process the actual analog voltages ρ_i. The normal practice is to quantize these voltages and to make available a set of numbers which represent the quantization level in which the received voltage occurred. If the quantization is fine enough, this obviously does not differ from the continuous case. From an engineering viewpoint, however, it is desirable to make the quantization relatively coarse. This minimizes the cost of the analog-to-digital converter and also the number of bits that are required to represent each quantization level. Other than binary, eight-level quantization is the most commonly used scheme. This channel was previously discussed in Section 1.2.7.

We recall that the binary input, M-ary output channel that results from quantizing the matched filter output is completely characterized by a set of transition probabilities $\Pr(j|x)$, where $\Pr(j|x)$ is the probability that the output voltage falls in level j given that the input is x. In the usual case the channel is symmetric and $\Pr(j|0) = \Pr(Q - 1 - j|1)$.

We define a distance function for this type of channel as follows. If a sequence $\mathbf{S}$ of n symbols is transmitted, then the probability of a particular received sequence $\boldsymbol{\rho}$ is written as the product of the appropriate transition probabilities. Thus,

$$\Pr(\boldsymbol{\rho}|\mathbf{S}_l) = \prod_{i=1}^{n} \Pr(\rho_i | S_{li})$$

Taking the logarithm of both sides we have

$$\log \Pr(\boldsymbol{\rho}|\mathbf{S}_l) = \sum_{i=1}^{n} \log \Pr(\rho_i | S_{li}) \tag{1-37}$$

As in our previous discussion, we note that the decoder will seek to maximize $\Pr(\boldsymbol{\rho}|\mathbf{S}_l)$, and that this function will be maximized whenever the negative sum occurring on the right-hand side of (1-37) is minimized. The right-hand side of (1-37) clearly has the properties we are seeking. Using (1-37) as a guide, we may now define a related function which is somewhat more convenient for computation. Let the *symbol metric* m_j be defined as

$$m_j = -A - B \log \Pr(j|0) \tag{1-38}$$

where $\Pr(j|0)$ is the transition probability given that a zero was transmitted. We normally choose the constants A and B such that the minimum m_j is zero and all other values of m_j lie in some convenient positive range. The use of A and B enables us to define a distance function that is more like our usual notion of distance in that it is never negative and is zero whenever the transmitted and received sequences "match." Different values of B simply correspond to taking different bases for the logarithm and thus provide a convenient means of adjusting the scale of the metric. In the case of the hard decision channel A and B can be adjusted so that the metric values are 0 and 1. Thus, the metric corresponds exactly to the number of errors, and choosing the code word which contains the smallest number of errors is indeed an optimum strategy.

One may observe that the ability to define a metric is independent of the underlying noise process. All that is required is that a set of transition probabilities be specified and that the probability of error be independent from symbol to symbol. Thus the notion of "metric" or distance is quite general and applies to a very broad class of communication channels.

One disadvantage to the use of the metric functions defined above is that they require that we know the transition probabilities. Further, in an optimum decoder we would have to adjust the metric values as the signal-to-noise ratio changed. An engineering solution to this problem is to pick a fixed set of metric values that can be easily specified and represent a good compromise over many different signal-to-noise ratios. A scheme that is often used is to let $m_j = j$. In an eight-level scheme, the metrics would assume the integer values 0 through 7. Surprisingly, this is an excellent approximation in many practical decoders and causes only a small degradation in performance over a broad operating range. This point is explored more thoroughly in the next section.

Using metrics defined in this fashion a soft decision maximum-likelihood decoder can be implemented. The decoder selects the code word at minimum Euclidean distance [as given by (1-36)] from the received signal. Performance of this decoder can be estimated using the union bound. Assume that coherent PSK is used with an unquantized demodulator output. Then to apply the union bound we must calculate $\Pr(B_j'' \,|\, A_0)$, the probability that the distance between the received sequence and a weight-j code word is less than the distance to the transmitted all-zero word. This probability is a function only of E_s/N_0 and the number of positions in which they differ. This is easily seen by observing that the two code words are of the form

$$(000 \cdots 000)$$

and

$$(\underbrace{11 \cdots 1}_{j} \; \underbrace{0 \cdots 00}_{n-j})$$

Obviously, the $n - j$ positions in which both code words are zero have the same contribution in the calculation of Euclidean distances. Thus, these positions do not affect probability of error, and $\Pr(B_j'' \,|\, A_0)$ is the probability of error in deciding between two possible transmitted words, either j zeros or j ones. In summing the demodulator output voltages for these j positions, we obtain a new random variable which has a mean of $\pm j E_s^{1/2}$ depending on whether a 1 or 0 was transmitted and a variance of $j N_0/2$. Since the demodulator is coherent and unquantized, this probability of error is identical to (1-8) but with a factor of j more energy. Thus,

$$\Pr(B_j'' \,|\, A_0) = Q\left[(2j E_s/N_0)^{1/2}\right] \tag{1-39}$$

and the union bound on sequence error probability is

$$P_s \le \sum_{j=1}^{n} n_j Q\left[(2jE_s/N_0)^{1/2}\right]$$

$$= \sum_{j=1}^{n} n_j Q\left[(2jRE_b/N_0)^{1/2}\right] \tag{1-40}$$

Note that the parameter $E_b = E_s/R$ is the received energy per information symbol. In a similar fashion the average bit error probability is bounded by

$$P_b \le \sum_{j=1}^{n} \frac{\delta_{0j}'}{k} n_j Q\left[(2jRE_b/N_0)^{1/2}\right] \tag{1-41}$$

1.3.5. Quantization Effects

We are now in a position to examine the effects of both quantization and choice of metric values on decoder performance. We recall that the union bound of Section 1.3.2 is a weighted sum of the probabilities of error when each individual choice is between the all-zero sequence and a weight-j sequence. This suggests that one might develop considerable intuition regarding the performance of decoders in general by solving a set of simple problems.

Suppose we have a communication system in which we transmit one of two equally likely sequences. One sequence consists of seven zeros, and the other sequence consists of seven ones. The receiving system is an optimum symbol detector (matched filter), followed by a quantizer, followed by a sequence decoder. We assume that the system can operate in one of several different modes. In mode (1) the decoder operates with the actual analog voltages from the matched filter (usually referred to as either unquantized or infinitely quantized). In mode (2) the decoder operates with eight-level quantized decisions where the quantizer thresholds are chosen to be uniformly spaced Δ volts apart as in Fig. 1-3 with $\Delta = (2/7) E_s^{1/2}$. The decoder first computes an optimum metric using (1-38), and then selects the sequence for which the sum of the metrics is smallest. In mode (3) the decoder operates with eight-level decisions as in mode (2), but in this case the uniform metric values 0 through 7 are used. In modes (4) and (5) the decoder operates with four-level decisions, where the spacing between thresholds is now $\Delta = (4/7) E_s^{1/2}$. The decoder assigns optimum metrics for mode (4) and linearly spaced metric values 0 through 3 for mode (5). In modes (6) and (7) we go to a three-level quantizer with the middle zone symmetrical about the origin. The middle zone width is $(4/7) E_s^{1/2}$ for mode (6) and $(8/7) E_s^{1/2}$ for mode (7). Finally, in mode (8) binary quantization is achieved by retaining only the quantizer threshold at 0. For each mode it is desired to compute the probability of error as a function of signal-to-noise ratio.

In mode (1) there is no information lost in the quantizer and the system is indistinguishable from one in which there is a filter matched to each sequence. Then, as in (1-39), the probability of error will be the same as making an individual symbol decision with 7 times more energy and

$$P_e = Q\left[\left(\frac{2 \cdot 7 E_s}{N_0}\right)^{1/2}\right]$$

In modes (2) through (8) it is first necessary to compute the individual transition probabilities and then compute the probability that the sum of the metrics associated with the erroneous sequence exceeds the sum of the metrics associated with the correct sequence. The transition probability $\Pr(j|0)$ is given by

$$\Pr(j|0) = \int_{l_j}^{l_{j+1}} \frac{1}{(\pi N_0)^{1/2}} \exp\left[-(\beta + E_s^{1/2})^2/N_0\right] d\beta$$

where the thresholds on either end (l_0 and l_Q) are $-\infty$ and $+\infty$. These transition probabilities are also symmetric so that for a Q-level scheme

$$\Pr(j|0) = \Pr(Q - 1 - j|1)$$

For ease of notation we define $P_j = \Pr(j|0)$.

The metric value (m_0, m_1, etc.) can be regarded as a random variable and for modes (2) and (3) has the discrete density function shown in Fig. 1-7. In order to compute the probability that the sum of the metrics for the "zeros" sequence exceeds the sum of the metrics for the "ones" sequence, it is advantageous to form the density function for the difference in metric values for the two sequences as shown in Fig. 1-8. The probability of error is then computed by convolving this density function with itself 6 times and then integrating (or summing) over the positive half. At this point one can foresee a computational problem. Unless the metric values are integers, the number of discrete terms involved in the computation increases by a

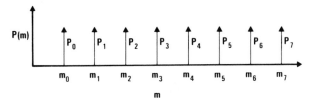

Figure 1-7. Probability density function for the metric value for the eight-level discrete channel.

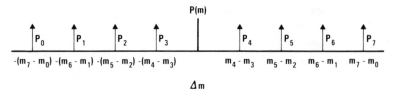

Figure 1-8. Probability density function for the metric differences for the eight-level discrete channel.

factor of 8 following each convolution. Thus, after the sixth convolution, one has a total of 8^7 discrete terms. For integer metric values many of the terms in each partial computation are summed together so that the number of discrete terms grows at a much smaller rate. A way around this difficulty is to compute the actual metric differences and then scale them to a range where they can be closely approximated by integers. For example, for $E_s/N_0 = 0.0$ dB the transition probabilities are given in Table 1-3. Also shown is the $\log(P_j/P_{Q-1-j})$ and an appropriately scaled version of the $\log(P_j/P_{Q-1-j})$ which can be rounded to the nearest integer and taken to be the metric differences $m_j - m_{Q-1-j}$. In this case the total number of terms in the sixfold convolution can never exceed $53 \times 7 = 371$. This is a much more manageable number than 8^7. Also, we observe that the terms associated with the center values are in the ratio of 1, 3, and 5 and are precisely the same as would have been produced by a linear metric. Only the end value is different. That is, the metric assignment 0, 1, 2, 3, 4, 5, 6, and 8.67 will produce this set of differences and is therefore nearly optimum.

One may also observe a very important property of metrics. Namely, that the performance is affected only by the difference in symmetrically spaced metric values and not by the individual metric values themselves. Thus, the three metric assignments $\{0, 1, 2, 3, 4, 5, 6, 7\}$, $\{0, 0, 1, 1, 2, 4, 5, 7\}$, and $\{0, 0, 0, 0, 1, 3, 5, 7\}$ will all produce an identical result when used with a decoder. This property can sometimes be exploited in order to simplify the actual decoder hardware. It is also interesting to note that for the three-

Table 1-3. Optimum Metric Computation for Eight-Level Channel at $E_s/N_0 = 0.0$ dB

j	P_j	$\log(P_j/P_{Q-1-j})$	$k\log(P_j/P_{Q-1-j})$	Δm
0	0.58	2.1299	26.09	26
1	0.148	1.2307	15.075	15
2	0.115	0.7384	9.045	9
3	0.0784	0.2449	3.00	3
4	0.0446	−0.2449	−3.00	−3
5	0.021	−0.7384	−9.045	−9
6	0.0087	−1.2307	−15.075	−15
7	0.0043	−2.1299	−26.09	−26

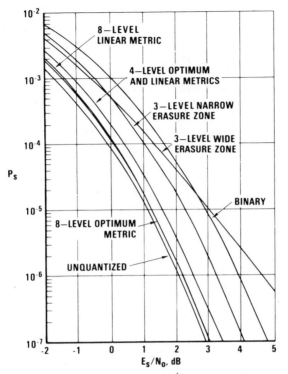

Figure 1-9. Probability of sequence error versus E_s/N_0 for a weight-7 sequence with various combinations of metric assignment and quantization.

level case any metric assignment in which the metric values are different is optimum and, therefore, the linear assignment $\{0, 1, 2\}$ is as good as any.

A set of performance curves for the eight different modes is shown in Fig. 1-9. One may make several interesting observations. First of all there is almost no advantage to using an optimum metric assignment as opposed to a linear assignment for the signal-to-noise ratios that are shown. Second, there is only a small difference between the eight-level quantization scheme and the unquantized case. Hence, it is not useful to quantize to more than eight levels. Further, on the eight-level channel there is little to be gained by going to a nonuniform spacing between levels or by changing the value of Δ relative to the signal mean. In the case of the three-level channel, a narrow erasure zone is better than a wide erasure zone except at very low signal-to-noise ratios. In fact, at high signal-to-noise ratios the choice of a wide erasure zone actually causes the performance to degrade over that obtainable with binary quantization. These results are qualitatively the same for sequences which differ in other than seven places. Thus, the curves are a good qualitative indication of the behavior of any decoding scheme whose performance is accurately predicted by a union bound.

1.3.6. Effect of Noncoherent Combining

We noted that in an unquantized coherent system one can obtain the effect of an increase in E_s by a factor of j by transmitting the same symbol j times. One might wonder if the same effect is observed in a noncoherent system. Unfortunately, the answer is no. The probability $\Pr[B''_j | A_0]$ can be calculated for binary orthogonal signaling and noncoherent detection. This probability is equal to the probability that the sum of j samples from the signal-absent envelope detector [with $p_n(\rho)$ given by (1-11)] is greater than the sum of j samples from the signal-present detector [with $p_{s+n}(\rho)$ given by (1-10)]. This probability has been derived by Marcum[7] as

$$\Pr[B''_j | A_0] = \frac{e^{-\rho/2}}{2} \sum_{i=0}^{j-1} \frac{(\rho/2)^i}{i!(j+i-1)!} \sum_{k=i}^{j-1} \frac{(k+j-1)!}{(k-i)!\,2^{k+j-1}} \quad (1\text{-}42)$$

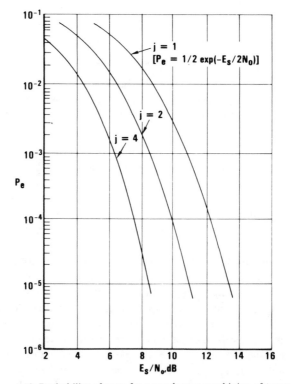

Figure 1-10. Probability of error for noncoherent combining of j samples.

where

$$\rho = \frac{jE_s}{N_0} \tag{1-43}$$

This result is shown in Fig. 1-10 for $j = 1$, 2, and 4. The reader should observe that when the number of samples is increased from 1 to 4 the value of E_s/N_0 can only be decreased by approximately 4.6 dB (at $P_e = 10^{-5}$) and still provide the same probability of error. This is in contrast to a coherent system in which increasing the number of samples from 1 to 4 would permit a 6-dB decrease in E_s/N_0. This loss of efficiency in combining multiple transmissions in a noncoherent system leads to different conclusions about how coding should be applied in such a system when compared with a coherent system. This aspect is explored in succeeding chapters.

1.3.7. Coding Gain

The usual figure of merit for a communication system is the ratio of energy per information symbol to noise spectral density (E_b/N_0) that is required to achieve a given probability of error. The term "coding gain" describes the amount of improvement that is achieved when a particular coding scheme is used. The usual method of determining coding gain is to plot the probability of error versus E_b/N_0 for both coded and uncoded operations and to read the difference in required E_b/N_0 at a specified error rate. As an example, the performance of the well-known (23, 12) Golay code (to be discussed in Section 2.3.2) with hard decision decoding is shown in Fig. 1-11. On the same plot we show the performance of the ideal demodulator operating alone. The coding gain at $P_e = 10^{-5}$ is 2.15 dB and at $P_e = 10^{-3}$ is 1.35 dB. At sufficiently low signal-to-noise ratios, one may observe that the coding gain actually becomes negative. This thresholding phenomenon is common to all coding schemes. There will always exist a signal-to-noise ratio at which the code loses its effectiveness and actually makes the situation worse.

Coding gain is a useful concept only in those situations where it is meaningful to obtain performance improvements by increasing the power. For instance, if one has a communication link that is subject to random erasures, then at high signal-to-noise ratios there is a floor on performance that cannot be overcome by increasing power. A coding scheme, however, might significantly reduce this floor or make it disappear entirely. One might be tempted to say that the coding gain is infinite. In fact, however, the desired performance could never have been obtained without coding, and such a statement would be meaningless.

Asymptotic coding gain is a quantity that is sometimes used as a figure of merit for a particular code. It depends only on the code rate and

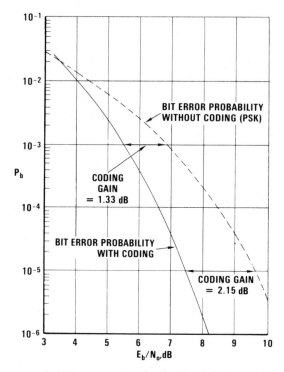

Figure 1-11. Bit error probability versus E_b/N_0 for the (23, 12) Golay code with hard decision decoding.

minimum distance and can be defined for both an unquantized channel and a binary quantized channel.

At high signal-to-noise ratios the probability of bit error for a rate R, t error-correcting code used with a binary quantized PSK demodulator is

$$P_b \approx K p^{t+1}$$
$$\approx K \left\{ Q \left[\left(\frac{2RE_b}{N_0} \right)^{1/2} \right] \right\}^{t+1} \tag{1-44}$$

An uncoded transmission will have a bit error rate of

$$P_b' = Q \left[\left(\frac{2E_b'}{N_0} \right)^{1/2} \right] \tag{1-45}$$

At high signal-to-noise ratios the Q function can be overbounded and approximated by

$$Q(\alpha) \le e^{-\alpha^2/2} \tag{1-46}$$

One may solve for the change in signal energy that is required to produce the same error rate by equating (1-44) and (1-45) and making use of (1-46). Thus,

$$K \left[\exp\left(-\frac{2RE_b}{N_0} \frac{1}{2} \right) \right]^{t+1} = \exp\left(-\frac{2E_b'}{N_0} \frac{1}{2} \right)$$

Taking the log of both sides and noting that $\log K$ is negligible for large E_b/N_0, we obtain

$$\frac{E_b'}{E_b} = R(t + 1)$$

Using this we define the asymptotic coding gain as

$$G_a = 10 \log \left[R(t + 1) \right]$$

Thus, a two-error-correcting rate-$1/2$ code has an asymptotic coding gain of 1.7 dB. One should note that this gain is achieved only in the limit as $E_b/N_0 \to \infty$. At modest signal-to-noise ratios the actual gain may be significantly less.

Using a similar argument one may show that for the unquantized channel (see Problems)

$$G_a = 10 \log \left[Rd \right]$$

where d is the minimum distance of the code. Although this result indicates that soft decision decoding is about 3 dB more efficient than hard decision decoding at very high E_b/N_0, a figure of 2 dB is more common at realistic signal-to-noise ratios.

1.4. Coding Bounds

A significant percentage of all the available coding literature deals with coding bounds in one form or another. There are essentially two kinds of coding bounds. These are bounds on minimum distance and bounds on performance. The bounds on minimum distance are obtained by examining certain of the structural aspects of codes. The principal useful bounds of this type are the *Hamming bound* (which was discussed earlier), the *Plotkin bound*, and the *Gilbert bound*. The Hamming and Plotkin bounds indicate the maximum possible minimum distance for a given code length and code rate, while the Gilbert bound is an achievable bound and thus provides a lower bound on the minimum distance of the "best" code. These bounds on minimum distance are often used when attempting to find new codes to indicate how close one is to the best that is possible.

The bounds on performance are all based on random coding arguments. They indicate that the average performance of all block codes exhibits a probability of error that decreases exponentially with code length. This, of course, implies the existence of specific codes which do better than average. Although random coding bounds show that it is possible to drive the error rate to zero in a manner that decreases exponentially with n they do not provide information useful in selecting such a code. Also, these bounds are not very useful for estimating the absolute performance of a code since good codes exhibit a probability of error that is considerably better than that predicted by the bound. An area where the bounds can be quite useful is in predicting the relative performance changes that occur when the noise process or channel model changes.

1.4.1. Bounds on Minimum Distance

In this section we will state each of the minimum distance bounds and discuss the conditions under which they are valid. No attempt will be made to provide rigorous proofs, which in some cases depend upon properties to be developed in Chapter 2. Rather, the goal will be to impart an understanding of what the bounds tell us and why they are applicable. The reader who is interested in a detailed discussion of the proofs and an exposition on some related bounds is referred to the excellent texts by Peterson and Weldon,[1] Berlekamp,[2] Gallager,[3] and MacWilliams and Sloane.[8]

The Hamming bound has already been discussed in connection with the construction of the decoding table. It can be stated as in (1-4), which expresses the maximum number of code words that can be provided for a given n and t or it can be expressed in terms of the amount of redundancy that is required for a fixed value of k and t as

$$n - k \geq \log_2 \left[\sum_{i=0}^{t} \binom{n}{i} \right]$$

For nonbinary codes this bound generalizes to

$$n - k \geq \log_q \left[\sum_{i=0}^{t} \binom{n}{i} (q - 1)^i \right] \qquad (1\text{-}47)$$

where q is the number of symbols in the code alphabet.

The Plotkin bound is also an upper bound on the minimum distance that can be achieved for a given n and k. The principle of the Plotkin bound is that the minimum weight code word of a group code is no larger than the average weight of all nonzero code words. It can be shown that if all the code words in an (n, k) binary group code are arranged as rows in a matrix, each nonzero element appears 2^{k-1} times in each column. With n

columns in the matrix, the sum of the weights of all code words in an (n, k) linear code is $n2^{k-1}$. Since there are $2^k - 1$ nonzero code words and since the minimum weight code word cannot have weight greater than the average weight, the following inequality holds:

$$d \le \frac{n2^{k-1}}{2^k - 1} \tag{1-48}$$

For nonbinary codes, the Plotkin bound generalizes to

$$d \le \frac{n(q-1)q^{k-1}}{q^k - 1} \tag{1-49}$$

Equations (1-48) and (1-49) are the simplest statements of the Plotkin bound. Although they readily provide the maximum possible value of d for a given n and k, they do not readily provide the maximum value of k for a given value of d and n. An alternate form of the Plotkin bound can be written

$$k \le n - 2d + 2 + \log_2 d \tag{1-50}$$

for the binary case, and

$$k \le n - \frac{qd - 1}{q - 1} + 1 + \log_q d \tag{1-51}$$

for the nonbinary case.

The Hamming bound tends to provide the tightest bounds for high-rate codes (i.e., large k/n) while the Plotkin bound is more appropriate for low-rate codes.

The Gilbert bound (also called the Varsharmov–Gilbert bound) states that it is always possible to find an (n, k) code with minimum distance at least d with at most $n-k$ parity checks provided

$$2^{n-k} > \sum_{i=0}^{d-2} \binom{n-1}{i} \tag{1-52}$$

A nonbinary version of the Gilbert bound can also be written as

$$q^{n-k} > \sum_{i=0}^{d-2} \binom{n-1}{i}(q-1)^i \tag{1-53}$$

An example of how these bounds might be used is as follows. Suppose it is desired to find a length-63 code with minimum distance of 5 and the largest possible value of k. A (63, 51) BCH code is an example of an algebraically constructed code with $d = 5$. (This class of codes will be discussed in Section 2.3.1.) The Hamming and Gilbert bounds will be used to estimate the "goodness" of this code relative to this measure. The Hamming bound states that

$$\sum_{i=0}^{2} \binom{63}{i} \leq 2^{n-k} \tag{1-54}$$

or

$$2017 \leq 2^{n-k}$$

Thus, the smallest number of parity checks for which (1-54) holds is 11. The Gilbert bound states that

$$\sum_{i=0}^{3} \binom{62}{i} > 2^{n-k} \tag{1-55}$$

or

$$39774 > 2^{n-k}$$

The smallest number of parity checks for which (1-55) holds is 16. Thus, the Hamming bound insures that no codes exist with $n-k < 11$ while the Gilbert bound guarantees the existence of codes with $n-k \leq 16$. Therefore, one must conclude that the (63,51) code is a good code and little improvement could be obtained by further searching.

Table 1-4 provides upper and lower bounds on k for different values of n and d. The values of k_{max} were obtained using the Hamming bound except those values marked with an asterisk where the Plotkin bound provides a tighter value. Lower bounds are obtained using the Gilbert bound. Also included in this table are parameters associated with some known constructable BCH codes.

Table 1-4. Maximum, Minimum, and Demonstratable Values of k for Various Length-31 Codes[a]

n	d	k_{min}	k_{max}	k_{BCH}
31	3	26	26	26
	5	19	22	21
	7	14	18	16
	9	10	15	—
	11	8	13	11
	13	4	10*	—
	15	3	6*	6
	17	1	3*	—

[a] Asterisk indicates values where the Plotkin bound provides a tighter value.

1.4.2. Random Coding Bound

The random coding bound is computed by calculating the average sequence error rate over the ensemble of all possible block codes with a specified rate and length. We will briefly outline the derivation of this bound for binary block codes utilizing PSK signaling on an unquantized additive Gaussian noise channel.

Consider the ensemble of all possible binary block codes of length n and rate $R = k/n$. Each code has $M = 2^k$ code words, and there are a total of 2^{Mn} possible codes. Each code has a set of code words $\{x_1, x_2, ..., x_M\}$ and an associated probability of sequence error, $P_s(x_1, x_2, ..., x_M)$. The average sequence error probability over the ensemble of all codes is

$$\bar{P}_s = 2^{-Mn} \sum_{x_1} \sum_{x_2} \cdots \sum_{x_M} P_s(x_1, x_2, ..., x_M) \tag{1-56}$$

where the sums range over all possible choices for the code words $x_1, x_2, ..., x_M$. (This ensemble of codes will include some very poor codes, i.e., those for which not all code words are distinct. Nevertheless, this technique provides some very useful insights into the fundamental behavior of coding schemes.)

Define the function $A_j(\{x_1, x_2, ..., x_M\} | x_i)$ as the total number of code words in the code $\{x_1, x_2, ..., x_M\}$ at distance j from the transmitted code word x_i. Then the average sequence error probability for any code may be written

$$P_s(x_1, x_2, ..., x_M) = \sum_{i=1}^{M} P_s(\{x_1, x_2, ..., x_M\} | x_i) \Pr(x_i)$$

$$= \sum_{i=1}^{M} \sum_{j=0}^{n} \frac{1}{M} A_j(\{x_1, x_2, ..., x_M\} | x_i) Q[(2jRE_b/N_0)^{1/2}]$$

$$\leq \sum_{i=1}^{M} \frac{1}{M} \sum_{j=0}^{n} A_j(\{x_1, x_2, ..., x_M\} | x_i) e^{-jRE_b/N_0} \tag{1-57}$$

where the bound $Q(\alpha) \leq e^{-\alpha^2/2}$ was used.

Now substituting (1-57) into (1-56) and interchanging summations we find that

$$\bar{P}_s \leq \sum_{j=0}^{n} e^{-jRE_b/N_0} \sum_{i=1}^{M} \frac{1}{M} \left[2^{-Mn} \sum_{x_1} \sum_{x_2} \cdots \sum_{x_M} A_j(\{x_1, x_2, ..., x_M\} | x_i) \right] \tag{1-58}$$

Since the term in the square brackets is averaged over all possible code words, it represents the "average" number of code words at distance j from the transmitted code word. This number is independent of $\mathbf{x}_i$ and is $(M - 1)\binom{n}{j}2^{-n}$. Thus,

$$
\begin{aligned}
\bar{P}_s &\le \sum_{j=0}^{n}(M - 1)\binom{n}{j}2^{-n}e^{-jRE_b/N_0} \\
&< 2^{-(n-k)}\sum_{j=0}^{n}\binom{n}{j}e^{-jRE_b/N_0} \\
&= 2^{-(n-k)}(1 + e^{-RE_b/N_0})^n
\end{aligned}
\tag{1-59}
$$

Defining the *exponential bound parameter* R_0 as

$$
R_0 = 1 - \log_2(1 + e^{-RE_b/N_0})
\tag{1-60}
$$

then the average sequence error probability may be bounded by

$$
\bar{P}_s < 2^{-n(R_0 - R)}
\tag{1-61}
$$

The parameter R_0 is also called the computational cutoff rate and denoted by R_{comp}. (This parameter is significant in the theory of sequential decoding as discussed in Chapter 7.) The R_0 parameter may be found in a similar fashion for discrete memoryless channels. Its value for these channels will be given shortly.

The bound (1-61) shows that as long as $R_0 > R$ it is possible to drive the error rate to zero provided that the code is long enough. Since this is the average performance of all possible codes, there must exist specific

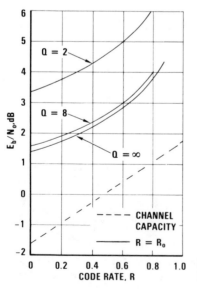

Figure 1-12. E_b/N_0 required to achieve $R = R_0$ for binary signaling.

schemes which will do even better. By letting $R = R_0 - \varepsilon$ with ε very small, we can show a bound on achievable coding gain. The result is shown in Fig. 1-12. This plot indicates the minimum possible value of E_b/N_0 for a given $R = k/n$ for which (1-61) can still be driven to zero. Since one can obtain a zero probability of error in the limit as $n \to \infty$, it follows that there is a finite value of n for which the probability of error can be any specified value. Hence, we have a bound on achievable coding gain. For example, it is known that with ideal PSK signaling and no coding an E_b/N_0 of 9.6 dB is required to produce a 10^{-5} error rate. Figure 1-12 shows that one can obtain an arbitrarily low error rate at 2.45 dB for an $R = 1/2$ code. Thus it is possible to obtain a coding gain of at least 7.15 dB at $P_s = 10^{-5}$ using an $R = 1/2$ code. In addition, observe that as the code rate goes to zero, the achievable coding gain approaches 8.2 dB. Thus we can argue that using codes in which the rate is significantly less than $1/2$ can only provide marginal improvements in performance. Finally, we note that as the code rate increases it becomes progressively harder to achieve coding gain with binary signaling.

No discussion of coding bounds would be complete without mentioning channel capacity. Using similar although considerably more involved arguments, one can show that in the limit of very long code lengths, it is possible to communicate with arbitrarily low error rates provided the code rate does not exceed the channel capacity C, which is given by

$$C = \frac{1}{2} \log_2 \left(1 + 2R \frac{E_b}{N_0} \right)$$

Further, it can also be shown that the error rate can never be made arbitrarily small for any code whose rate exceeds C. Thus, channel capacity provides an absolute limit on the minimum value of E_b/N_0 that is required to communicate with error-free or near-error-free performance. For comparison we have also shown the channel capacity bound in Fig. 1-12. From this we see that it might be possible to achieve an additional 2.4 dB of coding gain over what is predicted using R_0 at $R = 1/2$. To date, however, coding schemes which operate in this region have a high degree of complexity.

The bounding techniques used to derive (1-61) are somewhat crude in that they involve only the union bound. Using more sophisticated techniques one can bound the average sequence error rate over the ensemble of all codes by

$$\bar{P}_s \leq 2^{-nE(R)} \qquad (1\text{-}62)$$

where for a discrete memoryless channel the *error exponent*, $E(R)$, is given by

$$E(R) = \max_{0 \leq \rho \leq 1} \max_{\mathbf{Q}} \left[E_0(\rho, \mathbf{Q}) - \rho R \right] \qquad (1\text{-}63)$$

The so-called "Gallager function" is defined by

$$E_0(\rho, \mathbf{Q}) = -\log_2 \left\{ \sum_{j=0}^{J-1} \left[\sum_{k=0}^{K-1} Q(k) \left[\Pr(j|k) \right]^{1/(1+\rho)} \right]^{(1+\rho)} \right\} \quad (1\text{-}64)$$

This function is defined for the discrete memoryless channel with K inputs, J outputs, and transition probabilities $\Pr(j|k)$. The $\mathbf{Q}$ vector is the probability assignment over the input alphabet, $(Q(0), Q(1), \ldots, Q(K-1))$. After maximizing over $\mathbf{Q}$, the resulting Gallager function is denoted by $E_0(\rho)$. For *symmetric channels* (which are the ones of most interest), the optimizing $\mathbf{Q}$ distribution is uniform, i.e., $Q(k) = 1/K$.

The error exponent, $E(R)$, is nonnegative at all rates less than channel capacity, C. Thus, this form of the random coding bound demonstrates that for any desired P_s, one can find a value of n for which a code exists with this error rate provided that $R < C$. The error exponent for a typical case is shown in Fig. 1-13. This case is a BSC with $p = 0.0375$. Also shown is the error exponent of the form in (1-61) which is $R_0 - R$. Obviously, the bounds (1-61) and (1-62) are identical at low rates, but the bound (1-62) is substantially better at high rates. It turns out that on any DMC, R_0 is equal to the zero-rate exponent of the channel, i.e.,

$$R_0 = E_0(1) \quad (1\text{-}65)$$

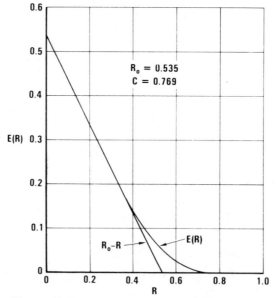

Figure 1-13. Error exponent at $E_s/N_0 = 2$ dB $(p = 0.0375)$.

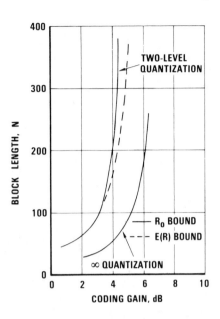

Figure 1-14. Coding gain predicted at $P_s = 10^{-5}$ by ensemble coding bounds for $R = 1/2$.

Therefore, for symmetric channels $[Q(k) = 1/K]$

$$R_0 = -\log_2 \left\{ \sum_{j=0}^{J-1} \left[\frac{1}{K} \sum_{k=0}^{K-1} [\Pr(j|k)]^{1/2} \right]^2 \right\} \tag{1-66}$$

which for the binary symmetric channel is

$$R_0 = 1 - \log_2 \{1 + 2[p(1-p)]^{1/2}\} \tag{1-67}$$

These expressions for R_0 were used in constructing the $Q = 2$ and $Q = 8$ curves in Fig. 1-12.

We can also use these bounds to compute the block length for which it is always possible to achieve a given coding gain. The result is shown in Fig. 1-14 for an $R = 1/2$ code. In this case the values of n are very conservative. For example, it is known that the (48, 24) quadratic residue code (to be discussed in Section 2.3.5) can provide a coding gain of approximately 5 dB with infinite quantization and that the (24, 12) Golay code can provide coding gains of approximately 2 dB and 4 dB with two-level and infinite quantization, respectively. The performance of both codes is significantly better than predicted by the bounds.

While these bounds may be conservative on an absolute basis, they are quite useful in comparing relative behavior. For example, the differences between the required E_b/N_0 for $Q = 2$, 8, and ∞ as predicted by Fig. 1-12 accurately reflect the differences that are typically observed in practice at

10^{-5} error rate. In addition, one can examine the R_0 parameter as a function of quantizer threshold settings (at the desired E_s/N_0 operating point) to find the optimum setting. Thresholds established in this manner exhibit near-optimum performance in a real system. Other examples of meaningful comparisons using these bounds are given in the succeeding chapters.

1.5. Remarks

We have observed that the two key ingredients of any coding scheme are redundancy and noise averaging. With linear block codes the redundancy is added by taking an input sequence of k information symbols and mapping it into a transmitted sequence of n symbols ($n > k$) which are linear combinations of the k information symbols. The mapping used defines the particular code. Several classes of good block codes are developed in the next chapter. Redundancy is added in a similar fashion for convolutional codes, and several classes of such codes are presented in Chapters 6 and 7. Noise averaging is accomplished by making the code block length sufficiently long. Some indication of achievable performance as a function of code block length and rate may be provided by the coding bounds presented previously, but direct performance analysis techniques are usually used for determining the required block length.

One of the principal topics addressed by this book is the formulation and implementation of decoding algorithms. The concept of maximum-likelihood decoding has already been introduced. This concept is important because maximum-likelihood decoders are used for several types of codes, and the performance of a maximum-likelihood decoder is often used as a benchmark with which to compare the performance of suboptimum decoding techniques. Frequently, one must use a code which is too long to be decoded by a maximum-likelihood decoder. Thus, many of the decoding techniques addressed in this book are suboptimum techniques chosen for their simplicity of implementation.

We also emphasize performance analysis of coding/decoding techniques in communication systems and various communication system applications. To this end, models for coherent and noncoherent communication systems were presented, and the effects of quantization to produce a discrete channel were discussed. In addition, a group of performance analysis techniques was developed which will be utilized throughout the book. Performance curves for many coding/decoding techniques are presented to facilitate comparisons among the various approaches.

The topics of code construction and coding bounds are treated in substantially more depth in several texts[1,2,3,8] than in this book. The classes of codes presented in this book are not all-inclusive, but we feel they are sufficient for an introduction to coding theory. Likewise, coding bounds are used sparingly in the remainder of the book. We feel that after developing an appreciation for the more practical aspects of coding theory, as presented in this book, the serious student could better understand the more difficult material in the classical texts.

Problems

1-1. Consider the code and decoding table of Table 1-1. Assume the code is used over a BSC with error rate p, and a complete decoding algorithm is used in decoding. Compute the probability of correct decision. Then, assume an incomplete decoding algorithm that classifies all received words that lie outside the dotted lines as deletions. Compute $P_{s_{ud}}$ and P_{s_d}.

1-2. Consider the $(5, 1)$ repetition code which consists of the two code words $0\,0\,0\,0\,0$ and $1\,1\,1\,1\,1$ corresponding to information bits of 0 and 1, respectively. Give the decoding table for this code. Is this a perfect code? What about the $(4, 1)$ repetition code? Determine all values of n for which $(n, 1)$ repetition codes are perfect. For the codes with n odd assume that all patterns of t or fewer errors are corrected $[t \le (n - 1)/2]$. Compute the probability of correct decision, P_{s_d}, and $P_{s_{ud}}$ as a function of p and t.

1-3. Assume a complete decoding algorithm for the $(5, 1)$ repetition code. Compute P_s exactly for this code and compare with a union bound.

1-4. Consider two sequences of length n, one with weight 0 and the other with weight j. Assuming a BSC with error rate p, compute the probability that if the all-zero sequence is transmitted, the received word is within distance t of the weight-j word. This result is given by (1-24).

1.5. Consider a block code with rate R and minimum distance d. Assuming an infinitely quantized matched filter output, find the asymptotic coding gain for PSK signaling over a Gaussian channel.

1-6. An $(n, n - 1)$ block code has a single overall parity symbol which is chosen to make each code word have even weight. This gives a code with $d = 2$. This code will allow correction of all single erasures. Assume a binary input, ternary output symmetric channel with a probability of erasure of s and a probability of error of p. Give a decoding algorithm that allows detection of single errors on all code words with no erasures, correction of all single erasures, and detection of all words with more than a single erasure. Find P_{s_d} and $P_{s_{ud}}$ for this algorithm.

1-7. For the binary input, ternary output channel of the previous problem, give the expression for the exponential bound parameter R_0 in terms of s and p. For PSK signaling with quantization thresholds at $\pm \alpha E_s^{1/2}$ the values of s and p are given by $p = Q[(1 + \alpha)(2E_s/N_0)^{1/2}]$ and $s = Q[(1 - \alpha)(2E_s/N_0)^{1/2}] - p$. Plot R_0 vs. $\alpha(E_s/N_0)^{1/2}$ to find the optimum value of α (for $E_s/N_0 = -5, 0, 5$, and 10 dB).

1-8. Consider the use of an $(n, 1)$ repetition code on the binary Z channel. This channel is unusual in that transmitted 1's cannot be received in error. Thus, the transition probabilities are:

$$\Pr(0|0) = 1 - p$$

$$\Pr(1|0) = p$$

$$\Pr(0|1) = 0$$

$$\Pr(1|1) = 1.$$

Find the proper metric to be used with a maximum likelihood decoder and give the decoding table. How many errors per code word can be corrected using this code? Calculate the sequence error rate.

2

Group Codes

The group codes constitute a vanishingly small percentage of all possible block codes. With very few exceptions, however, they are the only block codes of practical importance. They are often referred to by other names such as linear codes or generalized parity check codes. Within the set of all group codes there is a second major subdivision called polynomial generated codes. Examples of polynomial generated codes are the Bose–Chaudhuri–Hocquenghen (BCH) codes, Reed–Solomon codes, generalized Reed–Mueller codes, projective geometry codes, Euclidean geometry codes, and quadratic residue codes. Each of these classes of codes is described by a specific algorithm for constructing the code. The classes form overlapping sets so that a particular code may be a BCH code and also a residue code or it may be a generalized Reed–Mueller code and also a BCH code, etc. Polynomial generated codes are important for several reasons. First of all, the encoder can be implemented in hardware using a relatively simple feedback shift register. Second, this class contains many codes whose minimum distance is close to the best that can be found, especially for block lengths on the order of 100 or less. Third, there exist several decoding algorithms which enable one to decode certain of these codes using only moderate amounts of hardware.

In this chapter we will first develop some properties of group codes that are common to all members of the class. We will then concentrate principally on the polynomial generated codes. In order to discuss these codes it will be necessary to introduce certain elementary concepts from the theory of linear vector spaces and from the algebra of groups and fields. For the most part, the necessary mathematical properties will be introduced using an operational viewpoint and will be stated without proof. In lieu of proofs we will try to supply a sufficient number of examples to convince the average engineering reader of the validity of the particular property or operation.

This approach is not without its obvious pitfalls. It is the opinion of the authors, however, that one can gain a good manipulative understanding of coding theory using this approach. Afterwards the interested reader can go back and satisfy his curiosity regarding the "truth and beauty" aspects of the subject using any of several excellent texts.

We will assume that each code word in a group code can be divided into two portions. The first k-symbol portion is always identical to the information sequence to be transmitted. Each of the $n - k$ symbols in the second portion is computed by taking a linear combination of a predetermined subset of information symbols. Hence, they are referred to as generalized parity check symbols or, more simply, as parity symbols. Codes of this type in which the information symbols are transmitted in unaltered form were previously defined as *systematic* codes. It can be shown that any group code can be made to be systematic in some set of k positions by choosing the proper mapping between the input sequences and the code words. This point will become clearer as we proceed. Its significance is that we are not excluding any important group codes by considering only codes which are systematic. This does not imply that the average bit error rate is necessarily the same for both systematic and nonsystematic versions of the same code. Again, this will become clearer in the discussion which follows.

2.1. Generalized Parity Check Codes

A very simple binary group code is the $(n, n - 1)$ code that is formed by using a single overall parity check. A (4,3) code word, for example, would be written as a column vector $\mathbf{a}$, where

$$\mathbf{a}^T = (a_1, a_2, a_3, a_1 + a_2 + a_3)$$

the a_i's take on the values 0 or 1, and the symbol $+$ stands for modulo-2 addition. We observe that if a second code word

$$\mathbf{b}^T = (b_1, b_2, b_3, b_1 + b_2 + b_3)$$

is added symbol by symbol to the first code word, then the result is

$$\mathbf{c}^T = \mathbf{a}^T + \mathbf{b}^T$$
$$= (a_1 + b_1, a_2 + b_2, a_3 + b_3, a_1 + b_1 + a_2 + b_2 + a_3 + b_3)$$
$$= (c_1, c_2, c_3, c_1 + c_2 + c_3)$$

We note that the parity symbol in $\mathbf{c}$ is defined in exactly the same fashion as in $\mathbf{a}$ and $\mathbf{b}$. Consequently, $\mathbf{c}$ is also a code word. This illustrates what is probably the most essential characteristic of group codes. That is, the sum of any two code words is a code word. This property is called closure. The result extends in a trivial fashion to any group code.

As a second example, let us define a code word in a (6,3) code as

$$\mathbf{a}^T = (a_1, a_2, a_3, a_1 + a_2, a_2 + a_3, a_1 + a_2 + a_3)$$

Again, if a second code word $\mathbf{b}$ is added symbol by symbol to $\mathbf{a}$, the result is a word $\mathbf{c}$ with

$$c_1 = a_1 + b_1$$
$$c_2 = a_2 + b_2$$
$$c_3 = a_3 + b_3$$
$$c_4 = a_1 + b_1 + a_2 + b_2 = c_1 + c_2$$
$$c_5 = a_2 + b_2 + a_3 + b_3 = c_2 + c_3$$
$$c_6 = a_1 + b_1 + a_2 + b_2 + a_3 + b_3 = c_1 + c_2 + c_3$$

Thus, the three parity symbols in $\mathbf{c}$ are defined in precisely the same way as in $\mathbf{a}$ and $\mathbf{b}$ and consequently $\mathbf{c}$ is also a code word.

There are two important implications of this property. One is that it permits us to generate group codes in a simple fashion. The second is that it provides the mechanism for relating the distance properties of a group code to its weight structure. This feature greatly simplifies the problem of designing good group codes and, as we saw in Chapter 1, of computing their performance. For a moment we will postpone the problem of generating group codes and concentrate on their distance properties.

The *distance* $d(\mathbf{a}, \mathbf{b})$ between two code words $\mathbf{a}$ and $\mathbf{b}$ is defined as the number of positions in which they differ. The *weight* $w(\mathbf{c})$ of a code word is defined as the number of nonzero elements in the code word. It follows that if we combine two code words by modulo-2 addition of their elements, then the result will contain one's only in those positions where the code words are not the same. Thus, for any two code words $\mathbf{a}$ and $\mathbf{b}$,

$$d(\mathbf{a}, \mathbf{b}) = w(\mathbf{a} + \mathbf{b})$$

This result implies that the distance from a specific code word to all other code words is identically equal to the complete set of weights of all the code words. Another statement of the same phenomenon is that the distance between any two code words is the same as the distance from the all-zero code word to some other code word. Thus, if we are attempting to design a code with good minimum distance properties, it is sufficient to make sure that the weight of the nonzero words is as large as possible. Also, when computing the performance of a group code, it is sufficient to consider only the effects of transmitting the all-zero word, since the distances from from each of the other code words are identical.

The group codes are so named because the code words together with the operation symbol-by-symbol mod-2 addition, form a mathematical structure called a group. The essential features of a group are

1. the sum of any two elements in the group is always in the group (closure),
2. the associative law holds so that $(a + b) + c = a + (b + c)$,
3. the group always contains an identity element (all-zero element), and
4. for every element in the group there is an inverse element (in the binary code case each word is its own inverse) such that $a + (-a) = 0$.

It is obvious that the generalized parity check codes defined in this system satisfy all four of these properties. The example codes discussed in this chapter have been defined over the "field" of two elements $\{0,1\}$ using mod-2 arithmetic. However, nonbinary codes can be defined over a finite field of q elements $(q \neq 2)$. All of the principles that are being presented using binary codes also apply to these nonbinary codes. Arithmetic operations over nonbinary fields will be discussed in Section 2.2.1. An important class of nonbinary codes, the Reed–Solomon codes, will be presented in Section 2.3.

2.1.1. Parity Check Matrix

Rather than write out the parity check equations explicitly as was done in the example of the (6,3) code, it is often more convenient to use matrix notation. For example, the *parity check matrix*, **H**, defined below contains the same information as the sample code word **a** :

$$\mathbf{H} = \begin{bmatrix} 1 & 1 & 0 & 1 & 0 & 0 \\ 0 & 1 & 1 & 0 & 1 & 0 \\ 1 & 1 & 1 & 0 & 0 & 1 \end{bmatrix} \tag{2-1}$$

$$\mathbf{a}^T = (a_1, a_2, a_3, a_1 + a_2, a_2 + a_3, a_1 + a_2 + a_3)$$

Each column in the **H** matrix corresponds to a symbol in the code word. The first three columns correspond to the information symbols and the last three to the parity symbols. The first row expresses the relationship that the fourth symbol is the sum of a_1 and a_2. Likewise, the second row indicates that the fifth symbol is the sum of a_2 and a_3, etc. This matrix has been written in *echelon canonical form*. This means that the first k columns define the specific information symbols which enter into each parity equation while the last $n - k$ columns are always the identity matrix. As we shall see, other forms of the matrix are permissible and often useful.

We may formalize the above discussion by the statement that a sequence **a** is a code word if, and only if,

$$\mathbf{Ha} = \mathbf{0} \tag{2-2}$$

In this equation matrix multiplication is defined as usual, except that additions are modulo-2 with binary codes.

2.1.2. Minimum Distance Considerations

For any specific code word, (2-2) can be interpreted as requiring a subset of the columns of the **H** matrix to sum to zero. For instance, the sequence (1 0 0 1 0 1) is a code word in the (6,3) code in our previous example. In the matrix multiplication defined by (2-2), the nonzero elements of this sequence "sift" out the first, fourth, and sixth columns of **H**, viz.,

$$\begin{pmatrix} 1 \\ 0 \\ 1 \end{pmatrix} + \begin{pmatrix} 1 \\ 0 \\ 0 \end{pmatrix} + \begin{pmatrix} 0 \\ 0 \\ 1 \end{pmatrix} = \begin{pmatrix} 0 \\ 0 \\ 0 \end{pmatrix}$$

Since a relationship like this must hold for all code words, the following observation can be made. If the minimum weight and consequently the minimum distance of a code is d, then there must exist at least one subset of d columns of **H** that sum to zero. Further, there cannot exist any subsets containing $d - 1$ or fewer columns that also sum to zero. If we treat the columns of the **H** matrix as vectors, this is equivalent to saying that in a code of minimum distance d, all subsets of $d - 1$ columns of **H** must be linearly independent. This is one of the fundamental theorems of group codes. It provides a means of establishing the minimum distance for a specific group code when the **H** matrix is given. It also provides a means of constructing the **H** matrix to provide a guaranteed minimum distance. Examining the example (6,3) code, we see that all of the columns are distinct, and consequently no two can sum to zero. Further, there exists at least one set of three columns, 1, 2, and 5 for example, that sum to zero. This code must, therefore, have a minimum distance of 3.

The above property provides the mechanism by which the Gilbert bound was established in Chapter 1. Suppose we wish to construct an (n, k) code with minimum distance d. Choose the first two columns of the **H** matrix to be any set of ones and zeros. The only requirement is that they be different. Pick the third column so that it is not a linear combination of the first two. Continue in this manner choosing successive columns so that each new column is not a linear combination of any $d - 2$ or fewer previous columns. New columns may be added as long as the set of all linear combinations of $d - 2$ or fewer columns does not contain all possibilities. In the worst case every one of these linear combinations is a distinct nonzero element. Thus, just prior to the last step, the inequality

$$\binom{n-1}{1} + \binom{n-1}{2} + \cdots + \binom{n-1}{d-2} < 2^{n-k} - 1$$

must be satisfied and it will be possible to add one more column to $\mathbf{H}$.

2.1.3. Hamming Codes

At this point one might be led to ask the following question. For a specified number of parity symbols, how long can one make the code and still guarantee a minimum distance of 3? Since all sets of two columns must be linearly independent to guarantee distance 3, it is sufficient that all of the columns be distinct and nonzero. Thus, for three parity symbols there are seven distinct nonzero 3-tuples, for four parity symbols there are 15 distinct nonzero 4-tuples, etc. This implies a family of codes with (n, k) of the form $(2^p - 1, 2^p - 1 - p)$, where $p = n - k$. These are the so-called Hamming codes and were first described by Hamming in 1950.[9]

The Hamming codes have several unique properties. First, these codes are examples of the few known perfect codes. Note that a perfect code must satisfy (1-4), with equality, which for the Hamming codes is

$$2^{n-k} = 1 + n$$

Since $n = 2^{n-k} - 1$ for these codes, they are obviously perfect. Another unique property of the Hamming codes is that they are one of the few classes of codes for which the complete weight structure is known. Define the weight-enumerating function of a code as

$$A(x) = \sum_{i=0}^{n} A_i x^i$$

where A_i is the total number of code words in the code with weight i. Then the weight-enumerating function for the $d = 3$ Hamming codes of length $n = 2^p - 1$ is

$$A(x) = \frac{1}{n+1}\left[(1 + x)^n + n(1 + x)^{(n-1)/2}(1 - x)^{(n+1)/2} \right] \qquad (2\text{-}3)$$

The distance-3 Hamming codes can be converted to a minimum distance-4 code by appending one additional parity symbol that checks all of the symbols including the other parity symbols. This additional parity symbol causes all of the weight-3 code words to become weight-4, all of the weight-5 to become weight-6, etc. If, for instance, this procedure is used on the (7,4) Hamming code, the augmented parity check matrix becomes

$$\mathbf{H} = \begin{bmatrix} 1 & 1 & 1 & 0 & 1 & 0 & 0 & | & 0 \\ 1 & 1 & 0 & 1 & 0 & 1 & 0 & | & 0 \\ 1 & 0 & 1 & 1 & 0 & 0 & 1 & | & 0 \\ \hline 1 & 1 & 1 & 1 & 1 & 1 & 1 & | & 1 \end{bmatrix} \qquad (2\text{-}4)$$

The upper left-hand portion contained in broken lines is the original parity check matrix and contains all distinct 3-tuples. This new family of distance-4 codes that are generated by adding one additional parity symbol is sometimes referred to as extended Hamming codes. This results in a family of codes of the form $(2^p, 2^p - 1 - p)$.

One might have observed in the last example that the augmented $\mathbf{H}$ matrix for the distance-4 Hamming code was no longer in echelon canonical form. That is, the right half portion of the matrix is no longer the identity matrix and does not permit one to determine each of the parity symbols as a function only of the information symbols. This situation is remedied by adding each of the first three rows of this matrix to the last row. The resulting matrix $\mathbf{H}^1$ becomes

$$\mathbf{H}^1 = \begin{bmatrix} 1 & 1 & 1 & 0 & 1 & 0 & 0 & 0 \\ 1 & 1 & 0 & 1 & 0 & 1 & 0 & 0 \\ 1 & 0 & 1 & 1 & 0 & 0 & 1 & 0 \\ 0 & 1 & 1 & 1 & 0 & 0 & 0 & 1 \end{bmatrix}$$

We may verify that the last row is still a legitimate parity check by considering its effect on a code vector. If $\mathbf{h}_1$, $\mathbf{h}_2$, $\mathbf{h}_3$, and $\mathbf{h}_4$ are the four row vectors in $\mathbf{H}$, then the last row of $\mathbf{H}^1$ times a code vector $\mathbf{c}$ may be written

$$\mathbf{h}_4^1 \mathbf{c} = (\mathbf{h}_1 + \mathbf{h}_2 + \mathbf{h}_3 + \mathbf{h}_4)\mathbf{c}$$

$$= \mathbf{h}_1\mathbf{c} + \mathbf{h}_2\mathbf{c} + \mathbf{h}_3\mathbf{c} + \mathbf{h}_4\mathbf{c}$$

$$= 0$$

since each of the $\mathbf{h}_i\mathbf{c} = 0$.

In general, the rows of the parity check matrix form a linearly independent set. This is obvious when the matrix is written in canonical form because the identity matrix makes it impossible to find a linear combination that sums to zero. As should be obvious from the example, there are many other choices of rows for the $\mathbf{H}$ matrix that are acceptable. Any linear combination of the rows is a legitimate parity check equation, and any set of $n - k$ parity check equations formed in this fashion that are linearly independent may be used to form the **H** matrix. We will exploit this property at a later time to derive a simple decoding algorithm for group codes.

2.1.4. Generator Matrix

So far we have described the group codes in terms of the parity check matrix **H**. When written in canonical form, this matrix enables us to express each one of the parity check symbols as some linear combination of the information symbols. An alternate description is possible and often quite useful. We recall that the modulo-2 sum of any two binary code words is a code word. It follows by repeated application of this principle that any linear combination of code words (sum under modulo-2 addition) is also a code word. One might suspect that since the information symbols are specified independently, one could find different words, each one of which contains a single 1 in the information portion of the word. It would then be true that all 2^k possible code words could be formed by taking all 2^k possible linear combinations of these k *basis vectors*. The astute reader will recognize at this point that we are attributing to the code the properties of a *linear vector space*. Such is indeed the case and will be illustrated by the following example.

The parity check matrix for the (7,4) Hamming code is

$$\mathbf{H} = \begin{bmatrix} 1 & 0 & 1 & 1 & 1 & 0 & 0 \\ 1 & 1 & 0 & 1 & 0 & 1 & 0 \\ 1 & 1 & 1 & 0 & 0 & 0 & 1 \end{bmatrix}$$

We would like to find four different code vectors, each of which contains a single 1 in the first four symbol locations. Assume that the first of these vectors is given by

$$\mathbf{c}_1^T = (1\,0\,0\,0\,p_{11}\,p_{12}\,p_{13})$$

Taking the product of this vector with each of the rows of the **H** matrix, we find that p_{11}, p_{12}, and p_{13} assume the values of the first column of **H**. Similarly, with

$$\mathbf{c}_2^T = (0\,1\,0\,0\,p_{21}\,p_{22}\,p_{23})$$

then p_{21}, p_{22}, and p_{23} take on the values of the second column of **H**, etc. Thus, the parity portion of each of the four basis vectors is one of the first four columns of the **H** matrix. When these four vectors are arranged to form a matrix, the result is called the *generator matrix* for the code. When written in reduced echelon canonical form, it always consists of a $k \times k$ identity matrix adjoined to a $k \times (n - k)$ matrix of parity check symbols. The parity check portion of this matrix can be obtained by inspection from the **H** matrix (in canonical form) by transposing the submatrix consisting of the first k columns. Thus, for the example at hand

$$\mathbf{G} = \begin{bmatrix} 1 & 0 & 0 & 0 & 1 & 1 & 1 \\ 0 & 1 & 0 & 0 & 0 & 1 & 1 \\ 0 & 0 & 1 & 0 & 1 & 0 & 1 \\ 0 & 0 & 0 & 1 & 1 & 1 & 0 \end{bmatrix}$$

The general form of the relationship between the parity check matrix and generator matrix for any code, binary or nonbinary, should now be obvious. For a generator matrix of the form

$$\mathbf{G} = [\mathbf{I}_k, \mathbf{P}]$$

where **P** is the $k \times (n - k)$ matrix of parity check symbols, the corresponding parity check matrix is given by

$$\mathbf{H} = [-\mathbf{P}^T, \mathbf{I}_{n-k}]$$

Thus, with a generator matrix in echelon canonical form, one can write the parity check matrix from inspection, and vice versa.

2.1.5. Dual Codes

One interesting observation can be made about the **G** and **H** matrices. Both of these contain a set of linearly independent vectors. Thus, each matrix can be taken as the basis for a linear vector space. In addition, each of these vector spaces is a subspace of the vector space of all n-tuples. The dot product of any row in the **G** matrix with any row in the **H** matrix is always zero, i.e.,

$$\mathbf{H}\,\mathbf{G}^T = \mathbf{0}$$

Thus, one could interchange the two and use the **H** matrix to be the generator matrix and the **G** matrix to be the parity check matrix of another code. Codes that are so related are said to be *dual codes*. Thus, the vector space corresponding to the **H** matrix is said to be the *null space* of the vector space corresponding to the **G** matrix since all vectors in one space are orthogonal to all vectors in the other space. As an example, the Hamming codes are in the null space of the so-called "maximum-length" codes to be discussed later in this chapter.

There is also a very interesting relationship between the weight structure of a code and that of its dual code. This relationship has proved to be quite useful in determining the weight structure of many block codes. Consider an (n, k) binary group code with weight enumerator $A(x)$ and its dual code with weight enumerator $B(x)$. Then the weight enumerator of the $(n, n - k)$ dual code is given by the MacWilliams identity[10] as

$$B(x) = 2^{-k}(1 + x)^n A[(1 - x)/(1 + x)] \qquad (2\text{-}5)$$

At the present time one can determine the code weight structure through exhaustive search for values of k up to 25 or so using a computer. The value of (2-5) is that the weight structure of a code of any length with 25 or fewer parity symbols can be found by first finding the weight structure of the dual code and then applying (2-5).

2.1.6. The Syndrome

In the case of group codes the decoding table discussed in Chapter 1 takes on a special name and is called a standard array. As we noted in Chapter 1, this array has the property that an element in the ith column is equal to the sum of the code word which heads the ith column and the correctable error pattern which appears in the same row in the zeroth column. That is, each row in the array is formed by adding a fixed pattern to each of the possible code words. In group theory, the rows of this array are called *cosets* and the element in the left-most column is called the *coset leader*. A property of cosets that the reader can easily verify is that any element in the coset could be chosen to be the coset leader. The only effect of choosing a different coset leader is to rearrange the order of the elements within the coset. The usual case, of course, is to pick the minimum-weight element to be the coset leader.

As we discussed in Chapter 1, the decoding table or standard array forms the conceptual basis for the decoder. One method of implementing the decoder would be to store the table in a read-only memory. The received pattern forms the address and the code word the information that is stored at the address. The complexity of such a scheme could be significantly reduced if there was a simple way to determine in which coset a particular received n-tuple were located. We would then need a much smaller table in which we could look up the desired coset leader or error pattern. This pattern would then be added modulo-2 to the received sequence. Such an approach is indeed possible as will now be shown.

Suppose a received word $\mathbf{r}$ contains an error pattern $\mathbf{e}$. Then $\mathbf{r}$ may be written

$$\mathbf{r} = \mathbf{c} + \mathbf{e}$$

If the received word $\mathbf{r}$ is multiplied by the matrix $\mathbf{H}$, the resultant vector $\mathbf{s}$ (called the syndrome) is

$$\mathbf{s} = \mathbf{Hr}$$
$$= \mathbf{Hc} + \mathbf{He}$$
$$= \mathbf{He}$$

Table 2-1. Standard Array for a (6,3) Single-Error-Correcting Code

000000	100101	010111	110010	001011	101110	011100	111001
100000	000101	110111	010010	101011	001110	111100	011001
010000	110101	000111	100010	011011	111110	001100	101001
001000	101101	011111	111010	000011	100110	010100	110001
000100	100001	010011	110110	001111	101010	011000	111101
000010	100111	010101	110000	001001	101100	011110	111011
000001	100100	010110	110011	001010	101111	011101	111000
101000	001101	111111	011010	100011	000110	110100	010001

Thus, the parity check matrix permits us to determine a quantity which depends only upon the error pattern. Next, observe that any two n-tuples that have the same syndrome can differ at most by a code word. By the very construction of a standard array, elements appearing in the same row satisfy this requirement, whereas elements in different rows cannot differ by only a code word. Thus, each row in the standard array is characterized by a unique syndrome.

As an example, consider the standard array shown in Table 2-1. This array is for the (6,3) code defined by the **H** matrix in (2-1).

Note that the syndrome corresponding to the second row in the standard array (a single error in position 1) is simple the first column of the **H** matrix, the syndrome corresponding to the third row is the second column in **H**, etc. The syndrome corresponding to the last row in the standard array is the sum of columns 1 and 3 of the **H** matrix. The reader should observe that each of the syndrome patterns is unique, and that all possible 3-tuples are used.

This last example provides some more insight into the process of constructing good codes. We previously noted that for a code to have distance d, all sets of $d - 1$ or fewer columns of **H** must be linearly independent. Equivalently, for it to be t-error correcting, all linear combinations of t or fewer columns must be distinct. Since

$$t = \left\lfloor \frac{d - 1}{2} \right\rfloor$$

this latter formulation is sometimes easier to work with.

2.1.7. The Penny-Weighing Problem

At this point we have developed the principal concepts upon which the theory of binary group codes is based. It is worthwhile to pause and fix these concepts more firmly by way of some examples. A classical problem which relates directly to the theory of Hamming codes is the so-called "penny-weighing" problem.

Suppose you are given eight coins and a pair of scales and told that one of the coins is counterfeit. The counterfeit coin is perfect match for the other coins except that it differs slightly in weight. You are asked to find the counterfeit coin by using only three weighings. The solution, which is given quite simply by consulting the parity check matrix for the (7,4) Hamming code, is as follows.

First, number the coins 1 through 8 and let the first 7 correspond to the columns of the **H** matrix for the (7,4) Hamming code.

$$\begin{array}{ccccccc} 1 & 2 & 3 & 4 & 5 & 6 & 7 \end{array}$$
$$\mathbf{H} = \begin{bmatrix} 1 & 1 & 1 & 0 & 1 & 0 & 0 \\ 1 & 1 & 0 & 1 & 0 & 1 & 0 \\ 1 & 0 & 1 & 1 & 0 & 0 & 1 \end{bmatrix}$$

On the first weighing you select 1, 2, 3, and 5 (row 1 of the parity check matrix) and place two coins in each pan. It is immaterial which coins go in which pan. If the scales balance, you know that the counterfeit coin must be in the set 4, 6, 7, or 8, while if they do not balance, it must be in the set 1, 2, 3, or 5. Thus, by one weighing you have eliminated exactly half of the possibilities. On the second weighing you select coins 1, 2, 4, and 6 and repeat the experiment. Again, a balance condition means that the coin is in 3, 5, 7, or 8 and an unbalance means that it is in 1, 2, 4, or 6. This weighing in combination with the above has now eliminated two more possibilities. For example, if the scale had balanced on both weighings, then the counterfeit coin must be either 7 or 8 since it was not in the pan either time. Finally, on the last weighing, we select coins 1, 3, 4, and 7. The result of this weighing now determines the counterfeit coin uniquely. This process becomes clear if we consider the decision tree shown in Fig. 2-1.

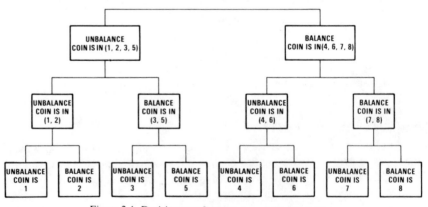

Figure 2-1. Decision tree for "penny-weighing" problem.

Upon consulting Fig. 2-1 we observe that each possible outcome corresponds to a column of the **H** matrix when we interpret an unbalance as a 1 and a balance as a 0. Thus, the set of weighings form a "syndrome" with each "syndrome" pattern corresponding to a possible counterfeit coin. The only difference between the penny-weighing problem and the decoding problem is the fact that one of the coins is known to be counterfeit, whereas in the decoding problem, the possibility exists that all of the symbols are correct. In either case we are trying to resolve one of eight possible events. If we had a channel in which it was guaranteed that a single error would always occur, then one could use a length-8 Hamming code and the analogy would be exact.

The above example illustrates a very important property of good codes. Each of the parity checks is designed to extract the maximum amount of information about the location of the error. Thus, the first check eliminates half of all the symbols, the second eliminates half of the remaining symbols, etc. Most codes, of course, will not be this efficient; however, the designer should attempt to apply this principle when designing a code for some special application.

2.1.8. A Simple Iterated Code

A simple coding system which has found considerable application can be constructed as follows. Suppose we wish to transmit nine information symbols. The nine symbols can be arranged in a square matrix as shown in Table 2-2 with a parity check appended for each row and each column. The one remaining symbol, P_7, is taken to be an overall parity check on all of the remaining symbols.

One may easily show that this code has minimum distance 4. First, note that any single error will cause both the row and column check containing the error to fail. Thus, a single error can be uniquely identified by taking the coordinates of the error to be the row and column in which failures occur. Since this pattern is unique for all distinct single-error patterns, the code must have a minimum distance of at least 3. Since P_7 will guarantee that all code words are of even weight, the minimum distance is at least 4. Actually, the distance is exactly 4 because in order for the distance

Table 2-2. Simple Iterated Block Code with Parity Checks on Rows and Columns

A_1	A_2	A_3	P_1
A_4	A_5	A_6	P_2
A_7	A_8	A_9	P_3
P_4	P_5	P_6	P_7

to be 5 one would have to resolve all double-error patterns uniquely. This is clearly not the case since A_5 and A_9 in error have exactly the same pattern of failures (syndrome) as A_6 and A_8 in error.

The parity check matrix for this code is given by

$$\mathbf{H} = \begin{bmatrix} 1 & 1 & 1 & 0 & 0 & 0 & 0 & 0 & 0 & 1 & 0 & 0 & 0 & 0 & 0 & 0 \\ 0 & 0 & 0 & 1 & 1 & 1 & 0 & 0 & 0 & 0 & 1 & 0 & 0 & 0 & 0 & 0 \\ 0 & 0 & 0 & 0 & 0 & 0 & 1 & 1 & 1 & 0 & 0 & 1 & 0 & 0 & 0 & 0 \\ 1 & 0 & 0 & 1 & 0 & 0 & 1 & 0 & 0 & 0 & 0 & 0 & 1 & 0 & 0 & 0 \\ 0 & 1 & 0 & 0 & 1 & 0 & 0 & 1 & 0 & 0 & 0 & 0 & 0 & 1 & 0 & 0 \\ 0 & 0 & 1 & 0 & 0 & 1 & 0 & 0 & 1 & 0 & 0 & 0 & 0 & 0 & 1 & 0 \\ 1 & 1 & 1 & 1 & 1 & 1 & 1 & 1 & 1 & 1 & 1 & 1 & 1 & 1 & 1 & 1 \end{bmatrix}$$

One should note there are several equivalent forms for the $\mathbf{H}$ matrix. One which has a distinct implementation advantage in some situations can be formed by adding rows 4, 5, and 6 to the last row. When this is done, the last row becomes

$$(0\ 0\ 0\ 0\ 0\ 0\ 0\ 0\ 0\ 1\ 1\ 1\ 0\ 0\ 0\ 1)$$

and P_7 is now computed as the sum of P_1, P_2, and P_3. By an identical argument, one could equally well take P_7 to be the sum of P_4, P_5 and P_6. The advantage to computing P_7 as the sum of P_1, P_2, and P_3 is that now all of the columns are computed in exactly the same fashion. The principle works independently of the size of the array. Thus, the technique is useful for data that occur naturally in array form such as characters on a data bus or paper tape when the length of the character string is not fixed. In many data transfer applications, it is usual to include a parity check on each character. It is usually a simple matter to compute an overall parity character by accumulating the character string in a single register using modulo-2 addition.

2.2. Polynomial Codes

In the previous sections an (n, k) code word was represented as an n-tuple of the form

$$(a_0, a_1, \ldots, a_{n-1})$$

An alternate method of representing the same code word is to let the elements $a_0, a_1, \ldots, a_{n-1}$, be the coefficients of a polynomial in x. Thus, we have

$$f(x) = a_0 + a_1 x + \cdots + a_{n-1} x^{n-1}$$

Using this notation, a polynomial code is specified as the complete set of polynomials of degree $n - 1$ or less that contains some specified polynomial $g(x)$ as a factor. The polynomial $g(x)$ is said to be the *generator polynomial* of the code. In order to multiply, factor, and otherwise manipulate these code polynomials, it is necessary to be able to add, subtract, multiply, and divide their coefficients. This is easily done if we require that the coefficients be elements of a finite field.

2.2.1. Finite Field Arithmetic

A finite field, also called a *Galois field* and designated GF(q), is a finite set of q elements for which we have defined some special rules for arithmetic. These rules are not very different from those which are used to do arithmetic with ordinary numbers. The principal difference is that there is only a finite set of elements involved and some modifications are necessary to account for this fact.

All finite fields have the following properties:

1. There are two operations that are defined for combining the elements: multiplication and addition.
2. The result of adding or multiplying two elements is always a third element that is contained in the field.
3. The field always contains the multiplicative identity element 1 and the additive identity element 0. Thus, $a + 0 = a$ and $a \cdot 1 = a$ for any element a.
4. For every element a there is an additive inverse element $(-a)$ and a multiplicative inverse element a^{-1} such that $a + (-a) = 0$ and $a \cdot a^{-1} = 1$. (Note 0^{-1} is not defined.) The existence of these elements permits us to use the customary notions of subtraction and division.
5. The usual associative $[a + (b + c) = (a + b) + c$ and $a \cdot (b \cdot c) = (a \cdot b) \cdot c]$, commutative $[a + b = b + a$ and $a \cdot b = b \cdot a]$, and distributive $[a \cdot (b + c) = a \cdot b + a \cdot c]$ laws apply.

Finite fields do not exist for any arbitrary number of elements. In general, they exist only when the number of elements is a prime number or is a power of a prime number. The former are called *prime* fields, while the latter are called *extension* fields over the prime field.

For each value of q there is one and only one unique field. That is, there is only one way that the rules for multiplication and addition can be defined so that all the other properties are satisfied. If q is a prime number, the field elements are the integers $0, 1, \ldots, q - 1$, and multiplication and addition are simply ordinary multiplication and addition modulo q. For example, the multiplication and addition tables for the prime field GF(5) are shown in Table 2-3. Subtraction and division can be accomplished by

Table 2-3. Multiplication and Addition Tables for GF(5)

·	0	1	2	3	4		+	0	1	2	3	4
0	0	0	0	0	0		0	0	1	2	3	4
1	0	1	2	3	4		1	1	2	3	4	0
2	0	2	4	1	3		2	2	3	4	0	1
3	0	3	1	4	2		3	3	4	0	1	2
4	0	4	3	2	1		4	4	0	1	2	3

using the tables to locate the appropriate inverse element and then adding and multiplying as usual. Thus

$$3 - 4 = 3 + (-4)$$
$$= 3 + 1$$
$$= 4$$

In a similar fashion

$$\tfrac{3}{4} = 3 \cdot 4^{-1}$$
$$= 3 \cdot 4$$
$$= 2$$

If q is a power of a prime number (say $q = p^m$), then the field elements are all possible polynomials of degree $m - 1$ where the coefficients are from the prime field $GF(p)$. The rules for multiplication and addition are obtained by multiplying and adding these polynomials in the usual way and then reducing the result modulo a special degree-m polynomial $p(x)$. This polynomial must have the property that it cannot be factored using only polynomials with coefficients from $GF(p)$. Such polynomials are said to be *irreducible* and are analogous to prime numbers. Like prime numbers, they are usually found by trial and error and have been tabulated in several references. (See Peterson and Weldon.[1])

The polynomial $p(x) = 1 + x + x^3$ is irreducible over $GF(2)$ and can be used as a basis for constructing $GF(8)$. Suppose that $\beta_1 = 1 + x + x^2$ and $\beta_2 = 1 + x^2$ are two elements in $GF(8)$. We may form their sum

$$\beta_1 + \beta_2 = (1 + x + x^2) + (1 + x^2)$$
$$= (1 + 1) \cdot 1 + x + (1 + 1) \cdot x^2$$
$$= x$$

and their product

$$\beta_1 \cdot \beta_2 = (1 + x + x^2) \cdot (1 + x^2)$$
$$= 1 + x + x^2 + x^2 + x^3 + x^4$$
$$= 1 + x + x^3 + x^4$$

To reduce this expression modulo $p(x)$, we make repeated use of the identity $p(x) = 0$. Thus,

$$x^3 = 1 + x$$
$$x^4 = x + x^2$$

and

$$\beta_1 \cdot \beta_2 = 1 + x + (1 + x) + (x + x^2)$$
$$= x + x^2$$

The reader should note that the reduction of a polynomial modulo $p(x)$ is identical to dividing by $p(x)$ and saving the remainder. This is easily verified for the previous example. A closely related concept is

$$a(x) \equiv b(x) \bmod p(x)$$

which is read "$a(x)$ is *congruent* to $b(x)$ modulo $p(x)$." This means that $a(x)$ and $b(x)$ have the same remainder after dividing by $p(x)$. Equivalently, $a(x) - b(x)$ is divisible by $p(x)$. This concept also applies with a, b, and p as integers.

In finite fields, the concept of logarithms apply just as they do with ordinary numbers. It is a property of all finite fields that there exists at least one element, called a *generator* or *primitive* element, such that every other nonzero element in the field can be expressed as a power of this element. For example, from Table 2-3, we see that 2 is a primitive element of GF(5) because

$$2^1 = 2$$
$$2^2 = 4$$
$$2^3 = 3 \qquad\qquad (2\text{-}6)$$
$$2^4 = 1$$

and

$$2^5 = 2^1 = 2$$

Thus, an alternate way of multiplying is to use the mapping defined by (2-6) to find the logarithms, add these numbers modulo 5, and then use (2-6) to find the inverse logarithm. Note that this requires that we define

$$2^0 = 1$$

just as in the ordinary case of real numbers.

Exactly the same concept can be used in GF(8). If we let $\alpha = x$, then it follows that

$$\alpha^2 = x^2$$
$$\alpha^3 = x^3 = 1 + x$$
$$\alpha^4 = x + x^2 \qquad\qquad (2\text{-}7)$$
$$\alpha^5 = x^2 + x^3 = 1 + x + x^2$$
$$\alpha^6 = x + x^2 + x^3 = 1 + x^2$$
$$\alpha^7 = x + x^3 = 1$$

etc.

Thus, we can multiply two numbers in $GF(8)$ by using (2-7) to define the logarithms, next add the logarithms modulo 7, and again use (2-7) to compute the inverse logarithm. In this example x conveniently turned out to be a primitive element and, consequently, it was a simple matter to generate a table of logarithms. Unfortunately, this is not always the case and locating a primitive element could prove to be a nontrivial problem. As a practical matter, x will always be primitive whenever the irreducible polynomial which specifies the field operations is also a primitive polynomial. The precise relationship between primitive polynomials and primitive elements will subsequently become clear. At this point it is sufficient to note that tables which list irreducible polynomials usually tell whether or not the polynomials are primitive. Hence, as a practical matter one can always select an irreducible polynomial which is also primitive and thus avoid the problem.

The existence of logarithms for finite fields means that there is a representation for field elements that is convenient for multiplication and a second representation that is convenient for addition. Once the mapping between these two representations has been established, it is no longer useful to explicitly write out the polynomial form of the field elements. A more convenient technique is to represent the field element as an m-tuple

Table 2-4. Mapping between the Ordinary and Logarithmic Representation of Elements in $GF(8)$

0	$\longleftrightarrow$ 0	0	0
$\alpha^0 = 1$	$\longleftrightarrow$ 1	0	0
α^1	$\longleftrightarrow$ 0	1	0
α^2	$\longleftrightarrow$ 0	0	1
α^3	$\longleftrightarrow$ 1	1	0
α^4	$\longleftrightarrow$ 0	1	1
α^5	$\longleftrightarrow$ 1	1	1
α^6	$\longleftrightarrow$ 1	0	1

when performing addition and as a power of the primitive element when performing multiplication. Thus, for GF(8), the representation in Table 2-4 is equivalent to the representation of equations (2-7). Using Table 2-4 one can readily do arithmetic in GF(8). For instance, a particular function $f(\alpha)$ might be evaluated as follows. Suppose that

$$f(\alpha) = (\alpha^2 + \alpha^5)\alpha + (1 + \alpha^3)\alpha$$

From Table 2-4 $\alpha^2 + \alpha^5 = \alpha^3$ and $(1 + \alpha^3) = \alpha$. Thus,

$$f(\alpha) = \alpha^3 \cdot \alpha + \alpha \cdot \alpha$$
$$= \alpha^4 + \alpha^2$$
$$= \alpha = (0\ 1\ 0)$$

In some cases it is not very convenient to carry out repeated conversions between the logarithmic representation and the representation as an m-tuple in performing finite field arithmetic. An alternate approach that is quite useful is to use the Zech logarithm, $z(n)$, as defined by

$$\alpha^{z(n)} = 1 + \alpha^n \tag{2-8}$$

Using this technique one can carry out finite field arithmetic operations entirely in the logarithm domain, without using an antilog table. The elements α^m and α^n can be added by

$$\alpha^m + \alpha^n = \alpha^m(1 + \alpha^{n-m})$$
$$= \alpha^{m + z(n-m)}$$

The table of Zech's logarithms for GF(8) is shown in Table 2-5. Using this table we note that

$$\alpha^2 + \alpha^5 = \alpha^{2 + z(3)}$$
$$= \alpha^3$$

Table 2-5. Zech's Logarithms in GF(8)

n	$z(n)$
$-\infty$	0
0	$-\infty$
1	3
2	6
3	1
4	5
5	4
6	2

2.2.2. Generation of Polynomial Codes

Earlier in this section we defined an (n, k) polynomial code as the complete set of polynomials of degree $n - 1$ or less that contain $g(x)$ as a factor. In general $g(x)$ is of degree $n - k$. For example, if

$$g(x) = (1 + x + x^3)$$

then the set of polynomials described by the equation

$$c(x) = (1 + x + x^3)(a_0 + a_1 x + a_2 x^2)$$

form a (6,3) code. We note that there are exactly eight distinct polynomials which correspond to the eight different ways of picking a_0, a_1, and a_2 from GF(2). One way of using this code would be to let the information symbols be the individual values of a_0, a_1, and a_2, and simply carry out the indicated multiplication. The result would be

$$c(x) = a_0 + (a_0 + a_1)x + (a_1 + a_2)x^2 + (a_0 + a_2)x^3 + a_1 x^4 + a_2 x^5$$
$$(2\text{-}9)$$

A circuit which performs this multiplication is shown in Fig. 2-2. Although this is a perfectly valid way of encoding, it has the disadvantage that the resulting code is usually nonsystematic. (In this particular instance the code is systematic in positions 1,5, and 6; however, this will usually not be the case.)

A property of (n, k) polynomial codes is that they can always be made to be systematic in the first k positions. For the example being discussed we define

$$b_0 = a_0$$
$$b_1 = a_0 + a_1 \qquad (2\text{-}10)$$
$$b_2 = a_1 + a_2$$

If we solve (2-10) for the a's and substitute into (2-9), the code polynomial becomes

$$c(x) = b_0 + b_1 x + b_2 x^2 + (b_1 + b_2)x^3 + (b_0 + b_1)x^4 + (b_0 + b_1 + b_2)x^5$$

This code polynomial is now in the familiar parity check form and the reader will notice that it would be identical to the (6,3) parity check code that was described earlier in (2-1) if the first and second parity symbols were interchanged. We could equally well have chosen a different set of positions to be independent message symbols. For example, choosing the last three results in

$$c(x) = (b_0 + b_2) + (b_0 + b_1 + b_2)x + (b_1 + b_2)x^2 + b_0 x^3 + b_1 x^4 + b_2 x^5$$
$$(2\text{-}11)$$

SHIFT REGISTER STAGE   MOD-2 ADDER

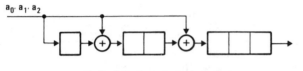

a_0, a_1, a_2

$a_0, a_0 + a_1, a_1 + a_2, a_0 + a_2, a_1, a_2$

Figure 2-2. Circuit for multiplying an arbitrary polynomial $a_0 + a_1 x + a_2 x^2$ by $g(x) = 1 + x + x^3$.

This latter choice is usually preferred, for reasons which will soon become clear.

It is obvious that one could build an encoder for either of these systematic versions of this code by first expressing the a's as the appropriate linear combination of the b's and then using the circuit of Fig. 2-2 to perform the required multiplication.

A second method, which in essence combines these two steps, is as follows. Suppose we multiply the polynomial $b_0 + b_1 x + b_2 x^2$ by x^3 giving a result $b_0 x^3 + b_1 x^4 + b_2 x^5$. If this result is divided by $1 + x + x^3$, a quotient $(b_0 + b_2) + b_1 x + b_2 x^2$ and a remainder $(b_0 + b_2) + (b_0 + b_1 + b_2)x + (b_1 + b_2)x^2$ are obtained. Now observe that if the remainder is added to the dividend, the result is the code word of (2-11) in systematic form. This procedure is simply an application of the Euclidean division algorithm, which states

$$\text{dividend} = (\text{quotient}) \cdot (\text{divisor}) + \text{remainder}$$

For the general case of an (n, k) code, the premultiplication is always by x^{n-k}. The degree of the remainder will always be less than the degree of the disivor, which is $n - k$. Since the sequence to be encoded was multiplied by x^{n-k}, the dividend will contain only terms whose exponents are greater than or equal to $n - k$. Consequently, adding the divisor to the quotient will not modify any of the terms in either expression. By choosing the divisor to be the generator polynomial, we ensure that the result is always a code word.

A circuit which implements this result for the (6,3) code is shown in Fig. 2-3. Using this circuit, the sequence b_2, b_1, and b_0 is applied to the input one symbol at a time. The contents of the various stages after each shift are as shown in Fig. 2-3. After three shifts, the remainder (three parity symbols) is stored in the three shift register stages. The reader can verify by actual

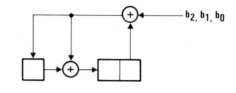

CONTENTS OF STAGES			SHIFT
0	0	0	0
b_2	b_2	0	1
b_1	$b_1 + b_2$	b_2	2
$b_0 + b_2$	$b_0 + b_1 + b_2$	$b_1 + b_2$	3

Figure 2-3. Circuit for simultaneously multiplying by x^3 and dividing by $1 + x + x^3$.

long division that this circuit duplicates the individual shift and subtract operations at each shift. The quotient, if it is desired, appears as successive terms in the feedback loop.

The circuit shown in Fig. 2-3 is very simple because all arithmetic operations are over $GF(2)$. In the general case of a code defined over $GF(q)$, multiplications must also be performed. Assume the generator polynomial has the form $g(x) = g_{n-k}x^{n-k} + \cdots + g_1 x + g_0$, where the g_i coefficients are elements of $GF(q)$. Then Fig. 2-3 must be modified in the following fashion. First, all addition operations are replaced by $GF(q)$ additions, and each shift register stage becomes a $GF(q)$ element. Second, the output of the adder that produces the feedback term must be multiplied by g_{n-k}^{-1}. Finally, for each nonzero g_i, $i < n - k$, the feedback term must be multiplied by $-g_i$. The resulting term added to the output of the preceding stage produces the input to the ith stage of the register ($i = 0$ through $n - k - 1$). This generalized encoding circuit is shown in Fig. 2-4. That this circuit produces the desired remainder is easily verified by long division.

A second viewpoint is also useful. Initialize the circuit of Fig. 2-3 so that 1 appears in the leftmost stage. If the circuit is now shifted several times with no input applied, the successive contents of the registers will be as shown in Fig. 2-5. We recognize that these are identical to the columns

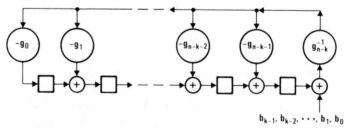

Figure 2-4. Circuit for multiplying by x^{n-k} and dividing by $g(x)$.

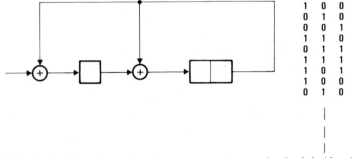

Figure 2-5. Feedback shift register and resultant sequence of states when loaded with a single 1.

of a parity check matrix, and if we stop after seven shifts, this matrix is the parity check matrix for the (7,4) Hamming code. By shifting fewer than seven times, the parity check matrix for a shortened version of this code is produced. We also note that by shifting more than 7 times, the sequence repeats itself. This is still a perfectly valid parity check matrix, but now the minimum distance has been reduced to 2 since the same column is contained more than once. One may also consider the operation of the circuit in Fig. 2-3 from the viewpoint of the generator matrix. This matrix for a (7,4) Hamming code is given by

$$G = \begin{bmatrix} 1 & 1 & 0 & 1 & 0 & 0 & 0 \\ 0 & 1 & 1 & 0 & 1 & 0 & 0 \\ 1 & 1 & 1 & 0 & 0 & 1 & 0 \\ 1 & 0 & 1 & 0 & 0 & 0 & 1 \end{bmatrix}$$

Now if a 1 is applied at the input and shifted once, the number 1 1 0 is stored in the respective stages. A second shift produces 0 1 1, a third 1 1 1, and a fourth 1 0 1. These four 3-tuples are identical with the four rows in the parity portion of the generator matrix. The effect of shifting 1 0 0 0 into the register (starting with the right-hand-most symbol) is to produce the

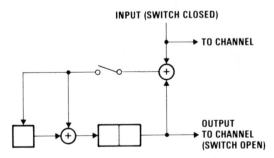

INPUT (SWITCH CLOSED)

TO CHANNEL

OUTPUT
TO CHANNEL
(SWITCH OPEN)

Figure 2-6. $(n - k)$-stage shift register encoder for the code generated by $g(x) = 1 + x + x^3$.

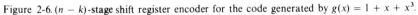

parity sequence 1 1 0. Similarly, the effect of shifting 0 0 0 1 into the register is to produce the parity symbols 1 0 1. Further, the device is completely linear. If we shift 1 0 0 1 into the register, the result will be 0 1 1. These are the parity symbols which are obtained by adding the first and last rows of the generator matrix.

In practice, the circuit of Fig. 2-3 is used as shown in Fig. 2-6. The sequence to be encoded is shifted simultaneously into the channel and into the encoder with the feedback enabled. Then the feedback is disabled, and the contents of the register are shifted into the channel.

2.2.3. Cyclic Codes

For certain values of n the polynomial codes exhibit a *cyclic* property. That is, a cyclic permutation of the symbols in a code word generates a new code word. The reason for this property can be illustrated by considering the generator polynomial, $1 + x + x^3$, from our previous example. We first note that $1 + x + x^3$ exactly divides $1 - x^7$. This can be verified by long division or by considering the circuit of Fig. 2-5. Recall that this circuit generates a cycle that repeats after every seven shifts. Thus, if we were to apply a single 1 to the input, shift seven times, and then apply a second 1 (i.e., divide $x^7 - 1$), the second input would cancel the contents of the register and we would be left with a remainder of zero.

Now, suppose we consider the effect of taking a code word from the (7,4) code generated by $1 + x + x^3$ and cyclically shifting it one place to the right. That is, we move every symbol one place to the right and take the right-hand-most symbol and move it around to the left end. Mathematically this is equivalent to multiplying by x and replacing the coefficient of x^7 by 1. That is, reducing the polynomial $\mod(x^7 - 1)$. Suppose we select a code word that has a 1 in the right-hand-most position. This code word may be written

$$c(x) = (1 + x + x^3)p(x)$$

where $p(x)$ is a degree-4 polynomial. Then

$$xc(x) \equiv xc(x) - (x^7 - 1)\mod(x^7 - 1)$$

Since $x^7 - 1$ is divisible by $1 + x + x^3$, the right-hand side of this equation is clearly divisible by $1 + x + x^3$. Consequently, cyclically shifting this code word produces a second code word. Note that if our code word did not have a 1 in the right-hand-most position, the only effect of a cyclic shift is to multiply by x, and the result still holds.

This result is easily extended to any (n, k) code for which $g(x)$ divides $(x^n - 1)$. The smallest value of n for which $g(x)$ divides $x^n - 1$ normally determines the largest useful length, $n_{\max}$, of the code. The reason for this is that two information symbols spaced $n + 1$ symbols apart would have exactly the same parity symbols and thus produce a code having minimum distance of 2. We also note that the largest possible value for n_{max} is $(2^m - 1)$, where m is the degree of the generator polynomial. This follows by noting that n_{max} is determined by the shortest cycle that the feedback shift register specified by $g(x)$ will produce when loaded with a single 1. Since this register has m stages, it can exist in no more than $2^m - 1$ different nonzero states. Hence, the largest possible value for n_{max} is $2^m - 1$. For some polynomials, the cycles may be significantly shorter than $2^m - 1$. For example, the polynomial $1 + x^3 + x^6$ generates a cycle of length 9, as can be easily verified by the reader.

Polynomial codes for which n is less than n_{max}, such as the (6, 3) code discussed earlier, are referred to as shortened cyclic or pseudocyclic codes. One can always find some degree-n polynomial for any value of n such that $g(x)$ divides this polynomial. Using this polynomial one could consider the class of permutations that result from multiplying a code word by x and then computing the result modulo this polynomial. Although this will obviously produce a code word, no one has as yet found any significant applications for this result. Normally, shortened cyclic codes are treated by considering them to be cyclic with the required end symbols set equal to zero.

2.2.4. Alternate Encoder for Cyclic Codes

We have shown that the encoder for an (n, k) polynomial code can be realized as an $(n - k)$-stage feedback shift register with feedback connections specified by the generator polynomial. It is also possible and sometimes more convenient to implement the encoder using a k-stage shift register. It turns out that the connections for this register are specified by the *parity check polynomial*

$$h(x) = \frac{x^n - 1}{g(X)}$$

We will illustrate this encoder using the same (7,4) code of our previous example.

In Section 2.2.2 we noted that the parity check matrix for the code generated by $1 + x + x^3$ could be produced by loading a single 1 into the feedback shift register specified by $1 + x + x^3$ and shifting seven times. The result is

$$\mathbf{H} = \begin{bmatrix} 0 & 0 & 1 & 0 & 1 & 1 & 1 \\ 0 & 1 & 0 & 1 & 1 & 1 & 0 \\ 1 & 0 & 0 & 1 & 0 & 1 & 1 \end{bmatrix} \qquad (2\text{-}12)$$

This operation is equivalent to dividing $x^7 - 1$ by $1 + x + x^3$, where the quotient appears as successive terms in the feedback loop. Thus, the first row of (2-12) is given by

$$h(x) = \frac{x^7 - 1}{1 + x + x^3}$$

$$= x^4 + x^2 + x + 1$$

If we add the first row of (2-12) to the last row, we obtain a new matrix

$$\mathbf{H}^1 = \begin{bmatrix} 0 & 0 & 1 & 0 & 1 & 1 & 1 \\ 0 & 1 & 0 & 1 & 1 & 1 & 0 \\ 1 & 0 & 1 & 1 & 1 & 0 & 0 \end{bmatrix}$$

in which each row is a shifted version of the first row. Thus, each parity symbol in the matrix is the same linear combination of the four symbols immediately to its right. This suggests that one could build the encoder using the circuit shown in Fig. 2-7. The feedback connections are determined by the parity check polynomial $h(x) = h_k x^k + \cdots + h_1 x + h_0$ (assume $h_k = 1$). In the general case for codes defined over $GF(q)$, the feedback term from the ith stage is multiplied by $-h_i$ ($i = 0$ through $k - 1$), and the adder is a $GF(q)$ adder. The convention is that the lowest-order stage is on the right side of the figure.

In Fig. 2-7 the encoder is operated by first loading the register with the four information symbols with the feedback disabled. The feedback is then enabled and the register shifted seven times. The first four symbols delivered to the channel are the four information symbols, and the next three are the parity symbols.

Another way of arriving at the same result is to observe that the product of $h(x)$ with any code word is zero when expressed modulo $(x^n - 1)$. That is,

$$h(x)\,g(x)\,p(x) = (x^n - 1)\,p(x) \equiv 0 \bmod (x^n - 1)$$

Since the entire resultant polynomial is zero, the coefficients of every power of x must also be zero. For the coefficient of x^6, we have

$$c_0 h_6 + c_1 h_5 + c_2 h_4 + \cdots + c_6 h_0 = 0$$

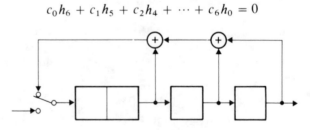

Figure 2-7. k-stage shift register encoder specified by $h(x) = (x^7 - 1)/(1 + x + x^3)$.

The other six coefficients will be similar with the index on h being a cyclic permutation of the above. Thus, the coefficients of $h(x)$ form a legitimate set of parity checks when written in reverse order and the $\mathbf{H}^1$ matrix could have been formed directly by noting this fact.

An important implication of these observations is that a code with generator polynomial $h(x)$ is the dual of a code with generator polynomial $g(x)$, where $h(x) = (x^n - 1)/g(x)$. Thus, in the present example, codes with generator polynomials $1 + x + x^3$ and $1 + x + x^2 + x^4$ are dual codes.

In addition to being useful for encoding low-rate codes (k/n small), the k-stage encoder is also useful when the encoder must perform a rate-buffering function as well as the encoding function. In most systems, information will be delivered to the encoder in a continuous serial stream at some data rate r_1. The encoder must divide this into k-symbol blocks, append $n - k$ parity symbols, and output the resultant n symbols at a continuous rate r_2, where $r_2 = (n/k)r_1$. One way of accomplishing this is to actually build a rate buffer which can be filled at rate r_1 and emptied at rate r_2 with $(n - k)$-symbol "holes" placed every k symbols. This data stream could then be fed directly to either the $(n - k)$-stage encoder or the k-stage encoder. A second method would be to operate two k-stage encoders in parallel. Encoder number (1) is filled at rate r_1 while encoder number (2) is being emptied at rate r_2. On the next cycle, the roles are reversed.

2.2.5. Additional Properties of Polynomials and Galois Field Elements

There is a further connection between polynomials and finite field elements that we have not yet developed and which provides significant insight into the properties of polynomial generated codes. The Galois field elements play precisely the same role for polynomials with finite-field coefficients that complex numbers do for polynomials with real number coefficients. Just as every polynomial with real coefficients can be factored if one permits the existence of complex numbers, similarly, every polynomial whose coefficients are from a finite field can be factored using elements from some extension field. The relationship between the factors and the roots of a polynomial with finite-field coefficients is precisely the same as the relationship between the factors and the roots of a polynomial with real number coefficients. For example, a given polynomial $f(x)$ can be factored as

$$f(x) = (x - \beta_1)(x - \beta_2) \cdots (x - \beta_j) \qquad (2\text{-}13)$$

where the j values of β are also the j roots of the polynomial $f(x)$. There are several properties of polynomials and Galois field elements that are useful both in describing codes and in structuring the decoder. These properties are listed below without proof. Following each property, one or more examples are given to illustrate the use of the particular property.

Property 1: Irreducible Polynomials. A polynomial $f(x)$ whose coefficients are contained in a particular field is said to be *irreducible* if the polynomial cannot be factored using only elements from that particular field. This same polynomial, however, can always be factored using elements from some extension field. Consequently, the polynomial will always have roots in the extension field. If $f(x)$ is an irreducible polynomial with coefficients from GF (p), and if α is a root, then α^p, α^{p^2}, α^{p^3}, etc. are also roots. Further, all of the roots of $f(x)$ can be found in this fashion. The polynomial $f(x)$ is called the *minimum function* of α. If α is a primitive element then $f(x)$ is a *primitive polynomial*.

Example. The polynomial $f(x) = 1 + x + x^3$ is irreducible over GF (2). The element α in Table 2-4 is a root since

$$f(\alpha) = 1 + \alpha + \alpha^3$$
$$= (1\,0\,0) + (0\,1\,0) + (1\,1\,0)$$
$$= (0\,0\,0) = 0$$

Using the above property α^2 and α^4 must also be roots. That is,

$$f(\alpha^2) = 1 + \alpha^2 + \alpha^6$$
$$= (1\,0\,0) + (0\,0\,1) + (1\,0\,1)$$
$$= (0\,0\,0) = 0$$

and

$$f(\alpha^4) = 1 + \alpha^4 + \alpha^{12}$$
$$= 1 + \alpha^4 + \alpha^5$$
$$= (1\,0\,0) + (0\,1\,1) + (1\,1\,1)$$
$$= (0\,0\,0) = 0$$

Alternately, if α, α^2, and α^4 are the roots of $f(x)$, then it must be possible to write $f(x)$ as

$$f(x) = (x - \alpha)(x - \alpha^2)(x - \alpha^4)$$
$$= x^3 - x^2(\alpha + \alpha^2 + \alpha^4) + x(\alpha^3 + \alpha^6 + \alpha^5) - (\alpha^7)$$
$$= x^3 + x + 1$$

Property 2: $f(x^p) = f^p(x)$. If $f(x)$ is any polynomial whose coefficients are from $GF(p)$, then it is always true that

$$f(x^p) = f^p(x)$$

This follows, because all of the cross terms generated in $f^p(x)$ are contained either p times or some multiple of p times and consequently must sum to zero.

Example. $(1 + x + x^2)^2 = 1 + x^2 + x^4 + 2(x + x^2 + x^3)$
$$= 1 + x^2 + x^4$$

The reader should note that Property 2 can be used to demonstrate that Property 1 is valid.

Property 3: Factors of $x^m - 1$. The *order* m of a finite field element β is the smallest value of m for which $\beta^m = 1$. By definition, β is a root of $x^m - 1$. If β is also a root of some irreducible polynomial $f(x)$, then $f(x)$ must be a factor of $x^m - 1$.

Property 4: $GF(p^m)$ and Roots of $(x^{p^m - 1} - 1)$. All of the roots of $x^{p^m - 1} - 1$ are all of the nonzero elements in $GF(p^m)$.

Example. (Properties 3 and 4). The polynomial $f(x) = x^7 - 1$ can be factored as

$$x^7 - 1 = (x^3 + x^2 + 1)(x^3 + x + 1)(x + 1)$$

From Table 2-4 one may verify that the roots of $x^3 + x + 1$ are α, α^2 and α^4, the roots of $x^3 + x^2 + 1$ are α^3, α^5, and α^6 and the root of $x + 1$ is 1. These seven roots are the seven nonzero elements in $GF(8)$.

Property 5: Divisibility of $x^m - 1$ by $f(x)$. The smallest value of m for which an arbitrary polynomial, $f(x)$, with no repeated roots divides $x^m - 1$ is given by the least common multiple of the order of the roots of $f(x)$. Consequently, m is the length of the shortest cycle generated by $f(x)$ when $f(x)$ is used to specify a feedback shift register.

Example. The order of the roots of $1 + x + x^3$ is 7, and the order of the roots of $1 + x^3 + x^4$ is 15. The least common multiple of 7 and 15 is 105. Thus,

$$g(x) = (1 + x + x^3)(1 + x^3 + x^4)$$
$$= 1 + x + x^5 + x^6 + x^7$$

will divide $x^{105} - 1$ and will generate a cycle of length 105.

Property 6: Divisibility of $x^n - 1$ by $x^m - 1$. $x^m - 1$ will divide $x^n - 1$ if and only if m divides n. This follows if we note that m must divide n in order for the roots of $x^m - 1$ to also be roots of $x^n - 1$.

Example. The reader can easily verify by long division that $x^3 - 1$ will divide $x^6 - 1$ but will not divide $x^5 - 1$ or $x^7 - 1$.

2.2.6. Codes Specified by Roots

An alternate method of specifying a polynomial code is as follows. A polynomial with coefficients from $GF(q)$ is a code word if and only if it has as roots $\beta_1, \beta_2, \ldots, \beta_r$, where β_i is an element in $GF(q^m)$. That is, $f(x)$ is a code word if, and only if, $f(\beta_1) = 0$, $f(\beta_2) = 0$, etc. As an example, the polynomials of degree 6 or less with coefficients from $GF(2)$ are code words if, and only if, α [an element from $GF(8)$] is a root. Since $x^3 + x + 1$ is the minimum function of α, this is equivalent to specifying that each code word is divisible by $x^3 + x + 1$.

We observe that imposing the condition $f(\beta) = 0$ is the same as specifying

$$f_0 \beta^0 + f_1 \beta^1 + f_2 \beta^2 + \cdots + f_{n-1} \beta^{n-1} = 0$$

For the code in our previous example

$$f_0 \begin{pmatrix} 1 \\ 0 \\ 0 \end{pmatrix} + f_1 \begin{pmatrix} 0 \\ 1 \\ 0 \end{pmatrix} + f_2 \begin{pmatrix} 0 \\ 0 \\ 1 \end{pmatrix} + f_3 \begin{pmatrix} 1 \\ 1 \\ 0 \end{pmatrix} + f_4 \begin{pmatrix} 0 \\ 1 \\ 1 \end{pmatrix} + f_5 \begin{pmatrix} 1 \\ 1 \\ 1 \end{pmatrix} + f_6 \begin{pmatrix} 1 \\ 0 \\ 1 \end{pmatrix} = \mathbf{0}$$

$$(2\text{-}14)$$

We recognize that the various powers of β in (2-14) are simply the columns in the parity check matrix. In general, when a code is specified by its roots, $\beta_1 \beta_2, \ldots, \beta_r$, one may write the parity check matrix in transposed form as

$$\mathbf{H}^T = \begin{bmatrix} \beta_1^0 & \beta_2^0 & & \beta_r^0 \\ \beta_1^1 & \beta_2^1 & & \beta_r^1 \\ \beta_1^2 & \beta_2^2 & & \beta_r^2 \\ . & . & \cdot \ \ \cdot \ \ \cdot & . \\ . & . & & . \\ . & . & & . \\ \beta_1^{n-1} & \beta_2^{n-1} & & \beta_r^{n-1} \end{bmatrix} \qquad (2\text{-}15)$$

When the parity check matrix is written in this form, it is only necessary to include one root from the set β, β^q, β^{q^2}, etc., since

$$f(\beta^q) = f^q(\beta)$$

by Property 2 of Section 2.2.5.

With the code defined in this fashion we note that the roots, $\beta_1, \beta_2, \ldots, \beta_r$ must also be roots of the code generator polynomial. Letting $m_i(x)$ denote the minimum function of β_i, we can define the code generator polynomial as

$$g(x) = \text{LCM}\left[m_1(x), m_2(x), \ldots, m_r(x)\right] \qquad (2\text{-}16)$$

where LCM refers to taking the least common multiple of the polynomials in the brackets. Many types of cyclic codes are specified in this fashion.

2.2.7. Syndrome Polynomials

A key step in most decoding algorithms for group codes is to compute the syndrome of the received sequence. We recall that this is simply the $(n - k)$-symbol vector that is produced by multiplying the received sequence by the parity check matrix. For polynomial codes, there are three closely related forms of the syndrome, each of which is useful in certain applications.

In the first form, we define the syndrome polynomial $S(x)$ as the degree-$(n - k - 1)$ polynomial that is formed by dividing the received sequence by the generator polynomial and saving the remainder. That is,

$$S(x) \equiv r(x) \bmod g(x)$$

Since all code words are divisible by $g(x)$, $S(x)$ will obviously be independent of the transmitted code word and will depend only on the error pattern. It is easy to show that this definition is identical to the definition in Section 2.1.6 by considering the circuit which is used to compute $S(x)$. We recall that the remainder can be computed by using a feedback shift register whose feedback connections are specified by the divisor. If the dividend is applied to the input, the quotient appears as successive terms in the feedback loop and the remainder is stored in the register. For instance, the circuit of Fig. 2-5 will compute any polynomial modulo $(x^3 + x + 1)$. As we have previously shown, this circuit also generates successive columns of the parity check matrix. When an arbitrary sequence is applied to the input, the circuit sifts out and accumulates the columns in the parity check matrix which corresponds to the nonzero elements in the input. Thus, it also accomplishes the necessary matrix multiplication.

A second form of the syndrome can be defined when the generator polynomial consists of two or more irreducible factors. Suppose that

$$g(x) = g_1(x) g_2(x) \cdots g_j(x)$$

Then we define

$$S_1(x) \equiv r(x) \bmod g_1(x)$$
$$S_2(x) \equiv r(x) \bmod g_2(x)$$

$$\vdots$$

$$S_j(x) \equiv r(x) \bmod g_j(x)$$

The fact that these two forms of the syndrome contain exactly the same information follows as a direct consequence of the Chinese remainder theorem for polynomials. This theorem states that there is a unique correspondence between the polynomial $S(x)$ and the set of polynomials $S_1(x)$, $S_2(x)$, etc. This viewpoint is developed in some detail in Chapter 2 in Berlekamp,[2] and the interested reader is referred to this text. Most engineering readers will be able to convince themselves of the correctness of this statement by working a specific example. This example also provides the key for transforming between the two different representations.

Suppose the generator polynomial is given by

$$g(x) = (x^3 + x + 1)(x^3 + x^2 + 1) = x^6 + x^5 + x^4 + x^3 + x^2 + x + 1$$

Since both polynomials have roots of order 7, $g(x)$ generates a code (in this case trivial) of length 7. The parity check matrix which corresponds to $S(x)$ can be found by considering the successive shifts of the feedback shift register specified by $g(x)$. It is given by

$$\mathbf{H}^T = \begin{bmatrix} 1 & 0 & 0 & 0 & 0 & 0 \\ 0 & 1 & 0 & 0 & 0 & 0 \\ 0 & 0 & 1 & 0 & 0 & 0 \\ 0 & 0 & 0 & 1 & 0 & 0 \\ 0 & 0 & 0 & 0 & 1 & 0 \\ 0 & 0 & 0 & 0 & 0 & 1 \\ 1 & 1 & 1 & 1 & 1 & 1 \end{bmatrix}$$

If the syndrome is specified as

$$S_1(x) \equiv r(x) \bmod (x^3 + x + 1)$$

and

$$S_2(x) \equiv r(x) \bmod (x^3 + x^2 + 1)$$

then the corresponding parity check matrix is

$$(\mathbf{H}')^T = \begin{bmatrix} 1 & 0 & 0 & 1 & 0 & 0 \\ 0 & 1 & 0 & 0 & 1 & 0 \\ 0 & 0 & 1 & 0 & 0 & 1 \\ 1 & 1 & 0 & 1 & 0 & 1 \\ 0 & 1 & 1 & 1 & 1 & 1 \\ 1 & 1 & 1 & 1 & 1 & 0 \\ 1 & 0 & 1 & 0 & 1 & 1 \end{bmatrix}$$

$S_1(x)$ and $S_2(x)$ can be computed independently by utilizing two separate feedback shift registers specified by $g_1(x)$ and $g_2(x)$. If the syndrome has been computed as $S(x)$ and one requires $S_1(x)$ and $S_2(x)$, then the necessary transform can be computed by taking the appropriate linear combination of the rows of $\mathbf{H}$ to create $\mathbf{H}'$. If we denote the ith row of $\mathbf{H}$ by $\mathbf{h}_i$ and the ith row of $\mathbf{H}'$ by $\mathbf{h}'_i$, then by inspection

$$\begin{aligned}
\mathbf{h}'_1 &= \mathbf{h}_1 + \mathbf{h}_4 + \mathbf{h}_6 \\
\mathbf{h}'_2 &= \mathbf{h}_2 + \mathbf{h}_4 + \mathbf{h}_5 + \mathbf{h}_6 \\
\mathbf{h}'_3 &= \mathbf{h}_3 + \mathbf{h}_5 + \mathbf{h}_6 \\
\mathbf{h}'_4 &= \mathbf{h}_1 + \mathbf{h}_4 + \mathbf{h}_5 + \mathbf{h}_6 \\
\mathbf{h}'_5 &= \mathbf{h}_2 + \mathbf{h}_5 + \mathbf{h}_6 \\
\mathbf{h}'_6 &= \mathbf{h}_3 + \mathbf{h}_4 + \mathbf{h}_5
\end{aligned} \qquad (2\text{-}17)$$

The transformation in the opposite direction can be found by inverting this set and is given by

$$\begin{aligned}
\mathbf{h}_1 &= \mathbf{h}'_1 + \mathbf{h}'_3 + \mathbf{h}'_6 \\
\mathbf{h}_2 &= \mathbf{h}'_1 + \mathbf{h}'_2 + \mathbf{h}'_3 + \mathbf{h}'_4 + \mathbf{h}'_6 \\
\mathbf{h}_3 &= \mathbf{h}'_1 + \mathbf{h}'_2 + \mathbf{h}'_4 + \mathbf{h}'_5 + \mathbf{h}'_6 \\
\mathbf{h}_4 &= \mathbf{h}'_2 + \mathbf{h}'_5 \\
\mathbf{h}_5 &= \mathbf{h}'_1 + \mathbf{h}'_4 \\
\mathbf{h}_6 &= \mathbf{h}'_2 + \mathbf{h}'_3 + \mathbf{h}'_5 + \mathbf{h}'_6
\end{aligned} \qquad (2\text{-}18)$$

In Fig. 2-8 a circuit is shown which first computes $S(x)$ and then uses

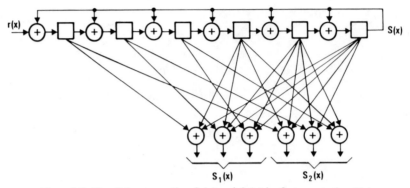

Figure 2-8. Circuit for computing $S_1(x)$ and $S_2(x)$ by first computing $S(x)$.

(2-17) to compute $S_1(x)$ and $S_2(x)$. In Fig. 2-9 the procedure is reversed so that $S_1(x)$ and $S_2(x)$ are computed first, and $S(x)$ is computed using (2-18). Actually, the transformation in this direction is rarely used and the purpose of the example is just to indicate that it is possible.

A third form of the syndrome is to define the vectors $\mathbf{S}_1, \mathbf{S}_2$, etc. as

$$\mathbf{S}_1 = r(\beta_1)$$
$$\mathbf{S}_2 = r(\beta_2)$$

etc.

where β_1 is a root of $g_1(x)$, β_2 is a root of $g_2(x)$, etc. This corresponds to the form of the parity check matrix given in (2-15). For the previous example,

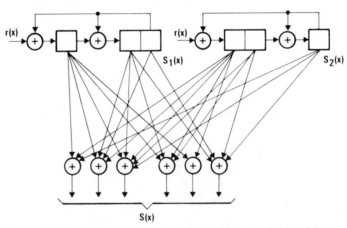

Figure 2-9. Circuit for computing $S(x)$ by first computing $S_1(x)$ and $S_2(x)$.

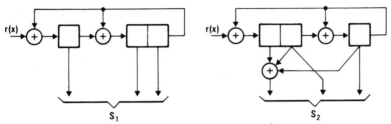

Figure 2-10. Circuit for evaluating $S_1 = r(\alpha)$ and $S_2 = r(\alpha^3)$ from $S_1(x)$ and $S_2(x)$.

the parity check matrix $\mathbf{H}''$ is given by

$$(\mathbf{H}'')^T = \begin{bmatrix} 1 & 0 & 0 & 1 & 0 & 0 \\ 0 & 1 & 0 & 1 & 1 & 0 \\ 0 & 0 & 1 & 1 & 0 & 1 \\ 1 & 1 & 0 & 0 & 0 & 1 \\ 0 & 1 & 1 & 1 & 1 & 1 \\ 1 & 1 & 1 & 0 & 1 & 0 \\ 1 & 0 & 1 & 0 & 1 & 1 \end{bmatrix} \qquad (2\text{-}19)$$

It is easy to show that $\mathbf{H}''$ can be computed from $\mathbf{H}'$ using

$$\mathbf{h}_1'' = \mathbf{h}_1'$$
$$\mathbf{h}_2'' = \mathbf{h}_2'$$
$$\mathbf{h}_3'' = \mathbf{h}_3'$$
$$\mathbf{h}_4'' = \mathbf{h}_4' + \mathbf{h}_5' + \mathbf{h}_6'$$
$$\mathbf{h}_5'' = \mathbf{h}_5'$$
$$\mathbf{h}_6'' = \mathbf{h}_6'$$

A circuit which realizes this computation is shown in Fig. 2-10. An alternate realization is to compute $\mathbf{H}''$ directly using the circuit shown in Fig. 2-11.

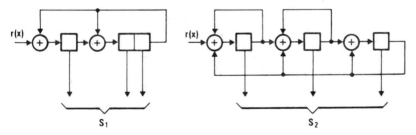

Figure 2-11. Circuit for computing S_1 and S_2 directly from $r(x)$.

The connections for the circuit which realizes S_2 can be obtained by inspection of the second set of columns in (2-19). From this set one may observe that a single shift causes the contents of register 1 to go to registers 1 and 2, the contents of register 2 to go to registers 2 and 3, and the contents of register 3 to go to registers 1, 2, and 3.

2.2.8. Modifications of Codes

Often in applying coding to a communication system there may be constraints on the code block length, n, or the number of information symbols, k, that require values other than those which the code designer would prefer. In such cases one can usually modify the desired code to fit the particular constraints on n and k. This can be done by increasing the block length by adding either more information or parity symbols, by decreasing the block length by omitting information or check symbols, or by increasing or decreasing the number of code words while holding the block length constant. These six basic code modifications are easily explained by using the (7,4) Hamming code shown in Fig. 2-12.

A code can be *extended* by annexing additional parity check symbols. The additional check symbols are usually carefully chosen to improve the code weight structure. The most common such modification is the addition of a single overall parity check. The value of the overall parity check is equal to the remainder obtained by dividing the original code word by the

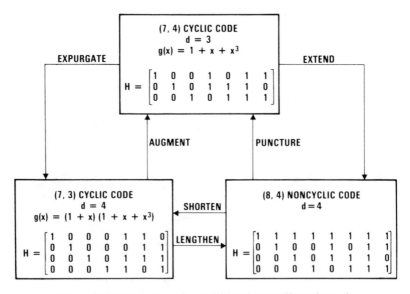

Figure 2-12. Codes derived by modifying the (7, 4) Hamming code.

polynomial $(x - 1)$. With an overall parity check the weight of every code word must now be even. Thus, the (7,4) $d = 3$ Hamming code becomes an (8,4) $d = 4$ code. In a similar fashion, all codes with an odd minimum distance will have their minimum distance increased by one by the addition of an overall parity check.

A code can be *punctured* by deleting parity check symbols. This process is the inverse of extending a code. Thus, the (8,4) extended Hamming code can be punctured to produce the (7,4) code. Unless the deleted parity check symbol is very carefully chosen, the minimum distance will decrease by one as a result of the puncturing operation.

A code may be *expurgated* by discarding some of the code words. This is easily done with cyclic codes by multiplying the generator polynomial by another factor of $x^n - 1$. The most common expurgated code is the one generated by $g(x)(x - 1)$, and the code words of this code are the even-weight code words of the original code. Thus, the (7,4) Hamming code can be expurgated to give a (7,3), $d = 4$ code with $g(x) = (1 + x)(1 + x + x^3)$.

A code may be *augmented* by adding new code words to the code. This process is the inverse of expurgating a code. Any cyclic code generated by the polynomial $g(x)$ may be augmented into another code of the same length using a generator polynomial that is a factor of $g(x)$. For example, if $g(x)$ contains $x - 1$ as a factor, then an augmented code is generated by $g(x)/(x - 1)$.

A code can be *lengthened* by adding additional information symbols. If one starts with a binary cyclic code which has $x - 1$ as a factor of $g(x)$, then the code is lengthened by a two-step process. First, the code is augmented into the code generated by $g(x)/(x - 1)$ (this adds the all-ones code word to the code). Second, the augmented code is then extended by adding an overall parity check. In most cases this process can be carried out without reducing the minimum distance of the code.

A code can be *shortened* by deleting information symbols. This is usually done in cyclic codes by making a segment of information symbols identically zero at the beginning of each code word.

2.3. Important Classes of Group Codes

In this section we will describe some of the more important classes of group codes. These codes are either cyclic or are derived by modifying cyclic codes. We will provide enough detail regarding the characteristics of each of these classes of codes to allow performance calculations and implementation of the practical decoding algorithms to be described in succeeding chapters. Additional properties of these codes as well as the relationships with other classes of codes may be found in the classical algebraic coding theory texts.[1, 2, 8]

2.3.1. BCH Codes

The BCH codes are a generalization of Hamming codes which allow multiple error correction. These codes are most easily defined in terms of the roots of the generator polynomial.

Definition. A *primitive t-error-correcting BCH code* over GF(q) of block length $n = q^m - 1$ has $\alpha^{m_0}, \alpha^{m_0+1}, \ldots, \alpha^{m_0+2t-1}$ as roots of the generator polynomial, $g(x)$ (for any m_0), where α is a primitive element of GF(q^m).

Thus, the generator polynomial of a BCH code may be written as

$$g(x) = \text{LCM}\left[m_{m_0}(x), m_{m_0+1}(x), \ldots, m_{m_0+2t-1}(x)\right] \qquad (2\text{-}20)$$

The codes with $m_0 = 1$ are called the *narrow-sense BCH codes*. The *nonprimitive codes* are defined in the same fashion except that α is replaced by a nonprimitive element, β, of GF(q^m), and the code block length becomes the order of β. Lists of code generator polynomials for both primitive and nonprimitive binary BCH codes are given in Appendix A.

Example. We wish to find the generator polynomial for the narrow-sense, 2-error-correcting BCH code of length 7. This requires that $\alpha, \alpha^2, \alpha^3$, and α^4 be roots of $g(x)$, where α is a primitive element of GF(8). Assume GF(8) is defined by Table 2-4. Then computing the minimum functions of the desired roots we see that

$$m_1(x) = m_2(x)$$
$$= m_4(x)$$
$$= (x - \alpha)(x - \alpha^2)(x - \alpha^4)$$
$$= x^3 + x + 1$$

and

$$m_3(x) = (x - \alpha^3)(x - \alpha^6)(x - \alpha^5)$$
$$= x^3 + x^2 + 1$$

Thus, the generator polynomial is

$$g(x) = (x^3 + x + 1)(x^3 + x^2 + 1)$$
$$= x^6 + x^5 + x^4 + x^3 + x^2 + x + 1$$

This $g(x)$ generates the trivial (7,1) repetition code. Actually, this code has $d = 7$ and can correct all triple errors. This is not particularly unusual. In a number of cases the actual error-correcting capability of BCH codes exceeds the value specified for the code.

The BCH codes provide a large class of easily constructed codes of arbitrary block length and code rate. They are important not only because of the flexibility in the choice of code parameters, but also because at block lengths of a few hundred or less many of these codes are among the best-known codes of the same block length and code rate. Further discussion of the code properties is deferred until Chapter 5, where a practical, algebraic decoding technique based on these properties is presented.

An important subclass of the BCH codes is the class with $m = m_0 = 1$. These are called the Reed–Solomon codes. For these nonbinary codes defined over $GF(q)$ the block length is $n = q - 1$, and the code generator polynomial is given by

$$g(x) = (x - \alpha)(x - \alpha^2) \cdots (x - \alpha^{2t}) \tag{2-21}$$

Note that $g(x)$ has degree $2t$ so that only $2t$ parity symbols are required for t-error correction. Typically, $q = 2^m$ is selected. This means that the code provides correction of 2^m-ary symbols and hence burst errors. These codes are very useful in two-level, "concatenated" coding schemes (described in detail in Chapter 8).

Example. We wish to construct the generator polynomial for the 2-symbol-error-correcting Reed–Solomon code of length 7. Again assume that $GF(8)$ is defined by Table 2-4. Then using (2-21) the generator polynomial is

$$g(x) = (x - \alpha)(x - \alpha^2)(x - \alpha^3)(x - \alpha^4)$$
$$= x^4 + \alpha^3 x^3 + x^2 + \alpha x + \alpha^3$$

Thus, $g(x)$ generates a $(7,3)$ code which can correct all double-error patterns. Note that the coefficients of this generator polynomial are typically elements of $GF(8)$.

Unfortunately, very little is known in general about the weight structure of BCH codes. The weight structure has been found by exhaustive search for some of the codes with either k or $n - k$ small. We previously gave the general form of the weight structure of the Hamming codes. However, the weight structure of an (n, k) Reed–Solomon code is known and is $A_0 = 1$, $A_j = 0\,(1 \leq j \leq n - k)$, and

$$A_j = \binom{n}{j} \sum_{h=0}^{j-1-(n-k)} (-1)^h \binom{j}{h} [q^{j-h-(n-k)} - 1] \tag{2-22}$$

for $n - k + 1 \leq j \leq n$.

2.3.2. Golay Codes

In searching for perfect codes, Golay noted that (1-4) is satisfied for $n = 23$, $k = 12$, and $t = 3$.[11] This implies that a perfect (23,12) code may exist. Such a code does exist and is a nonprimitive BCH code constructed in the following fashion.

Let α be a primitive element over $GF(2^{11})$, and let $\beta = \alpha^{89}$. Since $2^{11} - 1 = 89 \times 23$, β has order 23. The code is specified by requiring that β, β^2, β^3, and β^4 be roots of $g(x)$. The roots of the minimum function $m_1(x)$ of β are

$$\beta, \beta^2, \beta^4, \beta^8, \beta^{16}, \beta^{32} = \beta^9, \beta^{18}, \beta^{36} = \beta^{13}, \beta^{26} = \beta^3, \beta^6, \beta^{12}$$

Note that $m_1(x)$ contains all of the roots required to be in $g(x)$ so $g(x) = m_1(x)$. Depending upon which irreducible polynomial is used to construct $GF(2^{11})$ one obtains either

$$g(x) = x^{11} + x^9 + x^7 + x^6 + x^5 + x + 1$$

or

$$g(x) = x^{11} + x^{10} + x^6 + x^5 + x^4 + x^2 + 1$$

as the generator polynomial of the (23,12) Golay code. The generator polynomial was obtained by specifying a double-error-correcting BCH code. The BCH decoding algorithm described in Chapter 5 allows double-error correction. Actually, the minimum distance of this code is 7 and triple-error correction is possible by using a table look-up decoder or one of the techniques described in the next chapter. The extended (24,12) Golay code is quite useful because it has $d = 8$ and is exactly $R = 1/2$.

The weight structure of these codes is known. For the (23,12) code the weight enumerator is

$$A(x) = 1 + 253(x^7 + 2x^8 + 2x^{15} + x^{16}) + 1288(x^{11} + x^{12}) + x^{23}$$

The weight enumerator of the extended (24,12) code is derived directly from this since all odd-weight code words of the (23,12) code have their weight increased by one, i.e.,

$$A(x) = 1 + 759(x^8 + x^{16}) + 2576 x^{12} + x^{24}$$

The (23,12) code is the only known nontrivial multiple-error-correcting binary perfect code. The Hamming codes and the repetition codes with n odd are the only other binary perfect codes. Finally, the only known non-binary perfect code is a ternary [defined over $GF(3)$] Golay (11,6) code with $d = 5$.

2.3.3. Maximal-Length Codes

The *maximal-length codes* are the duals of the Hamming codes. Remember that a Hamming code of length $2^m - 1$ is obtained by taking a primitive polynomial, $p(x)$, as its generator polynomial. The dual code of the same length is obtained by letting the parity check polynomial $h(x) = p(x)$. This polynomial can be used as the feedback connection polynomial in the k-stage shift register encoder ($k = m$) shown in Fig. 2-7. This produces a $(2^m - 1, m)$ code.

The maximal-length codes have several interesting properties. Note that the m-stage encoder has only $2^m - 1$ nonzero states, which implies that the maximum possible cycle length is $2^m - 1$. Since this is also the code block length, each code word is an output sequence of maximum length. In addition, the encoder sequences through all $2^m - 1$ nonzero states in one cycle. This means that all $2^m - 1$ nonzero code words are cyclic shifts of a single nonzero code word. Therefore, all nonzero code words have the same weight; this code is a so-called *equidistant* or *simplex* code.

The weight structure of these codes is easily determined by examining the encoder structure. Note that for any nonzero code word, the encoder cycles through all $2^m - 1$ different nonzero states, and the weight of the code word is the total number of these states which have a most-significant bit equal to 1. Therefore, the weight of each nonzero code word is 2^{m-1} and the weight enumerator is

$$A(x) = 1 + (2^m - 1)x^{2^{m-1}}$$

This weight enumerator is related to that of the Hamming codes through the McWilliams identity (2-5). A union bound on sequence error rate for these codes with unquantized soft decision maximum-likelihood decoding and coherent PSK is given by

$$P_s \le (2^m - 1)Q\left[(2^m RE_b/N_0)^{1/2}\right] \tag{2-23}$$

with $R = m/(2^m - 1)$. Using (1-41) the union bound on average bit error probability is

$$P_b \le 2^{m-1}Q\left[(2^m RE_b/N_0)^{1/2}\right] \tag{2-24}$$

Some typical performance curves are shown in Fig. 2-13. Note that substantial coding gain is available with small m. At these values of m construction of a maximum-likelihood decoder is feasible.

Example. Consider the (15,4) maximal-length code. Any primitive polynomial of degree 4 can be used as $h(x)$, e.g., take

$$h(x) = x^4 + x + 1$$

Using this polynomial for the feedback connections, an encoder of the form

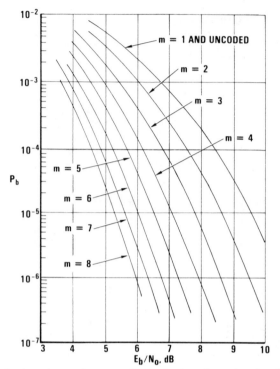

Figure 2-13. Union bound on performance of $(2^m - 1, m)$ maximum-length codes with unquantized soft decisions.

shown in Fig. 2-7 will generate the code. Note that the generator polynomial is

$$g(x) = (x^{15} - 1)/(x^4 + x + 1)$$
$$= x^{11} + x^8 + x^7 + x^5 + x^3 + x^2 + x + 1$$

This polynomial is a code word. The other 14 nonzero code words are the 14 cyclic shifts of this polynomial. Thus, the nonzero code words are the polynomials

$$c(x) = x^i g(x) \bmod (x^{15} - 1)$$

for $i = 0, 1, \ldots, 14$.

The encoder for these codes is sometimes called a maximal-length feedback shift register. It has a number of other uses. For m large, this shift register produces a sequence with excellent "randomness" properties. It has been used in applications requiring random number generation and also in applications requiring pseudorandom binary sequences such as in spread spectrum systems.

2.3.4. Reed–Muller Codes

The Reed–Muller codes are binary group codes that are equivalent to cyclic codes with an overall parity check added. These codes are defined in the following fashion.

Definition. Let v_0 be a vector whose 2^m components are all ones. Let $v_1, v_2, \ldots, v_m$ be the rows of a matrix with all possible m-tuples as columns. The rth order *Reed–Muller* code has as its basis vectors the vectors $v_0, v_1, \ldots, v_m$ and all of their vector products r or fewer at the time.

In this definition we define the vector product of $\mathbf{a} = (a_1, \ldots, a_n)$ and $\mathbf{b} = (b_1, \ldots, b_n)$ as

$$\mathbf{ab} = (a_1 b_1, \ldots, a_n b_n)$$

For any value of m the rth-order Reed–Muller code has parameters

$$n = 2^m$$

$$k = \sum_{i=0}^{r} \binom{m}{i}$$

and

$$d = 2^{m-r}$$

The rth-order code is the dual of the $(m - r - 1)$th-order code.

The first-order codes are closely related to the maximal-length codes. If we start with a maximal-length code and extend it by adding an overall parity check, we obtain an *orthogonal code*. This code has $n = 2^m$ and all nonzero code words have weight $d = 2^{m-1}$. Thus, each code word agrees in 2^{m-1} positions and disagrees in 2^{m-1} positions with every other code word. Using an antipodal signal set with this code will result in a set of 2^m orthogonal signals for the 2^m code words (hence the name orthogonal code). The first-order Reed–Muller code is obtained directly from the orthogonal code by augmenting it with the all-ones code word. In terms of the transmitted signal this produces the original orthogonal signal set plus the complement of each of these signals. For this reason, the resulting code is often called a *bi-orthogonal code*. Performance of both the orthogonal and bi-orthogonal codes is quite close to the performance of the maximal-length codes. However, the bi-orthogonal codes are often preferred in coherent systems because of implementation advantages. The most serious disadvantage of these codes is that they have very low code rates.

Generation of the higher-order Reed–Muller codes is demonstrated in Table 2-6 for $n = 8$. The first-order code is generated by using the vectors

Table 2-6. Basis Vectors for the Reed–Muller Codes of Length 8

$$
\begin{aligned}
v_0 &= (1 \quad 1 \quad 1 \quad 1 \quad 1 \quad 1 \quad 1 \quad 1) \\
v_1 &= (0 \quad 0 \quad 0 \quad 0 \quad 1 \quad 1 \quad 1 \quad 1) \\
v_2 &= (0 \quad 0 \quad 1 \quad 1 \quad 0 \quad 0 \quad 1 \quad 1) \\
v_3 &= (0 \quad 1 \quad 0 \quad 1 \quad 0 \quad 1 \quad 0 \quad 1) \\
v_1 v_2 &= (0 \quad 0 \quad 0 \quad 0 \quad 0 \quad 0 \quad 1 \quad 1) \\
v_1 v_3 &= (0 \quad 0 \quad 0 \quad 0 \quad 0 \quad 1 \quad 0 \quad 1) \\
v_2 v_3 &= (0 \quad 0 \quad 0 \quad 1 \quad 0 \quad 0 \quad 0 \quad 1) \\
v_1 v_2 v_3 &= (0 \quad 0 \quad 0 \quad 0 \quad 0 \quad 0 \quad 0 \quad 1)
\end{aligned}
$$

v_0, v_1, v_2, and v_3 as rows of the generator matrix. The second-order code is generated by augmenting this matrix with the rows $v_1 v_2$, $v_1 v_3$, and $v_2 v_3$. Finally, the third-order code is generated by adding the row $v_1 v_2 v_3$.

The Reed–Muller codes are important because these codes and some closely related codes, the Euclidean geometry and projective geometry codes, can be decoded using the threshold decoding algorithm presented in the next chapter. Another related class of codes, the difference-set cyclic codes, is presented in the next chapter after introducing the threshold decoding technique.

2.3.5. Quadratic Residue Codes

Let p be an odd prime number. Then the numbers $1^2, 2^2, 3^2 \ldots$, reduced mod p are called the quadratic residues mod p. One can find the complete set of *quadratic residues* mod p by considering only the squares of numbers from 1 to $p - 1$. In fact, since $(p - a)^2 \equiv a^2 \bmod p$, the complete set is generated by the squares $1^2, 2^2, \ldots, [(p - 1)/2]^2$. The remaining numbers mod p are called the *nonresidues*.

For example, if $p = 17$, the quadratic residues mod 17 are $1^2 = 1$, $2^2 = 4$, $3^2 = 9$, $4^2 = 16$, $5^2 = 25 \equiv 8$, $6^2 = 36 \equiv 2$, $7^2 = 49 \equiv 15$, and $8^2 = 64 \equiv 13$. The remaining numbers 3, 5, 6, 7, 10, 11, 12, and 14 are the nonresidues. Quadratic residue codes are specified by using a set of quadratic residues to specify the roots of the code generator polynomial.

Definition. Let R be the set of quadratic residues of the number $n = 8m \pm 1$, and let N be the set of nonresidues. Define the generator polynomials

$$
g_r(x) = \prod_{i \varepsilon R} (x - \alpha^i)
$$

and

$$
g_n(x) = \prod_{i \varepsilon N} (x - \alpha^i)
$$

where α is an element of order n in an extension field of $GF(2)$. Then the codes generated by $g_r(x), (x-1)g_r(x), g_n(x)$, and $(x-1)g_n(x)$ are referred to as *quadratic residue codes*.

The codes generated by $g_r(x)$ and $g_n(x)$ are equivalent codes in the sense that they have the same weight structure. [This statement is also true for $(x-1)g_r(x)$ and $(x-1)g_n(x)$.] Note that for $n = 8m \pm 1$ we have $x^n - 1 = (x-1)g_r(x)g_n(x)$, where $g_r(x)$ contains the residues and $g_n(x)$ contains the nonresidues. We can continue the previous example with $n = 17$. Let β be a primitive element of $GF(2^8)$. Since $255 = 15 \times 17, \alpha = \beta^{15}$ has order 17. Then the roots of $g_r(x)$ can be taken to be

$$\alpha, \alpha^2, \alpha^4, \alpha^8, \alpha^{16}, \alpha^{32} = \alpha^{15}, \alpha^{30} = \alpha^{13}, \alpha^{26} = \alpha^9, \quad \text{and} \quad \alpha^{18} = \alpha$$

This set of powers of α is the complete set of quadratic residues mod 17. Thus, these are all of the roots of $g_r(x)$. Using the polynomial $x^8 + x^4 + x^3 + x^2 + 1$ as the irreducible polynomial for constructing $GF(2^8)$, then $g_r(x)$ is the minimum function of $\alpha = \beta^{15}$, i.e.,

$$g_r(x) = x^8 + x^7 + x^6 + x^4 + x^2 + x + 1$$

In a similar fashion $g_n(x)$ is the minimum function of $\alpha^3 = \beta^{45}$, i.e.,

$$g_n(x) = x^8 + x^5 + x^4 + x^3 + 1$$

Both of the (17,9) quadratic residue codes have $d = 5$.

It can be shown that the minimum distance of quadratic residue codes satisfies the inequality

$$d^2 > n$$

This inequality can be strengthened to

$$d^2 - d + 1 \geq n$$

for $n = -1 \mod 4$. Unfortunately, these bounds are poor for large values of n and are not helpful in determining the minimum distance of most codes. However, the minimum distances of some quadratic residue codes of moderate lengths are greater than the minimum distances of BCH codes of comparable lengths. This observation has encouraged a substantial amount of research activity to find long quadratic residue codes and to determine their properties. Some of the more important theoretical results and an ample list of references may be found in MacWilliams and Sloane,[8] Berlekamp,[2] and Peterson and Weldon.[1]

Some of these codes are also examples of nonprimitive BCH codes [e.g., the (23,12) Golay code] and are listed in Table A-2. In many cases, these codes are more difficult to decode than BCH codes. However, decoding algorithms are presented in the next two chapters which can be applied to these codes.

2.3.6. Remarks

We have presented a discussion of some of the more important classes of block codes. These codes are attractive in terms of performance and implementation difficulty. The theory was presented in enough detail to allow an understanding of these issues in the succeeding chapters. In each of these cases substantially more detail on the theory may be found in one of the classical texts on algebraic coding theory. Finally, there are numerous classes of interesting codes which we have chosen not to present (such as Goppa codes, Srivastava codes, and generalized BCH codes) because they are beyond the scope of this book.

Problems

2-1. Consider the group code with the code word defined by

$$a = (a_1, a_2, a_3, a_4, a_5, a_1 + a_2 + a_4 + a_5, a_1 + a_3 + a_4 + a_5,$$

$$a_1 + a_2 + a_3 + a_5, a_1 + a_2 + a_3 + a_4)$$

Give the parity check matrix for this code and find n, k, and d for this code.

2-2. Given the parity check matrix

$$\mathbf{H} = \begin{bmatrix} 1 & 0 & 0 & 1 & 0 & 0 & 1 & 1 & 0 \\ 1 & 0 & 1 & 0 & 1 & 0 & 0 & 1 & 0 \\ 0 & 1 & 1 & 1 & 0 & 0 & 0 & 0 & 1 \\ 1 & 0 & 1 & 0 & 1 & 1 & 1 & 0 & 1 \end{bmatrix}$$

find the reduced echelon canonical form of this matrix.

2-3. Add an overall parity check to the code of Problem 2-1 and give the resulting parity check matrix. What is the minimum distance of this code?

2-4. Derive the generator matrices from the parity check matrices for the codes in Problems 2-1 and 2-3.

2-5. For the codes of Problems 2-1 and 2-3, give decoding tables that list coset leaders and the corresponding syndromes.

2-6. Give multiplication and addition tables for $GF(7)$ and find the multiplicative inverse of each nonzero element.

2-7. Derive the mapping between the ordinary and logarithmic representations of $GF(2^4)$ (as in Table 2-4) using the primitive irreducible polynomial $p(x) = 1 + x + x^4$.

2-8. A $(15, 5)$ binary BCH code has α, α^3, and α^5 as roots of the code generator polynomial, where α is a primitive element of $GF(2^4)$. Find the generator polynomial. Is this code a cyclic code? Find the parity check polynomial $h(x)$ of this code. Draw block diagrams of k-stage and $(n - k)$-stage encoders. Give the generator and parity check matrices of this code.

2-9. The polynomial $x^{63} - 1$ can be factored completely in $GF(2^6)$. If α is a primitive element of $GF(2^6)$, what is the order of α^7 and what is the smallest value of m for which α^7 is a root of $x^m - 1$? What is the degree of the minimum function over $GF(2)$ of α^7? If this minimum function was the generator polynomial of a cyclic code, what would be the length of the code?

2-10. Consider the (15, 5) BCH code of Problem 2-8. The parity check matrix, **H**, corresponding to $S(x) \equiv r(x) \bmod g(x)$ was found. By representing the minimum functions of α, α^3, and α^5 by $g_1(x)$, $g_2(x)$, and $g_3(x)$ one can form the syndromes

$$S_1(x) \equiv r(x) \bmod g_1(x)$$

$$S_2(x) \equiv r(x) \bmod g_2(x)$$

and

$$S_3(x) \equiv r(x) \bmod g_3(x)$$

Derive the form of the parity check matrix **H**′ corresponding to the syndromes $S_1(x)$, $S_2(x)$, and $S_3(x)$. Express the rows of **H**′ as linear combinations of the rows of **H**. This result provides the relationship between $S(x)$ and $\{S_1(x), S_2(x), S_3(x)\}$.

2-11. Repeat Problem 2-8 for the (15, 7) binary BCH code which has α and α^3 as roots of the code generator polynomial.

2-12. Consider the group code with the code word defined by

$$a = (a_1, a_2, a_3, a_3, a_3, a_2, a_1 + a_3, a_1 + a_3, a_1$$
$$+ a_2, a_1 + a_2 + a_3, a_1 + a_2 + a_3)$$

Give the generator and parity check matrices for this code and find n, k, and d. Add the overall parity check to the code and give the new generator and parity check matrices and the new code minimum distance.

2-13. Derive the weight enumerator, $A(x)$, for the orthogonal and bi-orthogonal codes. Using this function give a union bound on P_s and P_b for PSK signaling with coherent detection (infinitely quantized). Repeat for a BSC with transition probability p.

2-14. Consider a coded system that performs error detection but no error correction for a BSC with transition probability p. In terms of p and the weight enumerator, $A(x)$, give an exact expression for $P_{s_{ud}}$, the probability of undetected sequence error. Evaluate this expression for the following codes:
  (a) Hamming and extended Hamming codes
  (b) $(n, n - 1)$ codes with an overall parity check
  (c) $(n, 1)$ repetition codes
  (d) maximal-length codes

2-15. Derive the mapping between the ordinary and logarithmic representation of $GF(2^5)$ using the primitive irreducible polynomial $p(x) = x^5 + x^2 + 1$. Then find the generator polynomial, $g(x)$, for each of the following codes of length $n = 31$:

    (a) the Hamming code

    (b) the two- and three-error-correcting BCH codes

    (c) the maximal-length code

2-16. Find the quadratic residues mod 31. Letting β be a primitive element of $GF(2^{10})$ (note that $2^{10} - 1 = 3 \times 11 \times 31$), find the powers of β that are roots of the generator polynomial $g_r(x)$ for the $n = 31$ quadratic residue code. Repeat for $n = 73$. (Note that $2^9 - 1 = 7 \times 73$.)

3

Simple Nonalgebraic Decoding Techniques for Group Codes

Decoding techniques for group codes can be divided into two general categories, algebraic and nonalgebraic. The algebraic techniques basically involve the simultaneous solution of sets of equations for the location and values of the errors. The nonalgebraic techniques, while accomplishing the same goal, are based upon simple structural aspects of the codes which permit one to determine the error patterns in a more direct fashion. In this chapter three nonalgebraic techniques will be discussed. These are Meggitt decoders,[12] first introduced by Meggitt in 1961 for the correction of burst errors, threshold decoders[13] introduced by Massey in 1963, and information set decoding,[14] which was first introduced by Prange in 1962. The discussion will concern only binary codes except when it is specifically noted that the results apply to nonbinary codes as well.

3.1. Meggitt Decoders

Meggitt decoders can in principle be used to decode any cyclic code or shortened cyclic code. As a practical matter, the complexity of these decoders increases rather rapidly as the number of errors to be corrected grows. Hence, they are generally not useful for correcting more than 3 random errors or for correcting more than a single burst of errors. The syndrome is used to identify correctable error patterns. We will utilize the syndrome form, $S(x)$, defined in Section 2.2.7, but other forms could also be used. The basic idea involved in the Meggitt decoder is quite simple and relies on the following properties of cyclic codes:

1. There is a unique one-to-one correspondence between each member in the set of all correctable errors and each member in the set of all syndromes.
2. If $S(x)$ is the syndrome which corresponds to the error polynomial $e(x)$ then $xS(x) \bmod g(x)$ is the syndrome corresponding to $xe(x) \bmod (x^n - 1)$.

Property (2) states that if the error pattern is shifted cyclically one place to the right then the new syndrome is obtained by advancing the feedback shift register containing $S(x)$ one shift to the right. This implies that the set of error patterns can be divided into equivalence classes where each class contains all cyclic shifts of a particular pattern. For a block length n cyclic code, each class can be identified by advancing the feedback shift register containing $S(x)$ no more than n times and testing for a specific pattern after each shift. This property can be used to construct two slightly different versions of a Meggitt decoder. One version employs feedback while the other does not.

The considerations involved in building Meggitt decoders are illustrated by the following example. The polynomial $g(x) = 1 + x^4 + x^6 + x^7 + x^8$ generates a $(15, 7)$ BCH cyclic code. This code has minimum distance 5 and can correct all single errors, double errors, and approximately 30% of all triple errors. Alternately, it can also be used to correct burst errors and can uniquely identify all burst errors of length 4 or less. The basic decoder, when implemented without feedback, is shown in Fig. 3-1. The decoder is operated as follows. The received 15-symbol sequence is shifted

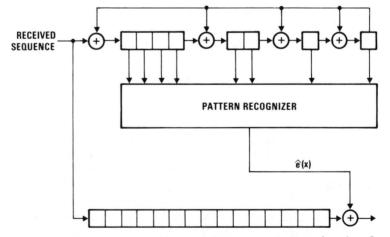

Figure 3-1. Basic decoder for the $(15, 7)$ code generated by $g(x) = 1 + x^4 + x^6 + x^7 + x^8$.

simultaneously into the feedback shift register and into a 15-symbol storage register. At the end of this sequence the syndrome polynomial $S(x)$ has been computed and is stored in the feedback shift register. The feedback shift register is now shifted 15 more times and the received sequence is delivered to the user. The pattern recognizer has been programmed to produce a 1 on the next shift any time the contents of the feedback shift register correspond to a correctable error pattern with an error in the last stage (on the right) of the received sequence register. If the decoder is to correct all single and double errors it must recognize 15 different patterns. One pattern corresponds to a single error in the last position, and 14 correspond to double errors with one error in location 15 and the other error in one of the locations 1 through 14. The patterns which must be identified are easily computed by examining the parity check matrix and computing the syndrome for each of the required error patterns. The transpose of this matrix is shown as cycle (a) in Table 3-1. Note that when a single error has advanced to location 15, the contents of the syndrome register are equal to the last row $(0\ 0\ 0\ 1\ 0\ 1\ 1\ 1)$. Similarly, for a double consecutive error, the syndrome register will contain the modulo-2 sum of rows 14 and 15 $(0\ 0\ 1\ 1\ 1\ 0\ 0\ 1)$ when the two errors are in locations 14 and 15. The other patterns are computed in a similar fashion.

The decoder in Fig. 3-1 does not take full advantage of the cyclic property of the code and consequently must recognize a larger number of patterns than is actually required. This fact is of absolutely no consequence if the pattern recognizer is being realized by a read-only-memory, but it might be quite significant if the pattern recognizer were to be implemented with standard logic functions. The decoder shown in Fig. 3-2 remedies this problem. It differs from Fig. 3-1 in two ways. After the received pattern has been loaded into the storage register and the initial syndrome polynomial $S(x)$ has been computed, the switch is placed in position (1). Then the contents of the storage register are circulated one or more times while simultaneously shifting the feedback shift register and looking for a correctable error pattern. Each time a correctable error is discovered it is corrected and its effect is removed from the feedback shift register by adding a 1 to the input. When the contents of the storage register have been circulated a number of times sufficient to reduce the number of remaining errors to no more than 1, the switch is moved to position (2). Then the pattern is delivered to the user with the final correction being made at this time. At the end of this cycle the contents of the syndrome register will have been set to zero if the designer has chosen to make all possible corrections. If, however, he has chosen to correct only a portion of the possible errors, then a nonzero syndrome may be present and this can be used as an error alarm to signal the user that the most recently delivered code word contains a detectable but uncorrectable error.

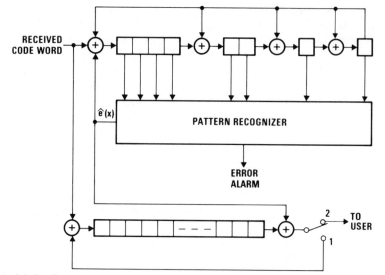

Figure 3-2. Feedback decoder for the (15, 7) code generated by $g(x) = 1 + x^4 + x^6 + x^7 + x^8$.

When the decoder is implemented in this manner, the pattern recognizer is programmed to recognize one member from each cyclic class of errors that also contains an error in position 15. Because the decoder can recognize only one member from each class, a particular symbol which is in error may have to be circulated one or more times before being ultimately corrected. Suppose, for example, that the decoder has been designed to correct all single and double errors and as many triple-error patterns as possible. If one of the correctable triple-error patterns occurs, then one of the three symbols that are in error will always be corrected on the first pass. As soon as this error and its effect on the syndrome has been removed, the decoder will contain a correctable double-error pattern. This pattern may require as many as 15 additional shifts to remove one more of the errors. Following this second cyclic shift the third error can be removed in the process of delivering the corrected word to the user.

Table 3-1 shows the complete set of cycles that can be generated by the feedback shift register for this example. The reader should observe that there are 19 distinct cycles, 16 of length 15 and 3 of length 5. As mentioned previously, cycle (a) is associated with a single error and the pattern recognizer must identify the last word in this cycle. Cycles (b), (c), (d), (f), (h), (i), and (1) are associated with the double-error patterns. The patterns which must be recognized can be obtained by locating an entry on each list that corresponds to an error in position 15. For example, cycle (d) corresponds to the class of error patterns $(1, 4), (2, 5), ..., (12, 15), (1, 13), (2, 14)$, and $(3, 15)$. One could perform the correction on recognizing the syndrome

Table 3-1. Cycles Generated by $g(X) = 1 + X^4 + X^6 + X^7 + X^8$ Generator Polynomial for the (15, 7) $d = 5$ Code

Cycle (a)	Cycle (b)	Cycle (c)	Cycle (d)	Cycle (e)	Cycle (f)
10000000	11000000	11100000	10100000	11110000	10010000
01000000	01100000	01110000	01010000	01111000	01001000
00100000	00110000	00111000	00101000	00111100	00100100
00010000	00011000	00011100	00010100	00011110	00010010
00001000	00001100	00001110	00001010	00001111	00001001
00000100	00000110	00000111	00000101	10001100	10001111
00000010	00000011	10010000	10001001	01000110	11001100
00000001	10001010	01000100	11001111	00100011	01100110
10001011	01000101	00100010	11101100	10011010	00110011
11001110	10101001	00010001	01110110	01001101	10010010
01100111	11011111	10000011	00111011	10101101	01001001
10111000	11100100	11001010	10010110	11011101	10101111
01011100	01110010	01100101	01001011	11100101	11011100
00101110	00111001	10111001	10101110	11111001	01101110
00010111	10010111	11010111	01010111	11110111	00110111

Cycle (g)	Cycle (h)	Cycle (i)	Cycle (j)	Cycle (k)	Cycle (l)
11010000	10110000	11111000	11011000	11101000	10011000
01101000	01011000	01111100	01101100	01110100	01001100
00110100	00101100	00111110	00110110	00111010	00100110
00011010	00010110	00011111	00011011	00011101	00010011
00001101	00001011	10000100	10000110	10000101	10000010
10001101	10001110	01000010	01000011	11001001	01000001
11001101	01000111	00100001	10101010	11101111	10101011
11101101	10101101	10011011	01010101	11111111	11011110
11111101	01010100	11000110	10100001	01111110	01101111
11110101	00101010	01100011	11011011	00111111	10111100
11110001	00010101	10111010	11100110	10010100	01011110
11110011	10000001	01011101	01110011	01001010	00101111
11110010	11001011	10100101	10110110	00100101	10011100
01111001	11101110	11011001	01011001	10011001	01001110
10110111	01110111	11100111	10100111	11000111	00100111

Cycle (m)	Cycle (n)	Cycle (o)	Cycle (p)
10100100	11110100	11010100	11111011
01010010	01111010	01101010	11110110
00101001	00111101	00110101	01111011
10011111	10010101	10010001	10110110
11000100	11000001	11000011	01011011
01100010	11101011	11101010	10100110
00110001	11111110	01110101	01010011
10010011	01111111	10110011	10100100
11000010	10110100	11010011	01010001
01100001	01011010	11100010	10100011
10111011	00101101	01110001	11011010
11010110	10011101	10110011	01101101
01101011	11000101	11010010	10111101
10111110	11101001	01101001	11010101
01011111	11111111	10111111	11100001

Cycle (r)	Cycle (s)	Cycle (t)
11001000	10101100	11111010
01100100	01010110	01111101
00110010	00101011	10110101
00011001	10011110	11010001
10000111	01001111	11100011

pattern corresponding to errors in locations 12 and 15 (row 13 — 0 1 0 0 1 0 1 1), or one could recognize the cyclic shift corresponding to errors in locations 3 and 15 (row 1 — 1 0 1 0 0 0 0 0). The remaining cycles can each be associated with a correctable triple-error pattern. Thus, a complete decoder must recognize 19 separate patterns while a double-error-correcting decoder must recognize 8 and a single-error-correcting decoder must recognize one.

One should especially notice that an error pattern that is contained in the first eight digits of the code word will be identical to its syndrome. An examination of Table 3-1 will reveal that all burst errors of length 4 or less lie on a separate cycle and consequently may be corrected. Thus, a burst-error-correcting decoder may be implemented in exactly the same manner as the random-error-correcting decoder. Alternately, by advancing the phase of the syndrome register by eight shifts with respect to the storage register (e.g., by multiplying $r(x)$ by x^8 before dividing by $g(x)$ as in Fig. 2-4 or by feeding back to the output of stage eight rather than stage 15 in the symbol storage register), the contents of the right-most four stages of the syndrome register will be identical to any correctable burst occurring in positions 12, 13, 14, and 15 of the cyclically shifted received sequence. This pattern can always be identified by looking for four consecutive zeros in the left-most four stages of the syndrome register. The reader may verify for himself that this is true by examining Table 3-1. In this case the decoder will make its correction by adding the error pattern in the syndrome register to the received sequence as it is delivered to the user.

3.2. Information Set Decoding

The use of information sets as the practical basis of a decoding algorithm for group codes was first proposed by Prange in 1962. Since that time the principle has been rediscovered by numerous others, and a family of closely related techniques has evolved. Although these algorithms are not as elegant as some of the algebraic techniques that were developed later, they have the property that they can be easily extended to the soft decision case. They also provide significant insight into the soft decision decoding problem and hence are quite useful for this purpose.

3.2.1. General Decoding Algorithm

In an (n, k) group code an *information set* is defined to be any set of k positions in the code word that can be specified independently. The remaining $n - k$ positions are referred to as the *parity set*. Obviously, if the generator matrix for the code can be written in echelon canonical form, then the first k positions form an information set. Any other set of positions can form an information set if it is possible to make the corresponding columns of the generator matrix unit weight through elementary row operations. For example, consider the (7, 4) Hamming code whose generator matrix is

$$
\mathbf{G} = \begin{bmatrix}
1 & 0 & 0 & 0 & 1 & 1 & 0 \\
0 & 1 & 0 & 0 & 0 & 1 & 1 \\
0 & 0 & 1 & 0 & 1 & 1 & 1 \\
0 & 0 & 0 & 1 & 1 & 0 & 1
\end{bmatrix}
$$

By adding the first row to the third and fourth rows, we may transform this matrix to

$$
\mathbf{G}^1 = \begin{bmatrix}
1 & 0 & 0 & 0 & 1 & 1 & 0 \\
0 & 1 & 0 & 0 & 0 & 1 & 1 \\
1 & 0 & 1 & 0 & 0 & 0 & 1 \\
1 & 0 & 0 & 1 & 0 & 1 & 1
\end{bmatrix}
$$

This has the effect of "interchanging" columns 1 and 5 and modifying column 6. We see that positions 2, 3, 4, and 5 now form an information set.

From this example we observe that a necessary and sufficient condition for being able to "interchange" any arbitrary column with one of the unit weight columns is that they both have a one in the same row. By this criterion we see that column 1 can be interchanged with 5 or 6 but not 7, that column 2 can be interchanged with 6 or 7 but not 5, etc.

Since the symbols contained in the information set can be specified independently, they uniquely define a code word. If there are no errors in these positions, then the remaining symbols in the transmitted code word can be reconstructed. This property provides the basis for all information set decoding algorithms. A general algorithm is as follows:

1. Select several different information sets according to some rule.
2. Construct a code word for each set assuming that the symbols in the information set are correct.
3. Compare each of the hypothesized code words with the actual received sequence and select the code word which is closest (smallest metric).

A flow diagram of this algorithm is shown in Fig. 3-3. The basic concept of the algorithm is the following. A collection of information sets is used to generate (from the received sequence) a corresponding collection of candidate code words. The algorithm then selects as its choice the candi-

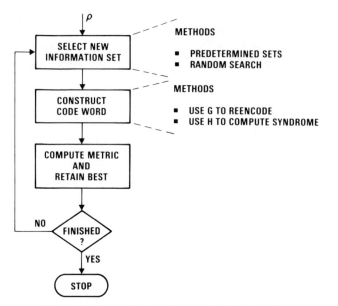

Figure 3-3. General information set decoding algorithm.

date code word with the best metric. This procedure ensures that if the received sequence contains no errors in one of the information sets, then the transmitted code word will be in the collection of candidate code words. Hence, if the actual error pattern is correctable by a maximum-likelihood decoder, then that code word will have the best metric and will be selected by the information set decoder. The required size of the collection of information sets is a strong function of the number of errors that must be corrected. (This statement will be quantified in Section 3.2.3.) There are several specific techniques that can be used in each of the first two steps of the algorithm, and the designer is at liberty to select any combination of these. In selecting an appropriate collection of information sets one can use either a "predetermined" collection of sets (as described in Section 3.2.3) or one can use a "random search" to construct the collection (as described in Section 3.2.4). In constructing a candidate code word, one can either reencode the selected information set using an appropriate $\mathbf{G}$ matrix (as described in this section), or one can simply compute a syndrome using an appropriate $\mathbf{H}$ matrix (as described in Section 3.2.2) to determine a candidate error pattern. The methods of using each of these techniques are developed through examples in the succeeding sections.

As an example of how this technique could be used, consider the (7, 4) Hamming code and let the information sets be $I_0 = \{1, 2, 3, 4\}$, $I_1 = \{4, 5, 6, 7\}$, and $I_2 = \{1, 2, 6, 7\}$. The three generator matrices corresponding to these sets are

$$G_0 = \begin{bmatrix} 1 & 0 & 0 & 0 & 1 & 1 & 0 \\ 0 & 1 & 0 & 0 & 0 & 1 & 1 \\ 0 & 0 & 1 & 0 & 1 & 1 & 1 \\ 0 & 0 & 0 & 1 & 1 & 0 & 1 \end{bmatrix} \qquad G_1 = \begin{bmatrix} 1 & 1 & 0 & 1 & 0 & 0 & 0 \\ 0 & 1 & 1 & 0 & 1 & 0 & 0 \\ 1 & 1 & 1 & 0 & 0 & 1 & 0 \\ 1 & 0 & 1 & 0 & 0 & 0 & 1 \end{bmatrix}$$

$$G_2 = \begin{bmatrix} 1 & 0 & 1 & 1 & 1 & 0 & 0 \\ 0 & 1 & 1 & 0 & 1 & 0 & 0 \\ 0 & 0 & 1 & 1 & 0 & 1 & 0 \\ 0 & 0 & 0 & 1 & 1 & 0 & 1 \end{bmatrix}$$

Suppose that the transmitted sequence is

$$C = (1\ 0\ 1\ 1\ 1\ 0\ 0)$$

and it is received with an error in the fourth position. That is,

$$\rho = (1\ 0\ 1\ 0\ 1\ 0\ 0)$$

Using G_0, G_1, and G_2, we generate three potential code words and compute the distance to ρ. Thus, we have for I_0

hypothesized code word	1 0 1 0 0 0 1
received word	1 0 1 0 1 0 0
error word	0 0 0 0 1 0 1

for I_1

hypothesized code word	0 1 1 0 1 0 0
received word	1 0 1 0 1 0 0
error word	1 1 0 0 0 0 0

and for I_2

hypothesized code word	1 0 1 1 1 0 0
received word	1 0 1 0 1 0 0
error word	0 0 0 1 0 0 0

Note that G_0 and G_1 produce code words that differ from ρ in two positions, whereas G_2 produces a code word that differs in one position. Since this is a distance-3 code, there can only be one code word that is within distance 1 and consequently the correct choice must be the word produced by G_2.

The three information sets specified in this example are always sufficient to correct any single symbol error. This becomes obvious if we note that any single error in positions 1, 2, or 3 will be corrected by I_1, that any error in 5, 6, or 7 will be corrected by I_0, and that the remaining error in position 4 will be corrected by I_2.

The above procedure can be modified slightly to work with soft decisions. Using eight-level quantization with a uniform metric 0 through 7, it is convenient to represent a code word with the symbols 0 and 7 (rather than 0 and 1). In this case the distance between a code word and an arbitrary received sequence is just the sum of the absolute values of the numerical difference on a component-by-component basis. Using this notion, assume that the transmitted sequence is

$$\mathbf{s} = (7\ 0\ 7\ 7\ 7\ 0\ 0)$$

and that it is received as

$$\boldsymbol{\rho} = (7\ 0\ 3\ 3\ 7\ 0\ 1)$$

In order to apply the information set decoding algorithm, we first make hard decisions in the positions corresponding to each information set, re-encode to produce three hypothesized code words, and then compute the difference pattern as before. Thus, for I_0

hypothesized code word	7 0 0 0 7 7 0
received word	7 0 3 3 7 0 1
error word	0 0 3 3 0 7 1

distance = 14

for I_1

hypothesized code word	0 7 7 0 7 0 0
received word	7 0 3 3 7 0 1
error word	7 7 4 3 0 0 1

distance = 22

and for I_2

hypothesized code word	7 0 7 7 7 0 0
received word	7 0 3 3 7 0 1
error word	0 0 4 4 0 0 1

distance = 9

Clearly, I_2 produces the code word which is closest to the received sequence and consequently is the maximum-likelihood choice.

This example also provides some insight for the process by which soft decision decoding produces an improvement in performance over hard decisions. Had we made hard decisions, the received sequence would have been

$$\rho = (1\ \ 0\ \ 0\ \ 0\ \ 1\ \ 0\ \ 0)$$

and a hard decision decoder would have decoded this as

$$c = (1\ \ 0\ \ 0\ \ 0\ \ 1\ \ 1\ \ 0)$$

which is clearly wrong.

Thus, a soft decision decoder allows correction of errors which could not have been corrected using a hard decision decoder. The extent to which this is true can be determined by noting that if the minimum Hamming distance between code words is 3, then the minimum distance using the soft decision metric is 21. Any sequence which differs from the transmitted one by 10 or less is uniquely decodable. Thus, many patterns of two hard errors are potentially decodable.

One final observation is in order. Since we reencode using the information set, there must be at least one information set that is error free on a hard decision basis. In our example, the hard decision process produced errors in positions 3 and 4. Had they occurred in positions 1 and 5 our algorithm would not have corrected them since none of the three information sets would have been error free. This problem can obviously be alleviated by adding more information sets. For the (7, 4) code, it can be shown that seven information sets are required to correct all potentially correctable eight-level patterns. This concept will be discussed in greater detail in Chapter 4.

3.2.2. Information Set Decoding Using the H Matrix

There is a second approach to information set decoding that enables us to focus more directly on the error mechanism. We recall that the basic concept is to assume the information set is error free, reencode using this set to produce a code word, and then to compare this resultant code word to the received sequence. If this comparison is done by adding the received sequence to the code word, then the result is the hypothesized error pattern. This pattern is obviously zero in the information set positions and contains nonzero terms only in the remaining or parity set positions. Further, the portion of the error pattern contained in the parity set is identical to the syndrome. That is, it is just the result of recomputing the parity checks and adding these new checks to the received parity symbols. Thus, we can consolidate the entire operation into one of computing the syndrome. One may, of course, make this observation directly from an examination of the H matrix. The appropriate form of the H matrix, which corresponds to a particular information set, is the one which contains unit weight columns in the parity set. If the error pattern is nonzero only in these positions, then it follows directly that the syndrome will be identical to this error pattern. For example, the **H** matrix for the Hamming code considered earlier is

$$\mathbf{H} = \begin{bmatrix} 1 & 0 & 1 & 1 & 1 & 0 & 0 \\ 1 & 1 & 1 & 0 & 0 & 1 & 0 \\ 0 & 1 & 1 & 1 & 0 & 0 & 1 \end{bmatrix}$$

This matrix corresponds to the information set $\{1, 2, 3, 4\}$. If a received sequence contains errors only in positions 5, 6, or 7, then the syndrome will exactly match this error pattern.

One may view the unit weight columns of the **H** matrix as providing a "window" by which we can look into the received word and see the error pattern. This concept is quite useful for it enables us to recognize a potentially correctable pattern by looking for some characteristics. For instance, in a situation where random errors are known to occur, we may test the weight of the potential error pattern and determine if it is within the error-correcting capability of the code. In a burst-error environment we might look to see if the burst length (number of positions between the first and last error) is consistent with the burst-correcting capability of the code. This fact was used to simplify the Meggitt decoder design in Section 3.1.

When considering information set decoding from the viewpoint of computing syndromes, the important observation to make is that an error pattern can be identified if one can find a parity set which completely contains it. Such a parity set is said to cover the error pattern and a collection of parity sets which covers all error patterns of a particular type is said to form a *covering*. The practical decoding problem is to locate the parity set which will cover a specific unknown error pattern.

There are two basic strategies which can be invoked. One is to preselect a fixed number of covering patterns which will correct a specified set of errors. One then searches through this set in a deterministic fashion to locate the minimum weight error pattern. The advantage to this procedure is that the computational effort required to find the parity check matrix for each information set can be done in advance and then stored in the decoder. A second strategy is to attempt to direct the search by selecting the next covering set to be examined by observing the effect of the previous calculation or by utilizing any reliability information that might be available. Although the work required for each iteration of this procedure may be greater, the number of iterations can sometimes be less. Each of these strategies will be discussed in some detail.

3.2.3. Predetermined Covering Sets

When using a predetermined covering set there are two questions of major importance. The first question is what is the minimum number of distinct parity sets that are required to correct a specified number of errors? The second question is how does one go about finding these sets? Unfortunately, there is no completely satisfactory solution to either of these questions. The classical covering problem, which omits the constraint that a given covering pattern also be a parity set, is, except for a few special cases, an unsolved problem of long standing. Thus, there is little hope of finding an optimum solution to the more difficult constrained problem. One can, however, find answers to all of these questions that are acceptable for engineering purposes. Since no constructive procedure is known for finding optimum covering sets, one is forced to choose an ad hoc procedure which seems appropriate for a given situation and to make some determination of the reasonableness of the result. The usual path through this problem is to first find a lower bound on the number of patterns that are required. If this number appears reasonable, then the next step is to find, by trial and error, or otherwise, a set of patterns which cover all desired errors when the patterns are not constrained to be parity sets. One would normally iterate this step until the answer is reasonably close to the bound or the number of sets becomes acceptable from a cost and complexity viewpoint. The last step is to impose the constraint that each pattern also be a parity set.

At this point one will usually find that all but a few of the covering sets also satisfy the parity set constraint. The choices available are (1) to find additional patterns which will correct the remaining uncovered set of errors, (2) to modify the existing set of patterns by permuting the symbol locations to reduce or eliminate the number of covering sets which are not parity sets, or (3) to modify the decoding algorithm to permit the use of those coverings which are not legitimate parity sets. The most satisfying solution is usually a combination of (2) and (3). To further complicate this problem there is a tradeoff between minimizing the number of parity sets that are required and minimizing the effort (computation plus memory) that is required to use the parity sets. This latter consideration can be particularly important for cyclic codes, where it is often possible to generate a sequence of different parity sets through some simple operation such as a cyclic shift of the code word.

In the next several sections we first give a lower bound on the minimum number of parity sets that are required to cover all t-error patterns in an (n, k) group code. We next give an example of an ad hoc procedure for finding a set of covering patterns for a two-error-correcting $(16, 8)$ code. This is followed by a constructive technique which utilizes a second code to generate the covering set and which sometimes provides a near optimum solution. Finally, we discuss a variation of the information set decoding algorithm which is appropriate for certain cyclic codes and a modification which permits one to use a covering set which is not simultaneously a parity set.

3.2.3.1. Lower Bounds for Covering Sets. A crude lower bound on the number of different parity sets (or information sets) that are required to correct a specified number of errors can be obtained as follows. Assume an (n, k) code and the desire to correct all patterns of t or fewer errors. Note that it is only necessary to concentrate on the t-error case since any set of less than t errors is also covered. In all n positions, the total number of distinct error patterns is $\binom{n}{t}$. Since the size of the covering set is $n - k$, the maximum number of unique patterns that can be covered by a specified set is $\binom{n-k}{t}$. Thus, the smallest number of coverings, N_{cov}, that one can possibly choose and correct all t-error patterns is bounded by

$$
\begin{aligned}
N_{cov} &\geq \binom{n}{t} \Big/ \binom{n-k}{t} \\
&\geq \frac{(n)(n-1)\cdots(n-t+1)}{(n-k)(n-k-1)\cdots(n-k-t+1)}
\end{aligned}
\tag{3-1}
$$

The above result can be refined somewhat. Schonheim[15] has shown that a tighter bound is given by

$$
N_{cov} \geq \left\lceil \frac{n}{n-k} \left\lceil \frac{n-1}{n-k-1} \cdots \left\lceil \frac{n-t+1}{n-k-t+1} \right\rceil \cdots \right\rceil \right\rceil
\tag{3-2}
$$

where the symbol $y = \lceil x \rceil$ denotes the smallest integer, y, such that $y \geq x$. The validity of (3-2) can be established by induction on t and is left as an exercise for the reader (see Problems).

3.2.3.2. Procedures for Finding Covering Sets. A possible procedure for obtaining a covering set is to generate in sequence a set of weight-$(n - k)$ patterns such that each new pattern adds (if possible) some minimum number of previously uncovered errors. For example, suppose we are given a $(16, 8)$ code and would like to determine an acceptably small number of coverings to correct all double-error patterns. Let us assume that any set of eight positions can be a legitimate parity set. Let the first two sets be positions 1 through 8 and 9 through 16. Since these do not overlap, they each cover 28 error patterns and leave 64 uncovered. We next add a pattern which covers as many additional errors as possible. Any pattern which covers four from positions 1 through 8 and four from positions 9 through 16 satisfies this requirement. Thus $\{1, 2, 3, 4, 9, 10, 11, 12\}$ is a reasonable choice. Continuing in this fashion we arrive at the coverings shown in Table

Table 3-2. Coverings for the (16, 8) Code

Parity set																No. of patterns covered	No. of patterns remaining
1	2	3	4	5	6	7	8									28	92
								9	10	11	12	13	14	15	16	28	64
1	2	3	4					9	10	11	12					16	48
1	2	3	4									13	14	15	16	16	32
				5	6	7	8	9	10	11	12					16	16
				5	6	7	8					13	14	15	16	16	0

3-2. Thus, we can cover all double-error patterns with exactly six parity sets. Evaluation of (3-2) yields $N_{cov} \geq 6$ and hence this solution is optimum. There is no guarantee, of course, for a given code that parity sets can actually be formed using the positions indicated. One solution to this potential problem is to modify the parity sets by interchanging positions. For example, if locations 4 and 10 are interchanged the proposed parity sets are as shown in Table 3-3, and this new set can be evaluated to determine if each covering can be chosen as a parity set. A second solution is provided in Section 3.2.3.

From the previous example one might observe that the largest number of new patterns is added at each step if each new pattern differs from all of the previous patterns in as many places as possible. This implies that the distance between patterns should be large, and that the problem of generating a covering set is analogous to the problem of generating a good code. Indeed, there are some situations where large minimum distance codes also make near optimum covering sets. Baumert, McEliece, and Solomon[16] have shown that the (48, 6) punctured Solomon–Stiffler code can be used to find a covering set for the five-error-correcting (48, 24) code. [This code is formed by puncturing a (63, 6) maximal-length code.] They first observed that any five positions of the (48, 6) code are zero in at least one nonzero code word. Second, the (48, 6) code has minimum distance 24. Any word whose weight exceeds this minimum distance can be reduced to weight 24 without destroying the zero covering property. Thus, the 63 nonzero words may be used to form a covering set by choosing the zeros

Table 3-3. Modified Covering for (16, 8) C

1	2	3		5	6	7	8		10						
			4					9		11	12	13	14	15	16
1	2	3	4					9	10	11	12				
1	2	3							10			13	14	15	16
			4	5	6	7	8	9		11	12				
				5	6	7	8					13	14	15	16

in the code word to denote the locations in the parity set. From (3-2), N_{cov} must be 62 or greater. For the case examined by Baumert *et al.*, not all of the 63 nonzero words formed a legitimate information set and it was necessary to add 29 additional patterns using the technique discussed in Section 3.2.3.3. Thus, the total number of patterns actually required is 92, which is quite reasonable in many situations.

The previous example depended on the property that for any five positions there exists at least one nonzero code word that is identically zero in these five locations. This property is true for any number of positions less than the number of information symbols. If the positions in question form part of an information set, then they can obviously be set to zero in a non-zero code word. If all of the positions cannot be specified independently, it must mean that the positions which cannot are linear combinations of those that can. Thus, setting the independent positions to zero forces all of the dependent positions to be zero. Since there must exist at least one additional independent position which can be specified as nonzero, then there must exist a nonzero code word in which any $k - 1$ positions are all zero.

This technique of using a second code to specify a covering set works well for a number of different rate $1/2$ codes. For example, there is a $(12, 4)$ code with minimum distance equal to 6 that can cover all triple-error patterns in a $(12, 6)$ code, with 15 different coverings. The bound is $N_c \geq 14$. There is also a $(32, 5)$ code and a $(32, 6)$ code each having a minimum distance of 16 which can be used to cover all four- or five-error patterns in a $(32, 16)$ code using 31 and 63 different covering patterns, respectively. The lower bound for each of these cases is 30 and 62. Thus, these represent very good potential solutions.

In addition to using the code directly as a set of covering patterns, one may also use the code to provide a list of potential patterns from which one can sequentially choose candidates and be assured that the minimum distance between successive patterns is large.

3.2.3.3. Coverings Generated by Code Permutations. In Prange's original paper on information set decoding he also suggested a technique for generating a sequence of p different information sets that is applicable to any cyclic code. This technique is based upon the fact that all cyclic codes with symbols in $GF(q^m)$ are invariant under the symbol position[†] permutations

$$i \rightarrow (i + v) \bmod n \tag{3-3}$$

[†]In this section it is convenient to designate the symbol positions with the integers 0 through $n - 1$ instead of 1 through n.

and

$$i \rightarrow (q^v i) \bmod n \qquad (3\text{-}4)$$

where v is any integer. The permutation given by (3-3) is simply a statement of the cyclic property. That is, every cyclic shift of a code word is a code word. The permutation given by (3-4) is a result of the fact that

$$f^q(x) = f(x^q)$$

Thus, if $f(x)$ is divisible by $g(x)$ and is consequently a code word then $f(x^q)$ is also a code word and represents a permutation of the t location indices of the code by

$$i \rightarrow qi$$

As an example, suppose that

$$f(x) = a_0 + a_1 x + a_2 x^2 + a_3 x^3 + a_4 x^4 + a_5 x^5 + a_6 x^6$$

is a cyclic code word of length 7 over $GF(2)$. Then

$$f^2(x) = a_0 + a_4 x + a_1 x^2 + a_5 x^3 + a_2 x^4 + a_6 x^5 + a_3 x^6$$

is also a code word. The polynomial $f^2(x)$ is simply $f(x)$ with the location index permuted by $i \rightarrow 2i$. Since the set of permutations described by (3-3) and (3-4) maps code word into code word, the new code word generated by this process will still have an all-zero syndrome. An error pattern associated with this code word will, however, be mapped into a new set of positions. Thus, we can compute a sequence of syndromes corresponding to a sequence of different parity sets by repeated application of (3-3) and (3-4) to the received word.

Instead of actually rearranging the symbols in the received word, it is equally effective to rearrange the columns in the parity check matrix. This can be done by physically rearranging them, by taking appropriate linear combinations of the rows of the parity check matrix, or by taking linear combinations of the elements of the syndrome after the initial matrix multiplication has been carried out. For implementation purposes, the last-mentioned technique is usually the most effective.

In addition to the two permutations that have been mentioned, there are other sets of permutations for which certain specific codes are also invariant. (See Berlekamp[2] pages 352–361 for a more complete discussion.) Very little work has been done, however, to exploit this property. Of all the different permutations the cyclic one is the most useful, and all of the examples discussed in this chapter use only this property. MacWilliams[17] has investigated decoders which make use of both (3-3) and (3-4) and the interested reader is referred to her paper for more detail.

A particularly effective way of using the cyclic permutation is to compute the syndrome polynomial in the normal fashion using a feedback shift register based on $g(x)$. This results in a parity set in the first $n - k$ locations of the code word. Consequently, any error pattern can be corrected which can be mapped into a "burst" of length $n - k$ or less through cyclic permutations. (This was done with the Meggitt burst-error-correcting decoder discussed previously.) In some special cases this will include all of the error patterns of interest. In other cases it is necessary to "spread out" the original parity set by applying one or more linear transformations directly to the elements of $S(x)$. By successively shifting the feedback shift register one can cause these transformed parity sets to be moved in a cyclic fashion past the entire code word. This will allow correction of certain error patterns that could not be corrected by the simple cyclic permutation.

An example for which only the cyclic permutation is necessary is given in Fig. 3-4. This decoder is very similar to the Meggitt decoder in Fig. 3-2. In this case the "pattern recognizer" has been reprogrammed to respond to any pattern of weight 2 or less. The circuit is operated in a fashion that is almost identical to the Meggitt decoder. After the initial syndrome calculation the syndrome register and storage register are shifted simultaneously until a weight-two syndrome is observed. At this point the feedback is disabled and the contents of the syndrome register are added symbol by symbol to the shifted version of the received pattern. The decoder is then shifted through the remainder of the cycle and the decoded word is de-

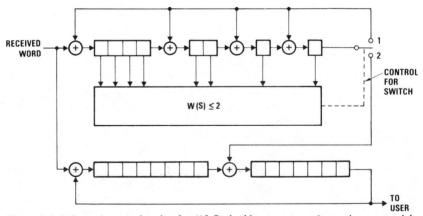

Figure 3-4. Information set decoder for (15, 7) double-error-correcting code generated by $g(x) = 1 + x^4 + x^6 + x^7 + x^8$.

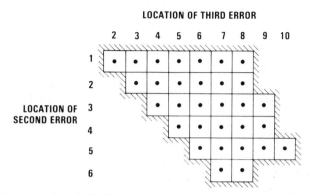

Figure 3-5. Representation of all triple-error patterns which must be covered in a cyclic code of length 15.

livered to the user. The reader should verify that any two-error pattern can always be mapped into positions 0 through 7, and consequently, this decoder will correct all such patterns. Note also that as with the burst-error-correcting Meggitt decoder, the phase of the syndrome register is advanced by eight shifts with respect to the storage register. This is done by performing the correction at the output of stage 8 of the storage register.

A somewhat more complicated decoder results when one attempts to correct all three-error patterns in a (15, 5) cyclic code. First one may verify that if each possible three-error pattern is cyclically shifted so that one of the errors is located in position zero, then all possible three-error patterns can be represented as one of the interior points of the polygon shown in Fig. 3-5. In this diagram the location of the error with the second lowest index is given by the vertical coordinate, and the location of the third error is given by the horizontal coordinate. By drawing two cycles of the pattern of the parity set on the edge of a piece of paper and sliding it by this diagram one can quickly verify which errors are covered by the parity set.

From Fig. 3-5 one quickly discovers that all errors except the one whose coordinates are 0, 5, and 10 are covered by the parity set $\{0, 1, 2, \ldots, 9\}$. At this point there are several options. One is to create a second parity set which together with the original set will cover all possible errors. A second option is to attempt to find a single parity set which will cover all possible errors. The third option is to use the original parity set and attempt to modify the decoding algorithm to account for the possibility of a single error outside this set. To explore these various options it is useful to refer to the parity check matrix for the code. This matrix is easily derived from the code generator

$$g(x) = 1 + x + x^2 + x^4 + x^5 + x^8 + x^{10}$$

and is given by

$$
H = \begin{bmatrix}
1 & 0 & 0 & 0 & 0 & 0 & 0 & 0 & 0 & 0 & 1 & 0 & 1 & 0 & 1 \\
0 & 1 & 0 & 0 & 0 & 0 & 0 & 0 & 0 & 0 & 1 & 1 & 1 & 1 & 1 \\
0 & 0 & 1 & 0 & 0 & 0 & 0 & 0 & 0 & 0 & 1 & 1 & 0 & 1 & 0 \\
0 & 0 & 0 & 1 & 0 & 0 & 0 & 0 & 0 & 0 & 1 & 1 & 0 & 1 \\
0 & 0 & 0 & 0 & 1 & 0 & 0 & 0 & 0 & 0 & 1 & 0 & 0 & 1 & 1 \\
0 & 0 & 0 & 0 & 0 & 1 & 0 & 0 & 0 & 0 & 1 & 1 & 1 & 0 & 0 \\
0 & 0 & 0 & 0 & 0 & 0 & 1 & 0 & 0 & 0 & 0 & 1 & 1 & 1 & 0 \\
0 & 0 & 0 & 0 & 0 & 0 & 0 & 1 & 0 & 0 & 0 & 0 & 1 & 1 & 1 \\
0 & 0 & 0 & 0 & 0 & 0 & 0 & 0 & 1 & 0 & 1 & 0 & 1 & 1 & 0 \\
0 & 0 & 0 & 0 & 0 & 0 & 0 & 0 & 0 & 1 & 0 & 1 & 0 & 1 & 1
\end{bmatrix}
$$

In order to realize the first option it is only necessary to require the second parity set to have elements in locations 0, 5, and 10. This can be accomplished by "interchanging" column ten of the parity check matrix with any of the first ten columns, excluding, of course, columns 0 or 5. One possible choice is column 8. The new matrix can be created by adding row 8 to rows 0, 1, 2, 4, and 5. The same transformation can be applied to the original syndrome to create the new syndrome.

Although we did not start out with that objective, it turns out that the parity set which was just formed not only covers the pattern {0, 5, 10} but also covers all of the other patterns indicated in Fig. 3-5. Thus, we can realize the second option by using this parity set alone.

The third option can be realized if we note that any time an error occurs in position 10 its effect on the syndrome is to add the tenth column of the parity check matrix to whatever syndrome pattern is created by the other errors. Thus, if we were to postulate a single error in position 10 we could remove its effect from the syndrome by adding column 10 to the original syndrome. The syndrome that is thus created will correspond to the remaining errors. In the current example this is two or fewer errors in locations 0 and 5. Thus, to use the third option one first computes both syndromes. The original syndrome is tested for a pattern of weight 3 or less and the second for a pattern of weight 2 or less. As an interesting side note, the modified syndrome will also cover all three-error patterns and can be used alone. This will usually not be the case and normally one will have to consider both the presence and absence of errors in the positions outside the parity set.

A general decoder for the (15, 5) code which can implement all three of these options is shown in Fig. 3-6. In this figure, f_1 is a ten-component vector which represents the syndrome pattern corresponding to the ith

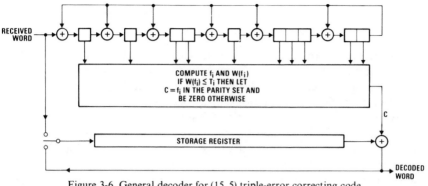

Figure 3-6. General decoder for (15, 5) triple-error correcting code.

parity set. The weight of this vector is $w(\mathbf{f}_1)$. This weight is compared with a threshold, T_i, to determine if the errors have been mapped into the parity set. $\mathbf{C}$ is a 15-component vector which represents the correction which is to be added to the received word. The components of $\mathbf{C}$ are equal to the appropriate corresponding elements of the syndrome in the parity set positions and are zero elsewhere. The pertinent data for the three different options are

Option 1:

$$\mathbf{f}_1 = (S_0, S_1, S_2, S_3, S_4, S_5, S_6, S_7, S_8, S_9)$$
$$T_1 = 3$$
$$\mathbf{C}_1 = (S_0, S_1, S_2, S_3, S_4, S_5, S_6, S_7, S_8, S_9, 0, 0, 0, 0, 0)$$
$$\mathbf{f}_2 = (S_0, S_1, S_2, S_3, S_4, S_5, S_6, S_7, S_9, S_0 + S_1 + S_2 + S_4 + S_5 + S_8)$$
$$T_2 = 3$$
$$\mathbf{C}_2 = (S_0, S_1, S_2, S_3, S_4, S_5, S_6, S_7, 0, S_9, S_0 + S_1 + S_2 + S_4 + S_5 + S_8,$$
$$0, 0, 0, 0)$$

Option 2:

$\mathbf{f}_1, \mathbf{C}_1, T_1$ not needed
$\mathbf{f}_2, \mathbf{C}_2, T_2$ same as Option 1

Option 3:

$\mathbf{f}_1, T_1, \mathbf{C}_1$ same as Option 1 (but not required)
$$\mathbf{f}_2 = (\bar{S}_0, \bar{S}_1, \bar{S}_2, S_3, \bar{S}_4, \bar{S}_5, S_6, S_7, \bar{S}_8, S_9)$$
$$T_2 = 2$$
$$\mathbf{C}_2 = (\bar{S}_0, \bar{S}_1, \bar{S}_2, S_3, \bar{S}_4, \bar{S}_5, S_6, S_7, \bar{S}_8, S_9, 1, 0, 0, 0, 0)$$

where

$$\bar{S}_i = 1 + S_i$$

The reader should observe that the technique used in Option 3 is the same technique that is required whenever a particular covering is desired, but all of the indicated positions do not form a parity set. In this case one assigns all possible combinations of errors to the positions outside the parity set and thereby generates the required patterns.

Other more complicated cases are possible as well. Once the number of errors exceeds 3 it becomes tedious to determine if the different error patterns are covered using hand computation. Patterns of 4, 5, and 6 errors, however, are easily handled using a computer. One may also modify the algorithm to decode beyond the minimum distance of the code. In this case enough additional coverings are selected to correct the additional errors. The decoder is operated in two phases. During the first phase the entire received word is scanned, and the location of the lowest-weight pattern is determined. In the second phase the syndrome register is advanced to this location, and the error is corrected.

There is an interesting class of codes for which the algorithm is particularly useful when the number of errors is not too large. These are the Reed–Solomon codes over $GF(2^m)$ which were introduced in Chapter 2 and are discussed in more detail in Chapter 5. For the moment it is sufficient to note that these codes are cyclic and have block lengths of $2^m - 1$. They have the property that any number of symbol errors up to $(2^m - 1)/2$ can be corrected, and each symbol to be corrected requires the use of two parity symbols. Thus, a t-symbol-correcting code will be a $(2^m - 1, 2^m - 1 - 2t)$ code. These codes also have the interesting property that any $2^m - 1 - 2t$ positions can be selected as a parity set. Thus, one can concentrate strictly

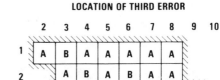

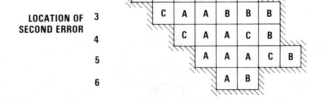

Figure 3-7. Triple-error patterns covered by at least one of the sets $A = \{0, 1, 2, 5, 7, 8\}$, $B = \{0, 1, 3, 5, 7, 10\}$, and $C = \{0, 3, 4, 5, 8, 9\}$ for the $(15, 9)$ RS code.

on the covering problem. As an example, there is a $(15, 9)$ RS code which will correct three or fewer symbol errors. One may verify that the parity sets $A = \{0, 1, 2, 5, 7, 8\}$, $B = \{0, 1, 3, 5, 7, 10\}$, and $C = \{0, 3, 4, 5, 8, 9\}$ will cover all three-error patterns given the appropriate cyclic shift. The sets were chosen sequentially so that each new set added as many new patterns as possible. Figure 3-7 shows all the patterns that are covered by A, all those covered by B and not A, and all those covered by C which are not covered by A or B.

3.2.4. Random Search Techniques

Rather than predetermine a set of coverings that are guaranteed to correct all t-error patterns, it is possible to decode by generating parity sets in a random fashion and testing to see if each set produces an acceptable solution. This random search can sometimes be aided by the use of soft decisions or from the results of the previous parity computation. The use of soft decisions will be deferred to the next chapter and we will concentrate, for the moment, on the hard decision case.

A random search algorithm for group codes was proposed by Omura[18] in 1969. In this algorithm one starts with a randomly chosen parity set and associated parity check matrix. The first step is to compute the syndrome and to assume that the actual error pattern is equal to the syndrome in the parity positions and is zero elsewhere. The next step is to modify the parity set by interchanging one of the elements in the parity set with one of the elements in the information set. The interchange is always done such that the weight of the new syndrome is less than or equal to the weight of the existing syndrome. One then assumes a new solution based upon the new syndrome. Thus, the change in the weight of the syndrome is used as a gradient function which will guide the search toward the correct solution. Unfortunately, although the desired solution is in fact the minimum-weight solution, attempting to minimize the weight of the syndrome at each iteration does not make it any more likely that one will minimize the number of steps required to find the correct solution. In some instances, in fact, this procedure may be counterproductive. One can show by simulations that the Omura algorithm behaves in the same fashion as an algorithm in which the positions are interchanged on a random basis. It is possible, however, to obtain some information about the location of the errors in the minimum-weight solution when certain chance events occur and thus cause the search to terminate faster.

In the next three sections we first discuss the details of the original Omura algorithm. Next, we examine the computational behavior of an algorithm which is identical to the Omura algorithm except that it is explicitly assumed that the interchanges are made on the basis of no information regarding the location of errors. This model is then used as a standard by which simulations of the Omura algorithm are compared. Finally, we discuss a modification to the Omura algorithm which improves the computational behavior under some circumstances. To make the discussion more relevant we assume that a $(48, 24)$ five-error-correcting quadratic residue code is to be decoded. This code is particularly interesting since there is no known method for decoding all five-error patterns except using information set techniques. In the previous section we noted that one could find the equivalent of 92 fixed parity sets that would completely cover all five-error patterns. For this code we will show that when a random algorithm is used it is possible to decode by examining, on the average, a small number of parity sets. However, there will be an occasional pattern which causes this number of parity sets to be quite large.

3.2.4.1. Some Details of the Omura Algorithm. The basic steps involved in the Omura algorithm are as follows:

(1) Start with a received sequence, ρ, an arbitrary parity set $\{P\}$, and the associated parity check matrix $\mathbf{H}$. With no loss in generality this initial $\mathbf{H}$ matrix can be the usual one in reduced echelon canonical form with the associated parity set $\{k, k + 1, ..., n - 1\}$.

(2) Compute the syndrome $\mathbf{S}$ and postulate an error vector $\mathbf{e}$ such that

$$\mathbf{He} = \mathbf{S}$$

Note that if

$$\mathbf{H} = [\mathbf{Y} \mathbin{\vert} \mathbf{I}]$$

and

$$\mathbf{S}^T = (S_0, S_1, ..., S_{n-k-1})$$

then

$$\mathbf{e}^T = (0, 0, ..., 0, S_0, S_1, ..., S_{n-k-1})$$

(3) Find an element in the information set such that if this element were interchanged with some element in the parity set and a new syndrome, $\mathbf{S}'$, were to be computed using this set, the weight of this new syndrome would be as small as possible.

(4) Carry out the indicated interchange, calculate a new matrix $\mathbf{H}'$ by elementary row operations, and postulate a new error vector $\mathbf{e}'$ such that $\mathbf{e}'$ is identical to $\mathbf{S}'$ in the new parity set and is zero elsewhere.

(5) If the weight of $\mathbf{S}'$ or equivalently of $\mathbf{e}'$ is less than or equal to t, the search is to be terminated. If not, go to step (2) and repeat using $\mathbf{H}'$ and $\mathbf{S}'$ as the starting point. If after a sufficiently large number of iterations the weight cannot be reduced to t or less, accept the latest $\mathbf{e}$ as the minimum-weight error vector.

The above steps are all self-explanatory with the exception of step (3). In order to implement step (3), one examines the effect of a postulated error in each of the positions corresponding to the current information set. That is, one determines what new error pattern would have to exist in the current parity set in order to account for the postulated error and still produce the current syndrome. This test can be done by simply adding the column of the parity check matrix corresponding to this postulated error location to the current syndrome. If the weight of this new error pattern (including the postulated error) is less than that of the currently assumed error pattern, then this must be a better solution. Thus, in step (3) one locates the best alternate solution to the parity equations that can be found by assuming a single error outside the parity set. In order to make the new parity set and new **H** matrix consistent with this alternate solution one must select an element from the existing parity set that (1) can be interchanged with the desired element in the information set and (2) also corresponds to a one in the current error vector. The last requirement follows readily if we recall that the error vector produces a syndrome by adding those columns of the **H** matrix that correspond to errors. The old error vector will still be a solution to the parity equations after the **H** matrix has been modified. Thus, the only way the selected column in the information set can actually modify the old syndrome is to "line up" with an existing error.

It may be, of course, that none of the alternate solutions are of lower weight than the existing solution. In this case one may either accept the alternate solution if it is of equal weight or keep the current solution. In order to keep the current solution, one simply makes any permissible interchange so that the element in the current parity set does not "line up" with a one in the current error pattern.

As an example, consider the (12, 4) two-error-correcting code with errors in positions 3 and 4. Then

$$\mathbf{H}\rho = \begin{bmatrix} 1 & 1 & 1 & 0 & 1 & 0 & 0 & 0 & 0 & 0 & 0 & 0 \\ 1 & 1 & 0 & 1 & 0 & 1 & 0 & 0 & 0 & 0 & 0 & 0 \\ 1 & 0 & 1 & 1 & 0 & 0 & 1 & 0 & 0 & 0 & 0 & 0 \\ 0 & 1 & 1 & 1 & 0 & 0 & 0 & 1 & 0 & 0 & 0 & 0 \\ 1 & 0 & 0 & 0 & 0 & 0 & 0 & 0 & 1 & 0 & 0 & 0 \\ 0 & 1 & 0 & 0 & 0 & 0 & 0 & 0 & 0 & 1 & 0 & 0 \\ 0 & 0 & 1 & 0 & 0 & 0 & 0 & 0 & 0 & 0 & 1 & 0 \\ 0 & 0 & 0 & 1 & 0 & 0 & 0 & 0 & 0 & 0 & 0 & 1 \end{bmatrix} \begin{pmatrix} \rho_1 \\ \rho_2 \\ . \\ . \\ . \\ \rho_{12} \end{pmatrix} = \begin{pmatrix} 1 \\ 1 \\ 0 \\ 0 \\ 0 \\ 0 \\ 1 \\ 1 \end{pmatrix}$$

$$\mathbf{e} = (0\ 0\ 0\ 0\ 1\ 1\ 0\ 0\ 0\ 0\ 1\ 1)$$

$$w(\mathbf{e}) = 4$$

Test column (1)

$$\begin{array}{ll} \mathbf{S} & 1\ 1\ 0\ 0\ 0\ 0\ 1\ 1 \\ \mathbf{h}_1 & 1\ 1\ 1\ 0\ 1\ 0\ 0\ 0 \\ \hline \mathbf{S} + \mathbf{h}_1 & 0\ 0\ 1\ 0\ 1\ 0\ 1\ 1 \end{array}$$

$\mathbf{e} = (1\ 0\ 0\ 0\ 0\ 0\ 1\ 0\ 1\ 0\ 1\ 1)$

$$w(\mathbf{e}) = 5$$

Test column (2)

$$\begin{array}{ll} \mathbf{S} & 1\ 1\ 0\ 0\ 0\ 0\ 1\ 1 \\ \mathbf{h}_2 & 1\ 1\ 0\ 1\ 0\ 1\ 0\ 0 \\ \hline \mathbf{S} + \mathbf{h}_2 & 0\ 0\ 0\ 1\ 0\ 1\ 1\ 1 \end{array}$$

$\mathbf{e} = (0\ 1\ 0\ 0\ 0\ 0\ 0\ 1\ 0\ 1\ 1\ 1)$

$$w(\mathbf{e}) = 5$$

Test column (3)

$$\begin{array}{ll} \mathbf{S} & 1\ 1\ 0\ 0\ 0\ 0\ 1\ 1 \\ \mathbf{h}_3 & 1\ 0\ 1\ 1\ 0\ 0\ 1\ 0 \\ \hline \mathbf{S} + \mathbf{h}_3 & 0\ 1\ 1\ 1\ 0\ 0\ 0\ 1 \end{array}$$

$\mathbf{e} = (0\ 0\ 1\ 0\ 0\ 1\ 1\ 1\ 0\ 0\ 0\ 1)$

$$w(\mathbf{e}) = 5$$

Test column (4)

$$\begin{array}{ll} \mathbf{S} & 1\ 1\ 0\ 0\ 0\ 0\ 1\ 1 \\ \mathbf{h}_4 & 0\ 1\ 1\ 1\ 0\ 0\ 0\ 1 \\ \hline \mathbf{S} + \mathbf{h}_4 & 1\ 0\ 1\ 1\ 0\ 0\ 1\ 0 \end{array}$$

$\mathbf{e} = (0\ 0\ 0\ 1\ 1\ 0\ 1\ 1\ 0\ 0\ 1\ 0)$

$$w(\mathbf{e}) = 5$$

No interchange is possible which does not increase the error weight. Therefore, we interchange any of the first four with a column which corresponds to a zero in the current syndrome. If we choose (1) and (7), the new parity equations become

$$
\mathbf{H_1 e} =
\begin{bmatrix}
0 & 1 & 0 & 1 & 1 & 0 & 1 & 0 & 0 & 0 & 0 & 0 \\
0 & 1 & 1 & 0 & 0 & 1 & 1 & 0 & 0 & 0 & 0 & 0 \\
1 & 0 & 1 & 1 & 0 & 0 & 1 & 0 & 0 & 0 & 0 & 0 \\
0 & 1 & 1 & 1 & 0 & 0 & 0 & 1 & 0 & 0 & 0 & 0 \\
0 & 0 & 1 & 1 & 0 & 0 & 1 & 0 & 1 & 0 & 0 & 0 \\
0 & 1 & 0 & 0 & 0 & 0 & 0 & 0 & 0 & 1 & 0 & 0 \\
0 & 0 & 1 & 0 & 0 & 0 & 0 & 0 & 0 & 0 & 1 & 0 \\
0 & 0 & 0 & 1 & 0 & 0 & 0 & 0 & 0 & 0 & 0 & 1
\end{bmatrix}
\begin{pmatrix} e_1 \\ e_2 \\ \cdot \\ \cdot \\ \cdot \\ \\ e_{12} \end{pmatrix}
=
\begin{pmatrix} 1 \\ 1 \\ 0 \\ 0 \\ 0 \\ 0 \\ 1 \\ 1 \end{pmatrix}
$$

Test column (2)

$$
\begin{aligned}
\mathbf{S} \quad & 1\ 1\ 0\ 0\ 0\ 0\ 1\ 1 \\
\mathbf{h_2} \quad & 1\ 1\ 0\ 1\ 0\ 1\ 0\ 0 \\
\hline
\mathbf{S + h_2} \quad & 0\ 0\ 0\ 1\ 0\ 1\ 1\ 1
\end{aligned}
$$

$\mathbf{e} = (0\ 1\ 0\ 0\ 0\ 0\ 1\ 0\ 1\ 1\ 1)$

$$w(\mathbf{e}) = 5$$

Test colum (3)

$$
\begin{aligned}
\mathbf{S} \quad & 1\ 1\ 0\ 0\ 0\ 0\ 1\ 1 \\
\mathbf{h_3} \quad & 0\ 1\ 1\ 1\ 1\ 0\ 1\ 0 \\
\hline
\mathbf{S + h_3} \quad & 1\ 0\ 1\ 1\ 1\ 0\ 0\ 1
\end{aligned}
$$

$\mathbf{e} = (1\ 0\ 1\ 0\ 1\ 0\ 0\ 1\ 1\ 0\ 0\ 1)$

$$w(\mathbf{e}) = 6$$

Test column (4)

$$
\begin{aligned}
\mathbf{S} \quad & 1\ 1\ 0\ 0\ 0\ 0\ 1\ 1 \\
\mathbf{h_4} \quad & 1\ 0\ 1\ 1\ 1\ 0\ 0\ 1 \\
\hline
\mathbf{S + h_4} \quad & 0\ 1\ 1\ 1\ 1\ 0\ 1\ 0
\end{aligned}
$$

$\mathbf{e} = (1\ 0\ 0\ 1\ 0\ 1\ 0\ 1\ 1\ 0\ 1\ 0)$

$$w(\mathbf{e}) = 6$$

Test column (7)

$$
\begin{array}{ll}
\mathbf{S} & 1\ 1\ 0\ 0\ 0\ 0\ 1\ 1 \\
\mathbf{h}_7 & 1\ 1\ 1\ 0\ 1\ 0\ 0\ 0 \\
\hline
\mathbf{S}+\mathbf{h}_7 & 0\ 0\ 1\ 0\ 1\ 0\ 1\ 1
\end{array}
$$

$$\mathbf{e} = (1\ 0\ 0\ 0\ 0\ 0\ 1\ 0\ 1\ 0\ 1\ 1)$$

$$w(\mathbf{e}) = (5)$$

Again, none of the potential choices are acceptable. The next logical step would be to interchange (2) and (8). If we did this, it would not permit us to produce a lower-weight error. To make the example shorter, we choose (4) and (8). This gives

$$
\mathbf{H}_2\mathbf{e} =
\begin{bmatrix}
0 & 0 & 1 & 0 & 1 & 0 & 1 & 1 & 0 & 0 & 0 & 0 \\
0 & 1 & 1 & 0 & 0 & 1 & 1 & 0 & 0 & 0 & 0 & 0 \\
1 & 1 & 0 & 0 & 0 & 0 & 1 & 1 & 0 & 0 & 0 & 0 \\
0 & 1 & 1 & 1 & 0 & 0 & 0 & 1 & 0 & 0 & 0 & 0 \\
0 & 1 & 0 & 0 & 0 & 0 & 1 & 1 & 1 & 0 & 0 & 0 \\
0 & 1 & 0 & 0 & 0 & 0 & 0 & 0 & 0 & 1 & 0 & 0 \\
0 & 0 & 1 & 0 & 0 & 0 & 0 & 0 & 0 & 0 & 1 & 0 \\
0 & 1 & 1 & 0 & 0 & 0 & 0 & 1 & 0 & 0 & 0 & 1
\end{bmatrix}
\begin{pmatrix}
e_1 \\
e_2 \\
\cdot \\
\cdot \\
\cdot \\
\\
\\
e_{12}
\end{pmatrix}
=
\begin{pmatrix}
1 \\
1 \\
0 \\
0 \\
0 \\
0 \\
1 \\
1
\end{pmatrix}
$$

Test column (2)

$$
\begin{array}{ll}
\mathbf{S} & 1\ 1\ 0\ 0\ 0\ 0\ 1\ 1 \\
\mathbf{h}_2 & 0\ 1\ 1\ 1\ 1\ 1\ 0\ 1 \\
\hline
\mathbf{S}+\mathbf{h}_2 & 1\ 0\ 1\ 1\ 1\ 1\ 1\ 0
\end{array}
$$

$$\mathbf{e} = (1\ 1\ 0\ 1\ 1\ 0\ 0\ 1\ 1\ 1\ 0)$$

$$w(\mathbf{e}) = 7$$

Test column (3)

$$
\begin{array}{ll}
\mathbf{S} & 1\ 1\ 0\ 0\ 0\ 0\ 1\ 1 \\
\mathbf{h}_3 & 1\ 1\ 0\ 1\ 0\ 0\ 1\ 1 \\
\hline
\mathbf{S}+\mathbf{h}_3 & 0\ 0\ 0\ 1\ 0\ 0\ 0\ 0
\end{array}
$$

$$\mathbf{e} = (0\ 0\ 1\ 1\ 0\ 0\ 0\ 0\ 0\ 0\ 0\ 0)$$

$$w(\mathbf{e}) = 2$$

This is the minimum weight solution, and the search is ended.

3.2.4.2. Behavior of a Completely Random Decoding Algorithm. We can get some further insight into the process by which the Omura algorithm converges to the correct error pattern by considering a completely random decoding algorithm. This will also provide a standard against which we can compare the behavior of the Omura algorithm. For concreteness, assume we have a (48, 24) five-error-correcting code and that the error pattern is known to be weight 5. Assume that the decoder functions exactly as with the Omura algorithm, except that we only use the weight of the syndrome to determine when the decoding operation is complete. That is, we compute the gradient function exactly as before, but only test to see if it is equal to 5. If it is equal to 5, we interchange the appropriate columns and decode the error. If it is not equal to 5, we ignore the value and interchange columns at random.

The decoder can exist in one of six different states depending on how many errors occur in the parity set and how many errors occur in the information set. We define state 0 to be no errors in the parity set and five errors in the information set, state 1 as one error in the parity set and four in the information set, etc. The object of the decoding algorithm is to reach state 5. We may describe the process by a set of initial occupancy probabilities for each state and a set of transition probabilities which indicate the probability of moving from one state to another. These are easily computed as follows. The initial occupancy probability of state j, $P_0(j)$, is just the probability of distributing t objects in n bins such that part of them fall in bins 1 through k and the remainder fall in bins $k + 1$ through n. In this example, we have $n = 48$, $k = 24$, and $t = 5$, and

$$P_0(j) = \binom{n-k}{j} \binom{k}{t-j} \bigg/ \binom{n}{t}$$

The probability $\Pr(i|j)$ of changing from state j to state i is also easy to compute. For example, if the decoder is in state 2, it means that two errors are covered and three errors are uncovered. In order to move to state 3, we must select an error-free position from the parity set and interchange it with a position containing an error in the information set. Using this line of reasoning, we may compute the transition probabilities as

$$\Pr(j-1|j) = \left(\frac{j}{n-k} \right) \left(\frac{k-t+j}{k} \right)$$

$$\Pr(j|j) = \left(\frac{n-k-j}{n-k} \right) \left(\frac{k-t+j}{k} \right) + \left(\frac{j}{n-k} \right) \left(\frac{t-j}{k} \right)$$

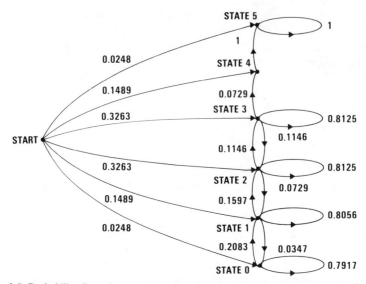

Figure 3-8. Probability flow diagram of random decoder when decoding a five-error pattern on a (48, 24) code.

and

$$Pr(j + 1 | j) = \left(\frac{n - k - j}{n - k} \right) \left(\frac{t - j}{k} \right)$$

with special cases

$$Pr(t | t - 1) = Pr(t | t) = 1$$

All other $Pr(i | j)$ are zero. $Pr(t | t - 1)$ is equal to 1 since the decoder will always decode correctly on the next iteration if it ever gets into state $t - 1$ (state 4 in this example). We may think of the decoder as being a sequential state machine and represent it diagramatically as in Fig. 3-8 The calculated initial occupancy probabilities and transition probabilities are shown in this figure. There are two figures of merit that one might want to determine about this decoder. One is the expected or average number of iterations to decode. The other is the probability of decoding in m or fewer iterations as a function of m. This is easily computed as follows. If we consider the occupancy probabilities at the mth iteration to be a vector $\{P_m(i)\}$ then we may compute the vector $\{P_m(i)\}$ at the end of m iterations as

$$\{P_m(i)\} = [P_{ij}]^m \{P_0(j)\}$$

where $[P_{ij}]$ is the matrix of transition probabilities $Pr(i | j)$. Thus, the probability of decoding in m or fewer iterations is obviously

$$\Pr \left\{ \begin{array}{c} \text{decoding in } m \text{ or} \\ \text{fewer iterations} \end{array} \right\} = P_m \quad (5)$$

and the expected number of iterations $\bar{m}$ to decode is

$$\bar{m} = \sum_{m=0}^{\infty} (m + 1) P_m \quad (4)$$

These are easily evaluated on a digital computer. $P_m(5)$ is plotted in Fig. 3-9 and $\bar{m}$ is given by

$$\bar{m} = 37.7$$

When the Omura algorithm is simulated on a computer and used to decode a large set of weight-5 error patterns, one finds that the computational performance is almost identical to the curve shown in Fig. 3-9. From this we can conclude that attempting to minimize the weight of the syndrome does not increase the speed with which the algorithm reaches the minimum values. This would seem to make sense in light of the fact that there is no mechanism by which minimizing the weight of S would tend to increase the upward transition probabilities or decrease the downward transition probabilities in Fig. 3-8.

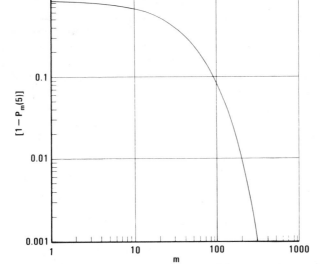

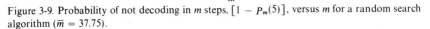

Figure 3-9. Probability of not decoding in m steps, $[1 - P_m(5)]$, versus m for a random search algorithm ($\bar{m} = 37.75$).

3.2.4.3. Modified Omura Algorithm. There is a mechanism by which one can increase the speed with which the decoder can find a minimum-weight error. First note that if e_{min} is the minimum-weight solution to

$$He = S$$

then any error vector of the form

$$e = e_{min} + C$$

is also a solution where C is any code word. Further, all 2^k possible solutions can be written in this fashion. Suppose in the course of decoding a weight-5 error pattern in the (48, 24) code, one were to discover a hypothesized error pattern of weight 7. This error pattern must differ from the actual weight-5 pattern by a code word. Since the minimum-weight code word is 12, the only way that this event can occur is for the weight-5 error and weight-7 error to be nonoverlapping. Thus, the weight-7 pattern identifies a set of positions that must be error free. The obvious strategy, then, is to map these positions into the information set as rapidly as possible and to never remove them.

One problem with the above approach is that one does not know *a priori* in a real decoding situation that the minimum-weight pattern is actually 5. This problem may be solved by assuming that the lowest-weight pattern that has been located will specify error-free positions and let the algorithm proceed on that basis. If a lower-weight pattern is found, then it replaces the previous pattern and all of the previous actions are disregarded.

Based on the above observation and empirical data that indicate that there is no particular value in actually generating a minimum-weight syndrome at each iteration, one may specify the following search procedure:

Modified Algorithm

(1) Start with the **H** matrix in reduced echelon canonical form and with the syndrome **S**. The associated information set is $I = \{0, 1, 2, ..., k - 1\}$, and the parity set is $P = \{k, k + 1, ..., n - 1\}$.

(2) Postulate an error vector based on the current syndrome. If the weight of this vector is t or less, then it must be the minimum-weight solution and the algorithm is stopped. If not, let the ones in this vector define an error-free set denoted M.

(3) Sequence through the information set, I, by adding each column of the parity check matrix individually to the current syndrome and for each case postulate an error-free pattern. If the weight of this error vector is equal to that of the initial error vector, then add the positions where this vector is one to the set M. If the weight of the postulated error is less than that of the initial error, then check to see if it is less than or equal to t. If so, the search terminates. If not, replace the set M by values determined by

the new error and then replace the original error vector with this vector.

(4) After scanning through the entire information set, choose two elements which can be interchanged such that, if possible, one element is in both P and M and the other is in I but not in M. If there are no elements in both P and M, then choose an arbitrary element from P. If there are no elements in I that are not in M, then set M to zero and choose an arbitrary element from I.

(5) Utilize elementary row operations to create a new H matrix consistent with the new parity set and information set. Compute a new syndrome and proceed to step 3.

(6) If the algorithm does not terminate naturally after some specified number of iterations, then choose the current hypothesized error to be the best solution.

In the above algorithm, steps 4 and 5 can be combined. In this case, one first accesses a given column of H and adds it to the existing S to obtain a hypothesized e. Then, before replacing it in memory, one must modify the column according to the particular elementary row operations that are to be performed. Thus, the creation of the new H matrix is done simultaneously with the determination of possible error patterns.

Whether or not the above algorithm represents any significant improvement over the Omura algorithm depends upon how often one is able to find a minimum-weight pattern that provides a legitimate "error-free" set. In the case of the (48, 24) code, the distribution of the number of interchanges required to decode a large number of weight-5 patterns is shown in Fig. 3-10. The average number of computations is approximately 16 and the overall shape of the curve is essentially the same as the random case. If the same algorithm is used to decode a good randomly generated (48, 24) code, the distribution of computation is virtually the same as the Omura algorithm. The difference can apparently be traced to the number of nearest neighbors and to the fact that a few weight-5 patterns could not be decoded with the random code. In the quadratic residue code there are 17,196 weight-12 code words. This is a large enough number that for a relatively large percentage of weight-5 errors, there will be 1, 2, or 3 error patterns of weight 7 that can be found very quickly and thus cause the true minimum-weight pattern to be located in a few iterations. In a nonnegligible percentage of the cases, the true error pattern was found without ever finding a weight-7 pattern. Although some of these were just chance events, in some cases it is probably true that there is no weight-12 code word which can be added to a given weight-5 error to produce the weight-7 pattern.

As an interesting point of comparison, one can compute a state transition diagram similar to Fig. 3-8 that does not allow for an error to be mapped out of the parity set once it has been included in this set. This removes all

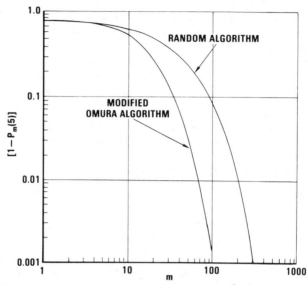

Figure 3-10. Probability of not decoding in m steps, $[1 - P_m(5)]$, versus m for the modified Omura algorithm with random algorithm being shown as a reference ($\bar{m} = 16.0$).

of the downward transitions and modifies all of the others slightly. When one computes the distribution of computation for this diagram, the result is nearly identical to that achieved for the modified algorithm. The modified algorithm will, of course, behave in this manner whenever one is able to locate error patterns of weight 7. The indication is that they occur frequently enough that this is a good approximation to its behavior.

These random search techniques are examples of algorithms with a variable decoding time. The distributions of computation shown in Figs. 3-9 and 3-10 are valid only for five-error patterns. Similar distributions for other error patterns can be computed using the same technique, and the required computational load is significantly smaller for the smaller-weight error patterns. One can then compute an average distribution of computation by weighting each individual distribution with the probability of occurrence of that size error pattern (this will be a function of channel error rate). Under typical operating conditions the average distribution of computation will be significantly better than that shown in Figs. 3-9 and 3-10.

The variability in the number of computations required to decode a code word presents a significant decoder design problem. (Similar behavior is observed with sequential decoding of convolutional codes. This behavior is discussed in detail in Chapter 7.) Obviously, the decoder must be capable, at least, of keeping up with the average computational requirement. However, because of the variability in computational demand, the decoder computational capability must be significantly in excess of the computational requirement. The effects of this variability can be lessened by providing a buffer several code words in length. Then, for a specified level of performance, there is a trade between the buffer size and the computational speed of the decoder.

3.3. Threshold Decoding

Threshold decoding is similar to Meggitt decoding; however, it requires the code to have a certain type of structure in order to work. For the codes on which it does work, threshold decoding is an extremely simple and powerful algorithm. Derivation of the general threshold decoding algorithm (which allows soft decision decoding) is deferred until Chapter 4. Here we present a simplified version of the algorithm (called *majority decoding*) which is applicable only to hard decision decoding. This approach will be developed through several examples. The discussion is restricted to binary codes since this case is of most practical interest. This decoding algorithm can also be used with convolutional codes as discussed in Chapter 7.

The basic idea involved in majority decoding can be illustrated by the following example. The (15, 7) BCH code discussed in Section 3.1 is one of the codes for which threshold decoding can be used. The parity check matrix for this code is given in transposed form by cycle (a) in Table 3-1. By taking columns 3, 7, 1 + 5, and 0 + 2 + 6, the following parity check equations can be written:

$$P_1 = r_3 + r_{11} + r_{12} + r_{14}$$

$$P_2 = r_7 + r_8 + r_{10} + r_{14}$$

$$P_3 + r_1 + r_5 + r_{13} + r_{14}$$

$$P_4 = r_0 + r_2 + r_6 + r_{14}$$

Since these are legitimate parity check equations, they depend only on the error terms, and consequently

$$P_1 = e_3 + e_{11} + e_{12} + e_{14}$$

$$P_2 = e_7 + e_8 + e_{10} + e_{14}$$

$$P_3 = e_1 + e_5 + e_{13} + e_{14}$$

$$P_4 = e_0 + e_2 + e_6 + e_{14}$$

Each error term is included in only one equation except for e_{14} which is included in all four. Parity equations having this property are said to be *orthogonal* on e_{14}. (Note that this definition of orthogonality is not the same as orthogonality of two vectors.) We observe that any single error will never cause more than one of the four equations to fail except when the error is e_{14}, in which case all four will fail. Similarly, double-error patterns can never cause more than two parity equations to fail except when one of the errors is e_{14}, in which case at least three parity equations will fail. Thus, the code can correct all double-error patterns using only these equations. A decoding rule is to compute the four indicated parity checks, take their arithmetic sum, and test whether it is 3 or larger. If so, the error in position 14 is to be corrected. The remaining errors are corrected by shifting the code word cyclically and repeating the operation. Note that this procedure will correct not only all one- and two-error patterns, but several of greater weight as well.

A majority decoder for this code is shown in Fig. 3-11. It also includes as an additional feature a feedback loop (from the majority gate). This step is not necessary for the decoder to function but can improve its performance (in this case marginally) at high signal-to-noise ratios. The decoder is operated exactly like a Meggitt decoder by first computing the syndrome and then shifting it at least one complete cycle to correct the errors. If the feedback is not used, then there is no benefit to using more than one cycle and the pattern may be delivered to the user as each symbol is corrected. If the feedback loop is enabled, then it is necessary to go through at least one more cycle in order to obtain all of the benefit of this circuitry. The use of the feedback loop actually extends the error-correcting range of the

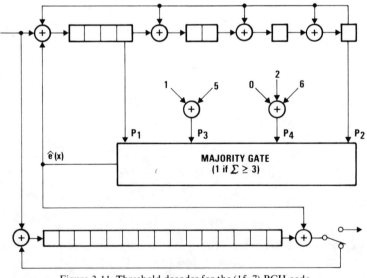

Figure 3-11. Threshold decoder for the (15, 7) BCH code.

decoder by including a few additional three- and four-error patterns. For instance, if the three-error pattern $\{9, 13, 14\}$ occurs, e_{14} can be corrected, but when the pattern is shifted so that e_{13} becomes e_{14} then the pattern looks like $\{0, 10, 14\}$ and the error in 14 would not be corrected. If, however, the feedback loop were used, the first correction also removes the effect of the error, and on the next shift, the error is the pattern $\{10, 14\}$ which is correctable.

Threshold decoding is most easily applied to cyclic codes. Note that the parity checks orthogonal on e_{14} are formed as linear combinations of syndrome symbols. However, for errors in other positions, a cyclic shift of the syndrome register is equivalent to a cyclic shift of the error pattern. Thus, eventually any pattern will be mapped into the e_{14} position and be corrected through a majority vote of the orthogonal estimates. The cyclic property of the code allows one to form estimates orthogonal on just one error symbol which can then be used at all other positions.

In the general case, one might be able to form J parity check equations orthogonal on the ith symbol. Any code word which contains a one in the ith position must also contain a one in one of the other positions checked by each parity check equation. Thus, the code word must have weight at least $J + 1$. For a cyclic code, orthogonal parity checks may also be formed for all other positions. Therefore, the code has minimum distance $d = J + 1$. Decoding is easily accomplished by using the majority decoding rule. That is, if one forms J estimates orthogonal on error symbol e_i, then set $e_i = 0$ when the sum of the estimates is less than or equal to $\lceil J/2 \rceil$ and set $e_i = 1$ when the sum is greater than $\lceil J/2 \rceil$. Using the arguments presented previously, this rule allows correction of all combinations of $\lfloor J/2 \rfloor$ or fewer errors.

The $(15, 7)$ code just discussed is a code which is *one-step orthogonalizable* (referring to the single majority gate required). In addition to the $(15, 7)$ code, the maximal-length codes can be orthogonalized in one step [it is always possible to find $d - 1 = (n - 1)/2$ parity checks orthogonal on any given position]. The other codes of most practical interest that are one-step orthogonalizable are the $(21, 11)$, $(73, 45)$, and $(273, 191)$ codes. These codes were introduced by Weldon[19] in 1966 and are referred to as difference-set codes since they can be constructed using perfect difference sets. For each of these codes one can find $d - 1$ parity equations (for d even) that are orthogonal on the last symbol. Then the simple majority decoding rule can be used. The generator polynomials and a set of parity checks orthogonal on the last digit are given in Table 3-4. The parity equations are given as linear combinations of the syndrome symbols. Also shown is the so-called "difference-set polynomial," $\theta^*(x)$. The polynomials $\theta^*(x), x\theta^*(x), \ldots,$ $x^{n-1}\theta^*(x) \bmod (x^n - 1)$ are in the null space of the code generated by $g(x)$ and contain all the orthogonal parity checks.

Table 3-4. Generator Polynomials and Parity Check Equations Which Are Orthogonal on the Last Symbol

Code	Generator polynomial and $\theta^*(x)$	Parity checks as a function of syndrome symbols
$(21, 11)$ $d = 6$	$g(x) = 1 + x^2 + x^4 + x^6 + x^7 + x^{10}$ $\theta^*(x) = x^{11} + x^9 + x^4 + x^3 + 1$	$A_1 = S_9$ $A_2 = S_1$ $A_3 = S_4 + S_6$ $A_4 = S_0 + S_5 + S_7$ $A_5 = S_2 + S_3 + S_8$
$(73, 45)$ $d = 10$	$g(x) = 1 + x^2 + x^4 + x^6 + x^8 + x^{12} +$ $\qquad x^{16} + x^{22} + x^{25} + x^{28}$ $\theta^*(x) = x^{45} + x^{43} + x^{35} + x^{21} + x^{20} +$ $\qquad x^{16} + x^9 + x^3 + 1$	$A_1 = S_{27}$ $A_2 = S_1$ $A_3 = S_7 + S_9$ $A_4 = S_{13} + S_{21} + S_{23}$ $A_5 = S_0 + S_{14} + S_{22} + S_{24}$ $A_6 = S_3 + S_4 + S_{18} + S_{26}$ $A_7 = S_6 + S_{10} + S_{11} + S_{25}$ $A_8 = S_5 + S_{12} + S_{16} + S_{17}$ $A_9 = S_2 + S_8 + S_{15} + S_{19} + S_{20}$
$(273, 191)$ $d = 18$	$g(x) = 1 + x^4 + x^{10} + x^{18} + x^{22} +$ $\qquad x^{24} + x^{34} + x^{36} + x^{40} + x^{48} +$ $\qquad x^{52} + x^{56} + x^{66} + x^{67} + x^{71} +$ $\qquad x^{76} + x^{77} + x^{82}$ $\theta^*(x) = x^{201} + x^{183} + x^{177} + x^{155} +$ $\qquad x^{151} + x^{134} + x^{98} + x^{89} +$ $\qquad x^{86} + x^{75} + x^{73} + x^{42} +$ $\qquad x^{35} + x^{34} + x^{15} + x^5 + 1$	$A_1 = S_{71} + S_{76}$ $A_2 = S_{17}$ $A_3 = S_5 + S_{23}$ $A_4 = S_{21} + S_{27} + S_{45}$ $A_5 = S_3 + S_{25} + S_{31} + S_{49}$ $A_6 = S_{16} + S_{20} + S_{42} + S_{48} + S_{66}$ $A_7 = S_{35} + S_{52} + S_{56} + S_{78}$ $A_8 = S_8 + S_{44} + S_{61} + S_{65}$ $A_9 = S_2 + S_{11} + S_{47} + S_{64} + S_{68}$ $A_{10} = S_{10} + S_{13} + S_{22} + S_{58} +$ $\qquad S_{75} + S_{79}$ $A_{11} = S_1 + S_{13} + S_{15} + S_{24} +$ $\qquad S_{60} + S_{77} + S_{81}$ $A_{12} = S_{30} + S_{32} + S_{43} + S_{46} +$ $\qquad S_{55}$ $A_{13} = S_6 + S_{37} + S_{39} + S_{50} +$ $\qquad S_{53} + S_{62}$ $A_{14} = S_0 + S_7 + S_{38} + S_{40} +$ $\qquad S_{51} + S_{54} + S_{63}$ $A_{15} = S_{18} + S_{19} + S_{26} + S_{57} +$ $\qquad S_{59} + S_{70} + S_{73}$ $A_{16} = S_9 + S_{28} + S_{29} + S_{36} +$ $\qquad S_{67} + S_{69} + S_{80}$ $A_{17} = S_4 + S_{14} + S_{33} + S_{34} +$ $\qquad S_{41} + S_{72} + S_{74}$

Example. Consider the (21, 11) code generated by $g(x) = x^{10} + x^7 + x^6 + x^4 + x^2 + 1$. The parity checks orthogonal on the first error symbol (represented by x^{20}) are found from

$$x^9 \theta^*(x) = x^{20} + x^{18} + x^{13} + x^{12} + x^9$$

$$x^{11} \theta^*(x) = x^{20} + x^{15} + x^{14} + x^{11} + x$$

$$x^{16} \theta^*(x) = x^{20} + x^{19} + x^{16} + x^6 + x^4$$

$$x^{17} \theta^*(x) = x^{20} + x^{17} + x^7 + x^5 + 1$$

and

$$x^{20} \theta^*(x) = x^{20} + x^{10} + x^8 + x^3 + x^2$$

Therefore, the corresponding parity check equations are

$$P_1 = r_{20} + r_{18} + r_{13} + r_{12} + r_9$$

$$P_2 = r_{20} + r_{15} + r_{14} + r_{11} + r_1$$

$$P_3 = r_{20} + r_{19} + r_{16} + r_6 + r_4$$

$$P_4 = r_{20} + r_{17} + r_7 + r_5 + r_0$$

and

$$P_5 = r_{20} + r_{10} + r_8 + r_3 + r_2$$

A set of computed performance curves for the difference-set codes is shown in Fig. 3-12. These curves show the probability of first decoding error and do not account for any improvement (or degradation) that might occur as a result of feedback. They were computed by enumerating all possible error events that would result in the first bit being decoded incorrectly. (See Problems.)

A more general procedure called *L-step orthogonalization* is illustrated by the following example. The (7, 4) Hamming code with $g(x) = x^3 + x + 1$ has a parity check matrix given by

$$\mathbf{H} = \begin{bmatrix} 0 & 0 & 1 & 0 & 1 & 1 & 1 \\ 0 & 1 & 0 & 1 & 1 & 1 & 0 \\ 1 & 0 & 0 & 1 & 0 & 1 & 1 \end{bmatrix} \qquad (3\text{-}5)$$

$$e_4 \; e_5 \; e_6 \; e_0 \; e_1 \; e_2 \; e_3$$

The error symbol corresponding to each column of $\mathbf{H}$ in (3-5) is indicated below the matrix. It is assumed that the syndrome will be calculated by a register that multiplies by x^3 and divides by $x^3 + x + 1$. This code can be

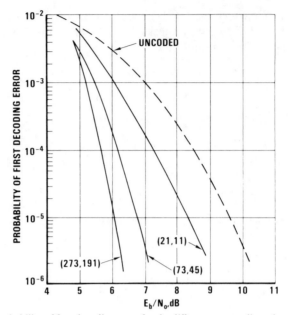

Figure 3-12. Probability of first decoding error for the difference-set cyclic codes using threshold decoding.

two-step orthogonalized. Denote the syndrome symbols corresponding to the first, second, and third rows of $\mathbf{H}$ by S_0, S_1, and S_2, respectively. Then the relevant orthogonal estimates are

$$P_1 = S_0 = e_1 + e_2 + e_3 + e_6$$

$$P_2 = S_0 + S_1 = e_0 + e_3 + e_5 + e_6$$

$$P_3 = S_0 + S_2 = e_0 + e_1 + e_4 + e_6 \tag{3-6}$$

and

$$P_4 = S_0 + S_1 + S_2 = e_2 + e_4 + e_5 + e_6$$

Note that P_1 and P_2 provide two estimates orthogonal on the sum $e_3 + e_6$. Thus, as long as no more than a single error occurs, this sum may be determined correctly. In a similar fashion, P_3 and P_4 allow calculation of the sum $e_4 + e_6$. Finally, note that the two sums calculated provide a pair of estimates orthogonal on e_6. Thus, a two-step majority decoder (referring to the number of levels of majority gates) for the $(7, 4)$ code has the structure shown in Fig. 3-13. Obviously, a simpler decoder for this particular code does exist, but this example serves to illustrate the concept of L-step decoding. In many cases, the L-step decoding algorithm is the simplest known algorithm for a particular code. Two well-known classes of cyclic codes

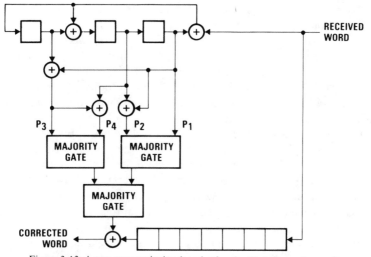

Figure 3-13. A two-step majority decoder for the (7, 4) Hamming code.

can be L-step decoded. These are the Euclidean geometry codes (some of which are equivalent to Reed–Muller codes with a symbol omitted) and the projective geometry codes. Both of these classes of codes are discussed in considerable detail in Peterson and Weldon.[1]

As a final note, threshold decoding is often useful in very-high-speed applications where a moderate amount of coding gain is desired. The usefulness in this situation stems from the fact that the algorithm will produce satisfactory results without any feedback. Thus, one can organize the hardware in a "pipeline" fashion such that the required computation is distributed over several symbol times, if necessary.

3.4. Remarks

In this chapter we have presented three nonalgebraic techniques for decoding block codes. In comparing these techniques we note that the Meggitt decoder is basically a table lookup method and is capable of providing maximum likelihood performance. In practice the Meggitt decoder is limited to codes for which $n - k$ and t are both relatively small. Hence the number of different codes which can be decoded using this technique is also rather small and the maximum practical coding gain which can be achieved is about 2.5 dB (at $P_b = 10^{-5}$). The (23, 12) Golay code is a typical example of a code which is easily decoded using the Meggitt decoder. Unfortunately, the Meggitt decoding concept does not extend directly to enable one to utilize soft decisions. As we will see in Chapter 4, however, the Meggitt decoder can be used as a component in a soft decision decoder.

Threshold decoding is also a remarkably simple technique but is limited to a small number of codes which have the necessary structural characteristics. Some of these codes have a relatively large minimum distance, and as a result it is possible to build threshold decoders that achieve a coding gain of 3.5 dB (at $P_b = 10^{-5}$). The threshold decoding algorithm also extends in a straightforward manner to include the use of soft decisions. This will be covered in Chapter 4.

In contrast, information set decoding techniques apply to a very large class of block codes and work well with both hard and soft decisions. For many codes they are the only known techniques for decoding out to the minimum distance of the code. A prime example is the (48, 24) extended quadratic residue code, which was treated in this chapter. Although the information set techniques are conceptually simple, they must be applied in an ad hoc fashion. There are several methods for using these techniques. The approach selected will depend on the data rate and desired hardware configuration. During the design process one needs to decide what is an acceptable level of performance degradation relative to maximum-likelihood decoding. This has a significant impact on algorithm complexity. In many cases the information set techniques are ideal for implementation in a microprocessor which can perform matrix operations. Obviously, there are some practical limitations on the codes that can be used. Equation (3-1) gives a bound on the number of covering patterns required for all possible t-error patterns. The number of decoding operations is proportional to this number, which is roughly $[n/(n - k)]^t$ for large n. This indicates that very long codes with large values of t are not practical.

The decoding techniques that have been presented in this chapter are the important hard-decision, block decoding techniques that are conceptually simple. We have deliberately deferred the presentation of the more difficult algebraic decoding techniques until Chapter 5. One reason the discussion of the algebraic techniques is deferred is that they do not generalize to use soft decisions. In the next chapter we show how the threshold decoding and information set decoding techniques can be modified to utilize soft decisions, and we show the relationship of these techniques to other soft decision decoding techniques.

Problems

3-1. The (21, 11) code has five orthogonal parity check equations and each equation contains five error terms.

(a) Show that Table P-1 describes the state of the parity check equations for various error patterns.

Table P-1. State of Parity Check Equations for Various Error Patterns

Total number of errors in word	Number of parity check failures when	
	One error is in common check position	No error is in common check position
1	5	1
2	4	0 or 2
3	3 or 5	1 or 3

(b) If the threshold is set so that a correction pulse is generated whenever four or five parity failures occur, what percentage of the time will an error in a three-symbol pattern be corrected?

(c) What is the effect of lowering the threshold to 3?

(d) With the threshold set at 3 compute the probability that a random number of errors will cause an error to be delivered to the user assuming there is an error in the common check position.

(e) Repeat (d) assuming that there is not an error in the common check position.

3-2. If a Meggitt decoder is used to decode a $(31, 21)$ cyclic code, how many distinct patterns must be recognized if

(a) all one- and two-error patterns are to be decoded and the decoder makes only one pass through the data, and

(b) the decoder makes two passes through the data and feeds back an error correction pulse to the syndrome computer.

3-3. Construct (by any means) a set of covering patterns that will cover all sets of three or fewer error patterns for a $(24, 12)$ code. How does this number compare with the Schönheim bound?

3-4. Find a set of three coverings that will cover all three-symbol error patterns in a $(23, 12)$ cyclic code if the patterns are cyclically shifted by the proper amount.

3-5. Show by induction on t that the number of patterns of $n - k$ parity symbols that are required to cover all t error patterns obeys the bound

$$N_{cov} \geq \left\lceil \frac{n}{n-k} \left\lceil \frac{n-1}{n-k-1} \cdots \left\lceil \frac{n-t+1}{n-k-t+1} \right\rceil \cdots \right\rceil \right\rceil$$

Hint: If $N(j)$ patterns are sufficient to cover j errors then every pattern of $j - 1$ errors is included in no fewer than

$$\left\lceil \frac{n-(j-1)}{n-k-(j-1)} \right\rceil$$

different covering sets.

3-6. A $(63, 50)$ distance-6 BCH code is generated by

$$g(x) = x^{13} + x^{12} + x^{11} + x^{10} + x^9 + x^8 + x^6 + x^3 + x + 1$$

Find a single parity set that will cover all two-error patterns if they are cyclically shifted by the proper amount. Find a linear transformation on the syndrome $S(x)$ that will generate the desired pattern. Draw a functional block diagram for the decoder which uses this transformation.

3-7. Consider the (15, 4) maximal-length code. What is the minimum distance d for this code? Find a set of $d - 1$ parity checks orthogonal on e_{14}. Give block diagrams for a shift register encoder and threshold decoder.

3-8. The (15, 5) BCH code is two-step majority decodable. Give the block diagrams of an encoder and a decoder that corrects all triple-error patterns.

4

Soft Decision Decoding
of Block Codes

The use of soft decisions with block codes provides the engineer with another degree of freedom when designing a communication system. The additional information provided by the soft decisions in most instances can provide about 2 dB of additional coding gain and can, therefore, significantly increase the usefulness of a particular code. Soft decision decoding is particularly effective when used with moderate length block codes and provides substantial performance benefits over a broad range of signal-to-noise ratios. It is also effective in concatenation schemes (see Chapter 8) which utilize two or more levels of encoding and decoding. In this case the use of soft decisions is restricted to the innermost coder/decoder, and it is necessary for the soft decision decoding method to provide near optimum performance at word error rates in the range of 10^{-2} to 10^{-3}. This is in contrast to the usual system requirement that the decoding scheme be effective at error rates less than 10^{-5}.

There are two optimum decoding rules that provide the conceptual basis for soft decision decoding. One rule is to select the code word which minimizes the "distance" from the code word to the received sequence. As was discussed in Chapter 1, this rule minimizes the probability of sequence error, and a decoder using this rule is a maximum-likelihood sequence estimator. An alternate rule is to decode each symbol in a code word so as to minimize the average symbol error rate. In this case the decoded sequence may not even be a code word. The performance associated with these two rules is virtually identical since minimizing symbol error rate also tends to minimize word error rate and vice versa. The importance of the two viewpoints is that each one provides insight into the soft decision decoding process that is not provided by the other. Unfortunately, neither rule can be implemented exactly except for very short codes. At this time, all of the practical soft decision decoding techniques for longer codes are essentially ad hoc schemes which attempt to approximate one or the other of these two rules.

Of those techniques which seek to minimize the symbol error rate, the most important are the Massey APP (a posteriori probability) decoding algorithm,[13] the Hartmann–Rudolph optimum symbol-by-symbol decoding algorithm,[20] and the Weldon algorithm.[21] Massey's APP decoding algorithm is simply the logical extension to soft decisions of the threshold decoding algorithm discussed in Chapter 3. For the limited class of codes to which it applies, it represents a nearly optimum decoding procedure. Although the mathematical operations that are required to carry out this technique are relatively complex, there are two special cases for which they can be simplified to a remarkable degree. One is when the three-level or binary erasure channel is used. The other is when only the likelihood information associated with the least reliable symbol is used. This latter technique, called approximate APP decoding, was proposed by Forney[22] and later by Davis.[23] It can be shown to be asymptotically optimum and at reasonable signal-to-noise ratios is only a few tenths of a dB inferior to the complete APP algorithm. The optimum symbol-by-symbol decoding algorithm proposed by Hartmann and Rudolph applies to any group code. It involves the computation of a decision function which is defined on each of the code words in the dual code (i.e., the set of all possible unique parity check equations). Thus the complexity of the algorithm increases with decreasing code rate. This algorithm forms the basis for two approximate procedures. One is to use only the minimum-weight terms in the decision function. When this is done for the case of a one-step orthogonalizable code, the algorithm bears a very strong resemblance to the APP threshold decoding algorithm. The second procedure, due to Greenberger,[24] involves first ordering the received symbols according to decreasing reliability. The Hartmann–Rudolph (HR) algorithm is applied successively to the k most reliable symbols, then the $k + 1$, $k + 2$, etc., until the decision functions associated with the first k symbols does not change appreciably. At low signal-to-noise ratios this procedure apparently requires only a small number of iterations and thus involves only a small number of the words in the dual code. The Weldon decoding technique, as it is normally applied, involves first creating m different sequences corresponding to each of the m bits in a demodulator that has been quantized to 2^m different levels. Each of these sequences is decoded using a hard decision decoder and each symbol is estimated using a weighted sum of the corresponding symbol in each decoded sequence. Although the Weldon technique is simple to apply, its performance is inferior to the other decoding methods.

Techniques which seek to minimize the sequence error rate rely heavily on the information set ideas developed in Chapter 3. These techniques all attempt to generate a small set of code words which, with high probability, contain the word which is at the minimum distance from the received sequence. The decoder then computes the distance to each of these words and chooses the word which is closest. A simple decoder of this type is the generalized minimum distance algorithm proposed by Forney.[25] This concept is closely related to a group of algorithms later proposed by Chase.[26] What Chase does is to locate a small set of the least reliable received symbols. Various trial sequences are generated by first making hard decisions on each of the received symbols and then assuming various combinations of low-weight error patterns in the least reliable symbol positions. Each trial sequence is decoded to produce a code word and the distance between each of these code words and the received sequence is computed. A second technique, which has been studied extensively by Baumert and McEliece,[27] is to arrange the received symbols in order of decreasing confidence and then to generate a number of different information sets with the highest-reliability symbols being included in most of these sets. Each set is then used to generate a trial code word to be compared with the received sequence. The two previous ideas can be combined in a single decoder so that for each information set that is generated various low-weight error patterns are assumed in the information set positions. This technique, which is a soft decision variant of the Omura algorithm, has also been investigated by Baumert and McEliece. Finally, there is a technique called partial syndrome decoding. This technique is used in conjunction with a confidence ordering of the received symbols and in essence uses the major portion of the parity equations to specify a "parity set" in the least reliable positions and the remaining portion to specify a partial syndrome which is used to locate various low-weight errors in the most reliable symbol positions. When this idea is combined with the idea of using several different parity sets (or information sets) the result is an extremely efficient soft decision decoder. These decoding techniques are discussed in more detail in the remainder of this chapter. The discussion is restricted exclusively to binary codes.

4.1. APP Threshold Decoding

The APP threshold decoding algorithm is an extension of the threshold decoding ideas discussed in Chapter 3. We will consider only the case of one-step orthogonalizable codes. In Massey's original work he considered two different variations of the decoding algorithm. The type I starts with a set of J parity check equations that are orthogonal on the mth received symbol. A decision function is defined on this set of parity checks and is compared to a threshold $T/2$ to estimate the value of the error in the mth symbol. Mathematically, this rule takes the form

$$e_m = 1$$

if and only if

$$\sum_{i=1}^{J} A_i w_i > \tfrac{1}{2}T \qquad (4\text{-}1)$$

In (4-1) A_i is the value of the ith parity check which is computed from hard decisions on the received sequence, w_i is a weighting term that is proportional to the reliability of the ith parity check, and T is the sum of the w_i's.[†] We see that if the w_i's are all equal to 1, this reduces to the majority decision rule given in Chapter 3. The type-II decoder starts with the same basic set of parity equations and computes a decision function which estimates the value of the mth code symbol directly. In this case, the decoding rule is

$$c_m = 1$$

if and only if

$$\sum_{i=0}^{J} B_i w_i > \tfrac{1}{2}T \qquad (4\text{-}2)$$

In (4-2) the B_i's are obtained from the A_i's by eliminating the mth symbol. Again, the weight w_i is a function of the reliability of the estimate B_i. The term B_0 is defined to be

$$B_0 = \hat{r}_m = c_m + e_m$$

where $\hat{r}_m$ is the hard decision associated with the mth received demodulator output r_m. Equations (4-1) and (4-2) can be obtained in a straightforward manner. The derivation of (4-2) and the development of the weighting terms, w_i, are given below while the derivation of (4-1) is left as an exercise for the reader (see Problem 4-1). In both cases the values of w_i and T turn out to be identical.

4.1.1. Derivation of the APP Decoding Rule

In order to minimize the average probability of symbol error we wish to assign to c_m a value, V, such that

$$\Pr(c_m = V \mid \{B_i\}) \qquad (4\text{-}3)$$

[†] Actually, T is defined as $T = \sum_{i=0}^{J} w_i$ and consequently includes a term w_0 which does not appear on the left-hand side of 4-1). The significance of w_0 will become clear upon solving Problem 4-1.

is maximized. Using Bayes' rule

$$\Pr(c_m = V | \{B_i\}) = \frac{\Pr(\{B_i\} | c_m = V) \Pr(c_m = V)}{\Pr(\{B_i\})}$$

Thus the decoding rule is as follows: Choose $c_m = 1$ if and only if

$$\Pr(\{B_i\} | c_m = 0) \Pr(c_m = 0) < \Pr(\{B_i\} | c_m = 1) \Pr(c_m = 1) \quad (4\text{-}4)$$

Since the parity check equations are orthogonal on the mth symbol the individual probabilities $\Pr(B_i | c_m = V)$ are all independent and (4-4) can be rewritten as

$$\Pr(c_m = 0) \prod_{i=0}^{J} \Pr(B_i | c_m = 0) < \Pr(c_m = 1) \prod_{i=0}^{J} \Pr(B_i | c_m = 1)$$

Finally, taking the logarithm of both sides and assuming that c_m is equally likely to be a 0 or 1, (4-4) becomes

$$\sum_{i=0}^{J} \log \left[\frac{\Pr(B_i | c_m = 0)}{\Pr(B_i | c_m = 1)} \right] < 0$$

Each of the B_i is computed by dropping the term corresponding to r_m from the ith parity equation. Thus, each of the B_i can be written as

$$B_i = c_m + \sum_{j=1}^{n_i} e_{ij}$$

where e_{ij} is the jth error term in the ith parity equation excluding, of course, e_m. (We have taken n_i to be the total number of terms in the ith parity equation excluding c_m.) Note that in the absence of errors each B_i is an independent estimate of c_m. We now observe that B_i takes on the value 0 when c_m is zero and there is an even number of errors or when c_m is one and there is an odd number of errors. Conversely, it takes on the value 1 when c_m is zero and there is an odd number of errors or when c_m is 1 and there is an even number of errors. If we let p_i denote the probability of an odd number of errors in the ith parity equation (excluding e_m) and $q_i = 1 - p_i$ be the probability of an even number of errors, it follows that

$$\Pr(B_i = 0 | c_m = 0) = \Pr(B_i = 1 | c_m = 1) = q_i$$

and

$$\Pr(B_i = 1 | c_m = 0) = \Pr(B_i = 0 | c_m = 1) = p_i$$

Thus, the decision function can be written as follows: Choose $c_m = 1$ if and only if

$$\sum_{i=0}^{J} (1 - 2B_i) \log\left(\frac{q_i}{p_i}\right) < 0 \qquad (4\text{-}5)$$

Finally one can show that the probability of an odd number of errors in a given set of n_i symbols is given by

$$p_i = \frac{1}{2}\left[1 - \prod_{j=1}^{n_i} (1 - 2\gamma_{ij})\right]$$

where γ_{ij} is the probability that the jth symbol in the ith parity equation is in error. For a coherent PSK system with an infinitely quantized demodulator output r_{ij},

$$\begin{aligned}
\gamma_{ij} &= \Pr(\text{error}|r_{ij}) \\
&= \frac{\exp\left[-(|r_{ij}| + E_s^{1/2})^2/N_0\right]}{\exp\left[-(|r_{ij}| + E_s^{1/2})^2/N_0\right] + \exp\left[-(|r_{ij}| - E_s^{1/2})^2/N_0\right]} \\
&= \frac{\exp\left[-4E_s^{1/2}|r_{ij}|/N_0\right]}{1 + \exp\left[-4E_s^{1/2}|r_{ij}|/N_0\right]}
\end{aligned} \qquad (4\text{-}6)$$

We may also write (4-5) as

$$\sum_{i=0}^{J} B_i \log\left(\frac{q_i}{p_i}\right) > \frac{1}{2}\sum_{i=0}^{J} \log\left(\frac{q_i}{p_i}\right) \qquad (4\text{-}7)$$

Although Massey originally formulated the decoding rule in the form given by (4-7), it is usually more convenient for implementation purposes to consider the form of the rule given by (4-5). In either case a key step in implementing the decoder is the evaluation of the weighting term

$$\begin{aligned}
w_i &= \log\left(\frac{q_i}{p_i}\right) \\
&= \log\left\{\left[1 + \prod_{j=1}^{n_i} (1 - 2\gamma_{ij})\right] \bigg/ \left[1 - \prod_{j=1}^{n_i} (1 - 2\gamma_{ij})\right]\right\}
\end{aligned} \qquad (4\text{-}8)$$

4.1.2. Computation of the Weighting Terms

From (4-8) we see that each individual weighting term is a nonlinear function of the n_i different γ_{ij}. Each γ_{ij} is simply the probability of error for an individual received symbol and depends only upon the magnitude of the soft decision output of the demodulator. Hence, the complexity associated with computing each individual weighting term depends strongly on the value of n_i and the number of bits required to represent the demodulator output. When the product of these two numbers is not too large, w_i can be computed using a read-only memory. For example, each orthogonal parity check in a $(21, 11)$ code involves four error terms and, consequently, each w_i is a function of three different γ_{ij}. If the demodulator is quantized to eight levels (three bits), then two bits are required to represent the absolute value of the output. Thus, w_i can be computed using a read-only memory with six address bits. For a $(73, 45)$ code each parity equation contains nine terms and, hence, each w_i depends upon eight different γ_{ij}. This would require a read-only-memory with 16 address lines for eight-level quantization or eight address lines for four-level quantization. As a practical matter, the difference between four-level and eight-level quantization represents only a few tenths of a decibel in performance. The design engineer must decide in a given situation whether this amount of additional gain is worth the added cost of the larger memory.

When the number of inputs required in a single read-only memory implementation becomes too large then it may be desirable to partition the process into two or more stages. In this case, an alternate form of (4-8) is useful. We define

$$\alpha_{ij} = -\log_e(1 - 2\gamma_{ij})$$

Then

$$1 - 2\gamma_{ij} = e^{-\alpha_{ij}}$$

and

$$\left[1 + \prod_{j=1}^{n_i} (1 - 2\gamma_{ij}) \right] \Big/ \left[1 - \prod_{j=1}^{n_i} (1 - 2\gamma_{ij}) \right]$$

$$= \left[1 + \exp\left(-\sum_{j=1}^{n_i} \alpha_{ij} \right) \right] \Big/ \left[1 - \exp\left(-\sum_{j=1}^{n_i} \alpha_{ij} \right) \right]$$

Using the identity

$$\coth x = \frac{e^x + e^{-x}}{e^x - e^{-x}}$$

we may write (4-8) as

$$w_i = \log\left[\coth\left(\frac{1}{2} \sum_{j=1}^{n_i} \alpha_{ij} \right) \right] \tag{4-9}$$

Like the γ_{ij}'s, the α_{ij}'s are functions only of the absolute value of the demodulator output. Using the form of the weighting function given by (4-9), the computation can be carried out by first computing the α_{ij}, next carrying out the summation, and finally performing the mapping indicated by

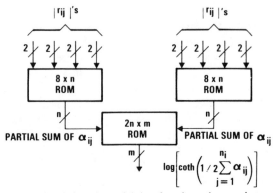

Figure 4-1. Circuit for computing the weighting function when each term contains eight entries.

$\log(\coth x)$. As we previously noted, a 16-input read-only memory is required for the computation of w_i when eight different γ_{ij} are involved and the demodulator is quantized to eight levels. Using the approach indicated above, w_i may be computed using two levels of read-only memories as shown in Fig. 4-1. At the first level four different values from the demodulator (indicated by $|r_{ij}|$) are converted to the partial sums of the α_{ij} in each of two read-only memories. These two memories each require eight input lines. At the second level the two partial sums are combined and the mapping $\log(\coth x)$ is performed. Obviously, many other variations on this idea are possible.

So far we have intentionally omitted any discussion as to how many bits are required to represent the w_i. In practice it turns out to be a surprisingly small number. The discussion in the next section sheds some light on this problem and provides a practical lower bound on the accuracy which is required.

4.1.3. Approximate Computation of w_i

Rather than attempt to compute each value of w_i exactly, it is possible to obtain a useful approximation by considering only the γ_{ij} which corresponds to the least reliable received symbol in each parity equation. Following a procedure suggested by Davis[23] let γ_l be the largest of the n_i different γ_{ij}'s in (4-8) and let $\gamma_{\min}$ be the smallest possible value of γ_{ij}. Then w_i is bounded above and below by

$$\log\left[\frac{1 + (1 - 2\gamma_l)^{n_i}}{1 - (1 - 2\gamma_l)^{n_i}}\right] \leq w_i \leq \log\left[\frac{1 + (1 - 2\gamma_l)(1 - 2\gamma_{\min})^{n_i - 1}}{1 - (1 - 2\gamma_l)(1 - 2\gamma_{\min})^{n_i - 1}}\right]$$

$$(4\text{-}10)$$

If γ_{min} is sufficiently small that $1 - 2\gamma_{min}$ is approximately one, (4-10) can be written as

$$\log \left[\frac{1 + (1 - 2\gamma_l)^{n_i}}{1 - (1 - 2\gamma_l)^{n_i}} \right] \le w_i \le \log \left[\frac{1 - \gamma_l}{\gamma_l} \right] \tag{4-11}$$

For high signal-to-noise ratios, γ_l becomes very small and in the limit (4-11) becomes

$$\log(1/\gamma_l) - \log n_i \le w_i \le \log(1/\gamma_l) \tag{4-12}$$

Thus the bounds differ only by a constant which is negligible for very small γ_l. Consequently, a scheme which uses only the largest value of γ_{ij} is as-symptotically equivalent to exact APP decoding. This suggests that we approximate w_i as

$$w_i \approx K \log \left[\frac{1 - \gamma_l}{\gamma_l} \right] \tag{4-13}$$

Using this approximation a practical set of weights for an eight-level quantization scheme might be obtained as follows: The transition probabilities for a uniformly quantized eight-level channel operating at an E_s/N_0 of 2.0 dB are given by

$$Pr(0|0) = 0.600 \qquad Pr(4|0) = 0.0265$$
$$Pr(1|0) = 0.177 \qquad Pr(5|0) = 0.00846$$
$$Pr(2|0) = 0.121 \qquad Pr(6|0) = 0.00210$$
$$Pr(3|0) = 0.064 \qquad Pr(7|0) = 0.000473$$

Thus the four possible values of γ_{ij} are

$$\gamma_{ij}^0 = \frac{Pr(4|0)}{Pr(3|0) + Pr(4|0)} = 0.2928$$

$$\gamma_{ij}^1 = \frac{Pr(5|0)}{Pr(2|0) + Pr(5|0)} = 0.0653$$

$$\gamma_{ij}^2 = \frac{Pr(6|0)}{Pr(1|0) + Pr(6|0)} = 0.0117$$

$$\gamma_{ij}^3 = \frac{Pr(7|0)}{Pr(0|0) + Pr(7|0)} = 0.00078$$

Table 4-1. Approximate Weighting Functions for
Eight-Level Channel at $E_b/N_0 = 2$ dB

Least reliable symbols in levels	γ_l	w_i	Normalized w_i
3.4	0.2928	0.8818	1.0 = 1.0
2.5	0.0653	2.661	3.017 ≈ 3.0
1.6	0.0117	4.436	5.03 ≈ 5.0
0.7	0.00078	7.15	8.1 ≈ 8.0

When these are used with (4-13) and suitably normalized, the result is as shown in Table 4-1. In practice the set of weights 1, 3, 5, and 7 works nearly as well as the set 1, 3, 5, and 8 and is suitable over a large range of signal-to-noise ratios. This set of weights establishes the minimum number of output lines for the read-only memory approach discussed in the previous section. The design engineer can use this as a starting point and include the effects of additional terms in each parity check so as to come as close as desirable to the exact APP algorithm.

There is one other approximation to computing the weight terms which is sometimes useful. If one observes that (4-8) is of the form

$$w_i = \log\left[\frac{1 + x_i}{1 - x_i}\right]$$

then w_i can be expanded in a power series in x and approximated by the leading term. Thus

$$w_i = 2\left(x_i + \frac{x_i^3}{3} + \cdots\right)$$

$$\approx 2x_i = 2\prod_{j=1}^{n_i}(1 - 2\gamma_{ij})$$

When this is used in (4-5) the decision function becomes

$$f_d = \sum_{i=0}^{J} 2(1 - 2B_i)\prod_{j=1}^{n_i}(1 - 2\gamma_{ij}) \tag{4-14}$$

This decision function has the same form as the decision function for the Hartmann–Rudolph algorithm to be discussed in Section 4.2.

4.1.4. Implementation of the APP Decoder

The most useful statement of the threshold decoding rule is given by

(4-5). We can think of (4-5) as specifying a decision function

$$f_d = \sum_{i=0}^{J} (1 - 2B_i)\, w_i \qquad (4\text{-}15)$$

which is greater than zero when $c_m = 0$ and is less than zero when $c_m = 1$. The w_i in this expression is always greater than zero whereas the coefficient $1 - 2B_i$ takes on the value $+1$ when B_i is zero and is -1 when B_i is one. Thus, this coefficient, in effect, determines the sign to be associated with each weight. Since each B_i is an independent estimate of c_m, each term in (4-15) will be positive whenever the individual estimate is $c_m = 0$ and will be negative whenever the individual estimate is $c_m = 1$.

For this algorithm it is useful to assign binary numbers to the various output levels of the demodulator such that the most significant bit (msb) specifies the sign and the remaining bits specify the magnitude. If this is done the msb is equivalent to a hard decision while the remaining bits specify the reliability of this decision. Using this convention, a general implementation for this algorithm is shown in Fig. 4-2. In this figure the orthogonal parity checks are computed directly from the hard decisions. Equivalently, one could replace this register with a feedback shift register and compute the orthogonal parity checks as was done in Chapter 3.

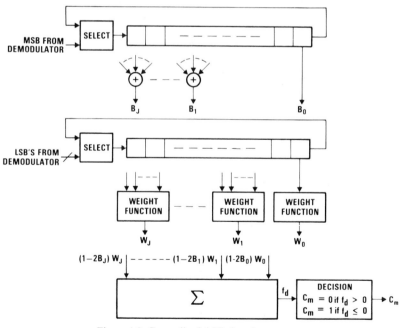

Figure 4-2. Generalized APP decoder.

In Chapter 3 it was noted that the performance of the threshold decoder could be improved by feeding back the decisions and making more than one pass through the decoder. This procedure can also be used with APP decoding, and the performance improvement is even more impressive. From the previous discussion it should be apparent that the magnitude of the decision function f_d will be large whenever the individual estimates collectively exhibit a large confidence value, and it will be small when they exhibit a small confidence value. Thus, the decision function, f_d, can be regarded as a soft decision estimate of c_m.

One may formalize this idea by observing that

$$f_d = \log \left[\frac{\Pr(c_m = 0 \,|\, \{B_i\})}{\Pr(c_m = 1 \,|\, \{B_i\})} \right]$$

Since the decoder chooses $c_m = 0$ whenever f_d is greater than zero and chooses $c_m = 1$ whenever f_d is less than zero it follows that

$$\Pr(\text{error} \,|\, f_d) = \Pr(c_m = 1 \,|\, \{B_i\}), \qquad f_d > 0$$

and

$$\Pr(\text{error} \,|\, f_d) = \Pr(c_m = 0 \,|\, \{B_i\}), \qquad f_d \leq 0$$

Thus, for $f_d > 0$

$$e^{f_d} = \frac{1 - \Pr(\text{error} \,|\, f_d)}{\Pr(\text{error} \,|\, f_d)}$$

and consequently

$$\Pr(\text{error} \,|\, f_d) = \frac{e^{-f_d}}{1 + e^{-f_d}}$$

Similarly for $f_d \leq 0$

$$\Pr(\text{error} \,|\, f_d) = \frac{e^{f_d}}{1 + e^{f_d}}$$

Thus, for all f_d

$$\Pr(\text{error} \,|\, f_d) = \frac{e^{-|f_d|}}{1 + e^{-|f_d|}} \tag{4-16}$$

We note that (4-16) has a form identical to expression (4-6) for the Gaussian channel. Thus, f_d can be treated in exactly the same manner as the demodulator output r_{ij}.

This suggests that we replace each received value r_m by the appropriate value of f_d as it is computed (properly scaled, of course) and recirculate the result. If this new value is, in fact, an improved estimate of the c_m then it in turn should contribute to improved estimates of the remaining code symbols. A general decoder which incorporates this feature is shown in Fig. 4-3. For very-high-speed operation, where the time required to perform the necessary computation is a limiting consideration, one might choose to cascade several levels of decoders rather than feed back to the original one. Alternately, one can feed back after several stages of delay to some interior point in the register and, hopefully, accomplish the same result. The value of this technique and the number of iterations that are required is best determined by simulation for a specific case.

One final simplification in the algorithm results when a three-level channel is used. In this case received symbols which fall in the erasure zone cannot be used to compute a parity check. The way out of this dilemma is to disregard any parity checks which contain erasures by setting the corresponding weight equal to zero. If the parity check does not contain an erasure, then the weight is assigned a value of 1. When this technique is used with the iterative scheme previously discussed, the erasures will tend to be filled in so that subsequent estimates will involve a larger number of parity check equations.

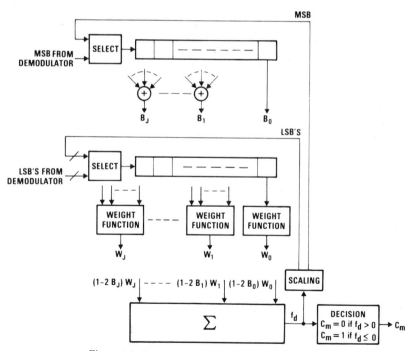

Figure 4-3. Generalized APP decoder with feedback.

4.2. Optimum Symbol-by-Symbol Decoding

In 1976 Hartmann and Rudolph[20] proposed an optimum decoding algorithm for group codes which minimizes the average symbol error rate. This algorithm bears a strong resemblance to the APP threshold decoding algorithm except that the decision function is computed using every possible unique parity check equation instead of using only those which are orthogonal on a given symbol. Consequently, the decoding rule can only be implemented exactly for codes having a small number of parity symbols. The formal statement of the Hartmann–Rudolph (HR) algorithm is quite simple. Let C_i be the ith code word in the dual code and let C_{ij} be the jth symbol in this code word. Define the likelihood ratio ϕ_j for the jth received symbol as

$$\phi_j = \frac{\Pr(r_j|1)}{\Pr(r_j|0)} \qquad (4\text{-}17)$$

Finally let

$$P_j = \frac{1 - \phi_j}{1 + \phi_j} \qquad (4\text{-}18)$$

The decoding rule can now be stated as follows: Set $c_m = 0$ if and only if

$$f_d = \sum_{i=1}^{2^{n-k}} \prod_{j=0}^{n-1} P_j^{C_{ij} \oplus \delta_{jm}} \geq 0 \qquad (4\text{-}19)$$

where δ_{jm} is equal to 1 when $j = m$ and is equal to zero otherwise. The symbol $\oplus$ stands for modulo-2 addition and is used to distinguish this operation from the ordinary arithmetic sum implied by the summation symbol.

As an example consider the (7, 4) cyclic Hamming code whose **H** matrix is given by

$$\mathbf{H} = \begin{bmatrix} 1 & 0 & 0 & 1 & 0 & 1 & 1 \\ 0 & 1 & 0 & 1 & 1 & 1 & 0 \\ 0 & 0 & 1 & 0 & 1 & 1 & 1 \end{bmatrix}$$

The eight words in the dual code are formed by taking all linear combinations of the rows of the **H** matrix and are (0 0 0 0 0 0 0), (0 0 1 0 1 1 1), (0 1 0 1 1 1 0), (0 1 1 1 0 0 1), (1 0 0 1 0 1 1), (1 0 1 1 1 0 0), (1 1 0 0 1 0 1), and (1 1 1 0 0 1 0). The decision function for the symbol in the last position (C_6)

is easily computed using these code words and is given by

$$f_d = P_6 + P_2P_4P_5 + P_1P_3P_4P_5P_6 + P_1P_2P_3 + P_0P_3P_5 + P_0P_2P_3P_4P_6$$
$$+ P_0P_1P_4 + P_0P_1P_2P_5P_6$$

Thus the symbol in position 6 is decoded as a zero whenever f_d is greater than zero and as a one whenever f_d is less than zero. The remaining symbols are decoded by cyclically shifting the received word and recomputing f_d.

4.2.1. Derivation of the Hartmann–Rudolph Algorithm

The derivation of the Hartmann–Rudolph decoding rule is straightforward and involves the following three steps. One first writes f_d as

$$f_d = \Pr(c_m = 0|\mathbf{r}) - \Pr(c_m = 1|\mathbf{r})$$

Next by applying Bayes' rule and enumerating all of the cases for which $c_m = 0$ and for which $c_m = 1$, one can show that

$$f_d = \sum_{i=1}^{2^K} \Pr(\mathbf{r}|\mathbf{c}_i)(-1)^{c_{im}}$$

That is, the decision function is the sum of all of the conditional probabilities of $\mathbf{r}$ given each code word in which c_m is a zero minus the sum of the conditional probabilities of $\mathbf{r}$ given each code word in which c_m is a one.

Finally, one invokes the following result from finite Fourier transform theory. Let $g(\boldsymbol{\xi})$ be any function which is defined on the set of all 2^N binary vectors of dimension N and let $G(\mathbf{v})$ be the finite Fourier transform of $g(\boldsymbol{\xi})$ which is given by

$$G(\mathbf{v}) = \sum_{\text{all } \boldsymbol{\xi}} g(\boldsymbol{\xi})(-1)^{\boldsymbol{\xi}\cdot\mathbf{v}}$$

Then it can be shown that

$$\frac{1}{|C'|} \sum_{\mathbf{v}\in C'} G(\mathbf{v}) = \sum_{\boldsymbol{\xi}\in C} g(\boldsymbol{\xi})$$

where $|C'|$ is just the number of vectors in C'. This result says that if one has any function $g(\boldsymbol{\xi})$ defined on a code C then the sum of this function over the entire set of code words is directly proportional to the sum of the Fourier transform of $g(\boldsymbol{\xi})$ over the set of all code words in the dual code. By using this result and by making use of the definitions for ϕ_j and P_j, (4-19) follows directly. The details of this proof are left as an exercise for the reader (see Problems).

4.2.2. An Alternate Form of the HR Algorithm

One can gain further insight into the nature of the Hartmann–Rudolph algorithm by writing (4-19) in a slightly different form. First, by substituting (4-17) into (4-18) we obtain

$$P_j = \frac{\Pr(r_j|0) - \Pr(r_j|1)}{\Pr(r_j|0) + \Pr(r_j|1)}$$

If we had made hard decisions, $\hat{r}_j$, on each of the received symbols, r_j, then

$$\Pr(r_j \text{ is in error}|r_j) = \frac{\min[\Pr(r_j|0), \Pr(r_j|1)]}{\Pr(r_j|0) + \Pr(r_j|1)}$$

It follows that the magnitude of P_j is

$$|P_j| = 1 - 2\Pr(\text{error}|r_j)$$

and consequently

$$P_j = (-1)^{\hat{r}_j}[1 - 2\Pr(\text{error}|r_j)] \tag{4-20}$$

We now make the following observations: First, only those values of P_j for which $C_{ij} + \delta_{jm} = 1$ are included in each product term in (4-19). Second, the nonzero elements of each C_i define a unique parity check equation. Thus if, for each value of i, we form the modulo-2 sum of $\hat{r}_j$ over those values of j for which $C_{ij} + \delta_{jm} = 1$, then each of these sums defines a different estimate, B_i, of c_m. Each estimate is of the form

$$B_i = c_m + \sum_{j=1}^{n_i} e_{ij}$$

where the e_{ij} are the error terms corresponding to $C_{ij} + \delta_{jm} = 1$. Finally, if we let γ_{ij} denote the probability that $e_{ij} = 1$ then (4-20) can be written

$$P_j = (-1)^{\hat{r}_j}(1 - 2\gamma_{ij})$$

When this is used in (4-19), the result is

$$f_d = \sum_{i=1}^{2^{n-k}} (-1)^{B_i} \prod_{j=1}^{n_i} (1 - 2\gamma_{ij})$$

or equivalently

$$f_d = \sum_{i=1}^{2^{n-k}} (1 - 2B_i) \prod_{j=1}^{n_i} (1 - 2\gamma_{ij})$$

We recall that p_i, the probability of an odd number of errors, is

$$p_i = \frac{1}{2} \left[1 - \prod_{j=1}^{n_i} (1 - 2\gamma_{ij}) \right]$$

and thus

$$q_i - p_i = \prod_{j=1}^{n_i} (1 - 2\gamma_{ij})$$

Hence f_d is given by

$$f_d = \sum_{i=1}^{2^{n-k}} (1 - 2B_i)(q_i - p_i) \tag{4-21}$$

Equation (4-21) provides an alternate statement of the HR algorithm, namely, the following

1. Using the parity check matrix compute every possible estimate of c_m using the $\hat{r}_j$'s.
2. Weight each estimate by the probability that the estimate is correct minus the probability that the estimate is incorrect.
3. Sum all the weighted estimates, compare the result to zero, and chose c_m accordingly.

From the form of the HR algorithm given by (4-21) it is obvious that this algorithm is simply a generalization of the APP threshold decoding algorithm. Instead of using only those parity checks which are orthogonal on a given symbol, one uses all possible parity check equations. In addition, the weight terms are computed in a slightly different fashion.

Since the APP algorithm uses a much smaller number of estimates, this has led to speculation[28] that the number of estimates for a general (n, k) code could also be reduced significantly if the weighting functions were modified. At this time, however, no positive results in this direction have been found except for L-step orthogonalizable codes. Interestingly enough, if one attempts to decode a one-step orthogonalizable code by using only the orthogonal parity checks and computing the weights via the HR algorithm, then the resulting decision function is identical to that given by (4-14). Thus, in this particular case, using only a small subset of the parity check equations results in an excellent decoding rule.

Whether or not one uses a complete set of dual code words or a partial set of dual code words, all of the modifications and implementation techniques discussed for the APP threshold decodable codes also apply to the HR algorithm. Thus, one can assign a weighting function based on the least reliable symbol in each estimate (à la Forney and Davis) or one can iterate the algorithm one or more times using the newly computed (and scaled) f_d to replace the original soft decisions.

4.2.3. A Design Example

The ideas developed in this section and in 4.1 will now be used to design a decoder for the (21, 11) code. The basic hardware configuration is assumed to be as shown in Fig. 4-3. Factors which can affect the complexity of this design are the number of estimates that are used and the degree of quantization. As a first step it is desirable to establish some baseline performance curves using the unquantized channel. Four such curves are shown in Fig. 4-4. The curve labeled "union bound" was obtained by computing a bound on bit error rate using (1-41). The other three curves were obtained from computer simulations. The curve for the complete HR algorithm was obtained by Chase et al.,[29] while the other two were provided by the authors. Both the APP algorithm and the simplified HR algorithm use the same set of five orthogonal parity equations to construct the estimates. In both cases, the scaling in the feedback loop was adjusted to minimize the error rate and the algorithms were iterated until no further performance improvement was obtained. The rate at which multiple itera-

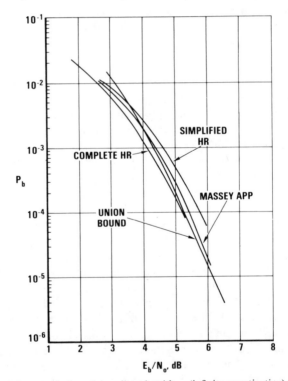

Figure 4-4. Performance of selected decoding algorithms (infinite quantization) with the (21, 11) code.

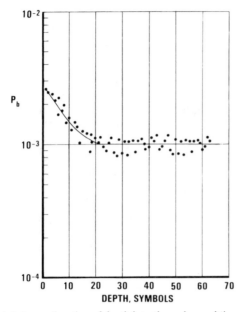

Figure 4-5. P_b as a function of depth into the code word (in symbols).

tions provided an improvement is typified by the curve shown in Fig. 4-5 for the APP algorithm. This curve shows bit error rate as a function of symbol position and was based on the decoding of 10^5 received sequences. From this curve one may conclude that two passes through the decoder are sufficient to obtain all of the improvement that is available.

Since the complete HR algorithm is optimum, it represents the best performance that can be obtained. The reader should observe that the curve for the complete HR algorithm is closely approximated by the union bound and that the two predict identical performance at high signal-to-noise ratios. One should also note that the APP algorithm does almost as well as the complete HR algorithm and is about 0.3 dB better than the simplified HR algorithm. Since the difference between the APP algorithm and the simplified HR algorithm is only in the weighting function, the choice is dictated by the relative difficulty of implementing this term. For the (21, 11) code, one would probably use a read-only-memory, and consequently the choice is clearly in favor of the APP algorithm.

The next step in the design is to determine the degree of quantization and the placement of quantization levels. Since demodulators which provide three-bit quantization with uniform spacing are commonly available, this case was chosen as a starting point. The weighting function was implemented on the computer as a look-up table where the entries in the table were computed assuming an input signal-to-noise ratio (E_s/N_0) of 2 dB.

Thus, the table would be optimum at an output symbol error rate somewhere between 10^{-3} and 10^{-4}. The entries in the table used the full accuracy of the computer, which was 24 bits. The quantizer in the feedback loop was adjusted so that the conditional probabilities of error given by (4-16) matched the conditional probabilities of error associated with the input quantized signal assuming a 2-dB signal-to-noise ratio. Several different spacings of the input quantization levels were chosen to determine the relative sensitivity of the algorithm to threshold spacing. The best performance was obtained for a spacing of $0.3E_s^{1/2}$; however, the degradation was very small for values in the range of $0.2E_s^{1/2}$ to $0.5E_s^{1/2}$. The performance curve for the $0.3E_s^{1/2}$ spacing is shown in Fig. 4-6. A similar exercise was carried out for the case of four-bit quantization. The best threshold spacing was determined experimentally to be $0.2E_s^{1/2}$ and the performance curve corresponding to this case is also shown in Fig. 4-6. From these curves it seems clear that there is little point in using more than three-bit quantization.

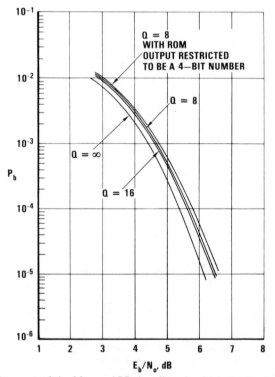

Figure 4-6. Performance of the Massey APP decoding algorithm for the (21, 11) code with several values of Q.

The one remaining task is to determine the degree of quantization that is required at the output of the weighting functions. This will then enable the designer to size the adding circuitry. From a practical viewpoint this is not of major importance since the complexity of this portion of the decoder will not be a strong function of the number of bits that are used. Since read-only-memories which provide four-bit outputs are commonly available as well as adders which can accept four-bit inputs, one additional simulation was carried out with the look-up table restricted to four-bit numbers. The resulting performance curve is also shown in Fig. 4-6.

4.2.4. Greenberger Approximation to the HR Algorithm

Greenberger[24] has proposed an alternate method of reducing the number of parity check equations used in the HR algorithm. The basis of his technique is that symbols that are received with low reliability do not contribute much information and therefore parity equations which contain these symbols can be disregarded. His procedure is as follows:

1. Order the received symbols according to their reliability with the most reliable symbols appearing at the head of the list.
2. Rearrange the columns of the parity check matrix so that the column corresponding to the most reliable symbol appears in the left-most position, the second most reliable symbol in the next position, etc.
3. Perform elementary row operations on the reordered H matrix to create all zeros above a main diagonal in the right-hand most positions. The first row of the modified matrix is now all zero in the $n - k - 1$ least reliable positions. The second row is all zero in the $n - k - 2$ least reliable positions, etc.
4. Using only the first row of the modified H matrix, apply the HR algorithm to estimate each of the received symbols. This operation will obviously use only two words in the dual code, the all-zero word and the word corresponding to the first row of H.
5. Now consider the first two rows in the modified H matrix. These two rows will generate four words in the dual code. Two of those will be the ones previously generated, and two will be new and will contain contributions from the $(k + 2)$th most reliable symbol. Use the two new words to update the estimates of each of the received symbols.
6. Continue by adding new rows one at a time and updating each estimate. Stop when the estimates tend to stabilize or when a predetermined number of iterations has occurred.
7. Based on the most recent set of estimates, choose the k most reliable symbols which form an information set and encode these to regenerate the code word.

8. Reorder the symbols and deliver this word to the user.

Greenberger has used this procedure to decode the (23, 12) Golay code. He found that the technique was particularly effective at low signal-to-noise ratios (output error rates of approximately 1 %) and that six iterations (64 dual code words) produced a performance that was almost identical to using the full algorithm (1024 dual code words). He also found that while the algorithm might not always provide a correct estimate for all of the required symbols, it more often than not would cause the erroneous symbols to have sufficiently low reliability that they would not appear among the k most reliable and would therefore not influence the choice of decoded word. The algorithm is definitely oriented toward a microprocessor-based implementation. It also has the disadvantages that the particular form of the **H** matrix cannot be computed in advance and that one cannot take advantage of any cyclic properties.

4.3. The Weldon Algorithm

Like both the APP algorithm and HR algorithm, the Weldon algorithm[21] obtains several different estimates of each transmitted symbol, weights these according to their reliability, and computes a decision function similar to (4-5) or (4-21). Unlike either of these algorithms, however, the initial estimates are computed using several hard decision decoders. The Weldon algorithm does not perform as well as other soft decision decoding techniques but is somewhat appealing because of its simplicity. Only a brief outline of the algorithm is included here. The reader is referred to Weldon's original paper for more details. The basic procedure proposed by Weldon is as follows:

1. Assume the decoder has available a sequence of n soft decisions and that the demodulator has been quantized to Q levels. Normally Q will be 2^m but any other choice can be accommodated. Assign a weighting function $\sigma_i (i = 0$ to $Q - 1)$ to each level such that $\sigma_0 = 0$,

$$\sigma_0 \leq \sigma_1 \leq \sigma_2 \leq \cdots \leq \sigma_{Q-1}$$

and

$$\sigma_i + \sigma_{Q-1-i} = \sigma_{Q-1}$$

For the usual case where $Q = 2^m$, a good but suboptimum choice is to let the σ_i be the integers 0 through $2^m - 1$.

2. The next step in the algorithm is to map each of the received σ_i into an r-bit binary number and thus create r different binary sequences of length n. This procedure requires some explanation. With the constraints on σ_i it is always possible to find a set of binary numbers a_{ij} and a set of

nonnegative weights V_j such that each value σ_i can be written

$$\sigma_i = \sum_{j=1}^{r} a_{ij} V_j$$

where V_j must satisfy the constraint

$$\sum_{j=1}^{r} V_j = \sigma_{Q-1}$$

in order that the a_{ij} all be nonnegative. For reasons which will soon be apparent it is desirable to make r as small as possible. The reader can easily verify that it is always possible to choose

$$\sigma_1 = V_1$$

$$\sigma_2 = V_1 + V_2$$

$$\vdots$$

$$\sigma_{\lfloor (Q-1)/2 \rfloor} = V_1 + V_2 + \cdots + V_{\lfloor (Q-1)/2 \rfloor}$$

and

$$\sigma_{Q-1} = V_1 + V_2 + \cdots + V_{\lfloor (Q+1)/2 \rfloor}$$

The remaining σ_i can then be determined using the constraint $\sigma_i + \sigma_{Q-1-i} = \sigma_{Q-1}$. Thus r need never be greater than $\lfloor (Q + 1)/2 \rfloor$ and in many cases may be made smaller by using a different assignment. In the usual case where $Q = 2^m$ and the σ_i are the integers less than $Q - 1$, it follows immediately that if V_j takes on the values $1, 2, 4, \ldots, 2^{m-1}$ then the a_{ij} are just the bits in the binary representation of σ_i and $r = m$.

3. Using a hard decision decoder, decode each of the r different sequences to obtain r different code words. Each symbol in each of the code words is treated as if it were an independent estimate of the transmitted symbol.

4. Combine the r estimates on a symbol-by-symbol basis using the decision function

$$f_d = \sum_{i=1}^{r} (1 - 2B_{ij}) w_i \tag{4-22}$$

and choose $C_j = 0$ if f_d is greater than zero and $C_j = 1$ otherwise. In this expression B_{ij} is the jth symbol in the ith decoded sequence and w_i is a weighting term that is proportional to the reliability of the estimate.

The w_i in (4-22) are computed as follows. Let

$$w_i = V_i R_i$$

where V_i is a fixed number associated with each decoder and R_i depends on the number of errors corrected by the decoder. In the usual case

$$V_i = 2^{i-1}$$

where $i = 1$ corresponds to the least significant bit, $i = 2$ to the next least significant bit, etc. Thus, for an eight-level scheme the values of V_i would be 1, 2, and 4. The value of R_i is given by

$$R_i = \max(0, d - 2n_i)$$

where n_i is actual number of hard decision errors made by the decoder.

It can be shown that the values of R_i and V_i given above will enable the Weldon algorithm to correct any soft error pattern whose weight is less than $(Q - 1)d/2$ assuming the error weight is computed using the σ_i given above. For example, suppose one has a length-15 distance-5 code which is used with an eight-level demodulator. Suppose that the all-zero sequence is transmitted and the received sequence $\mathbf{r} = (000000044540000)$. The actual weight of the error sequence is 17, which should be correctable since $(Q - 1)d/2 = 35$. The three hard decision sequences to be decoded (starting with the *lsb*) are

$$\mathbf{S}_1 = (0\ 0\ 0\ 0\ 0\ 0\ 0\ 0\ 0\ 1\ 0\ 0\ 0\ 0\ 0)$$

$$\mathbf{S}_2 = (0\ 0\ 0\ 0\ 0\ 0\ 0\ 0\ 0\ 0\ 0\ 0\ 0\ 0\ 0)$$

and

$$\mathbf{S}_3 = (0\ 0\ 0\ 0\ 0\ 0\ 0\ 1\ 1\ 1\ 1\ 0\ 0\ 0\ 0)$$

At worst, the decoder will decode $\mathbf{S}_3$ as some weight-5 sequence and admit to a single error being corrected. $\mathbf{S}_2$ will obviously be decoded as a weight-0 sequence with zero errors corrected and $\mathbf{S}_1$ as a weight-0 sequence with one error corrected. The w_i's for the three sequences are

$$w_1 = 1(5 - 2) = 3$$

$$w_2 = 2(5 - 0) = 10$$

$$w_3 = 4(5 - 2) = 12$$

When these weights are used in (4-22) w_1 and w_2 will always be multiplied by 1; however, w_3 will sometimes be multiplied by -1. Thus the worst case is

$$f_d = 3 + 10 - 12 = 1$$

Since f_d is always greater than zero the all-zero sequence will be decoded.

Unfortunately, the Weldon decoder will not decode correctly many of the soft error patterns that a true maximum-likelihood sequence estimator would decode. Assume, as before, that the all-zero pattern was transmitted but this time the received sequence is $\mathbf{r} = (1\,1\,1\,0\,0\,0\,0\,4\,4\,4\,4\,0\,0\,0\,0)$. The weight of this error sequence is 19; however, it is at least distance 20 from the nearest weight-5 code word. [Assume this to be $\mathbf{c}' = (7\,0\,0\,0\,0\,0\,0\,7\,7\,7\,7\,0\,0\,0\,0).$] The three hard decision sequences are

$$S_1 = (1\ 1\ 1\ 0\ 0\ 0\ 0\ 0\ 0\ 0\ 0\ 0\ 0\ 0\ 0)$$

$$S_2 = (0\ 0\ 0\ 0\ 0\ 0\ 0\ 0\ 0\ 0\ 0\ 0\ 0\ 0\ 0)$$

and

$$S_3 = (0\ 0\ 0\ 0\ 0\ 0\ 0\ 1\ 1\ 1\ 1\ 0\ 0\ 0\ 0)$$

S_1 and S_3 will both decode erroneously to the weight-5 word, $\mathbf{c}'$. Thus, $w_1 = 1(1)$, $w_2 = 2(5)$, and $w_3 = 4(3)$. Whenever B_{3j} is a one, W_3 will be included in the decision function as a negative quantity and will always cause f_d to be negative. Thus, the Weldon decoder will not decode this sequence correctly. This inability of the Weldon decoder to correctly decode many patterns whose weight exceeds $(Q - 1)\,d/2$ even extends to the case where the pattern would have been correctly decoded by a hard decision decoder. This latter effect can be ameliorated, to some extent, by adjusting the system gain so that the quantization levels are crowded closer to the origin. This has the effect of reducing the soft decision weight associated with some error patterns and can have a beneficial effect if not carried too far.

In order to provide an example of the performance that can be obtained, the Weldon algorithm was simulated using the $(21, 11)$ code discussed in the previous section. A three-bit soft decision demodulator was used and the σ_i were assigned the values $0, 1, \ldots, 7$. Each of the three hard decision decoders used a syndrome look-up table which provided correction of all one- and two-symbol errors, 490 of the 1330 possible three-symbol errors, 301 of the 5985 possible four-symbol errors, and one five-symbol error. The threshold spacing was established by noting that the decoding procedure guarantees correction of all patterns of weight 21 or less and using this fact to compute an upper bound on performance. The threshold spacing was varied until the bound was minimized. At a 4-dB value for E_b/N_0 the best spacing using this bound turned out to be $0.1E_s^{1/2}$. This is in contrast to an optimum spacing of approximately $0.3E_s^{1/2}$ that was found for the APP decoding example. This value was then used in the simulation. One slight modification was made in the algorithm. It was observed that a large number of errors occurred when all three of the decoders had a

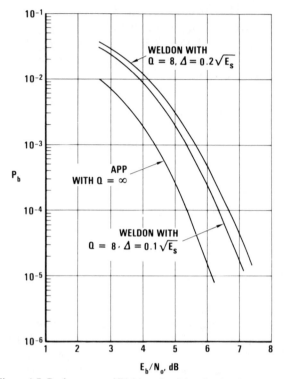

Figure 4-7. Performance of Weldon algorithm for the (21, 11) code.

weight, w_i, of zero on every symbol in the sequence. This condition was detected and the decoder then used only the results of the estimate on the sequence which corresponded to the most significant bit in the representation of σ_i. This resulted in almost all of these cases being decoded correctly. The resulting performance is shown in Fig. 4-7. For comparison the performance is also given for the same algorithm using a threshold spacing of $0.2E_N^{1/2}$ and for the APP algorithm using infinite quantization and feedback.

Since the structure of the Weldon algorithm is similar to both the APP algorithm and the HR algorithm, one might think that the use of feedback could also result in a performance improvement for the Weldon algorithm. This does not turn out to be the case. One possible explanation is that the initial application of the algorithm almost always produces a code word whereas the initial application of both the HR algorithm and APP algorithm often produces a sequence which is not a code word but is mapped into a code word on subsequent applications of the algorithm.

4.4. The Chase Algorithm

The Chase algorithm[26] is a technique for approximating the performance of a maximum-likelihood sequence estimator. Like the Weldon algorithm, the Chase technique makes use of several hard decision decoders but in an entirely different manner. The decoders generate a set of candidate code words, and the most likely of these is selected as the Chase decoder output.

4.4.1. Standard Chase Algorithms

In his original paper Chase proposed three different variations of the algorithm. Several others are possible and will undoubtedly occur to the reader. The basic procedure of Chase is as follows:

1. Make hard decisions on each symbol in the received sequence to produce the vector $\mathbf{Y}$.
2. Deliberately introduce various patterns of errors, $\mathbf{T}$, to generate test sequences, $\mathbf{S} = \mathbf{T} + \mathbf{Y}$. Decode each sequence created in this manner using a hard decision decoder.
3. Compute the distance from each of the code words to the (soft decision) received sequence, and select the code word which is closest.

The three variations proposed by Chase have to do with the method of generating the test sequences. Method (1) produces the largest number of sequences and also provides the best performance. Methods (2) and (3) use successively smaller numbers and, accordingly, have somewhat inferior performance.

Method (1): Generate as the list of vectors, $\mathbf{T}$, the all-zero pattern, and all error patterns with weight $\lfloor d/2 \rfloor$.

Method (2): The $\mathbf{T}$ vectors take on all $2^{\lfloor d/2 \rfloor}$ combinations of values in the $\lfloor d/2 \rfloor$ least reliable positions and zero in all other positions.

Method (3): Determine the $d - 1$ least reliable symbols. The $\mathbf{T}$ vectors have ones in the i least reliable positions and zeros elsewhere (for d odd, $i = 0, 2, 4, \ldots, d - 1$ while for d even, $i = 0, 1, 3, \ldots, d - 1$).

The basic idea of the Chase algorithm is to generate a list of code words that will almost always contain the word which is closest to the received sequence. It takes advantage of the fact that when the usual hard decision decoded word is not the correct choice, then one of the nearest neighbors usually is. The three algorithms proposed by Chase are essentially three ways of generating various sets of these nearest neighbors. We recall that any distance-d code is always capable of correcting a soft decision error whose weight is less than $(Q - 1)d/2$. Thus, the hard decision version of the received sequence will often contain as many as $d - 1$ errors. If one is able to remove $\lfloor d/2 \rfloor$ of these by using a **T** sequence of weight $\lfloor d/2 \rfloor$, then the hard decision decoder will always be capable of correcting the remaining $d - 1 - \lfloor d/2 \rfloor$. That is, if d is even, the test sequence will contain $(d - 2)/2$ errors, whereas if d is odd, the test sequence will contain $(d - 1)/2$ errors. Method (1) is obviously an overkill since it will produce the desired trial pattern $\binom{d}{\lfloor d/2 \rfloor}$ times. It also generates a large number of other patterns which are very improbable by changing highly reliable symbols. One should observe that whenever there are more than $d/2$ hard errors then it is highly probable that some of them will be very low reliability symbols. Thus by restricting the positions in which the **T** vectors are one to the least reliable symbols one will still find $\lfloor d/2 \rfloor$ of the errors a large portion of the time. Thus, method (2) provides a level of performance that is almost identical to method (1) with far fewer patterns. Method (3) also provides surprisingly good performance using a very small number of test patterns. Performance curves from Chase's original paper for the three methods used on a (24, 12) code are shown in Fig. 4-8.

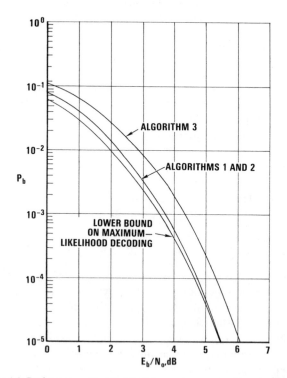

Figure 4-8. Performance of a (24, 12) Golay code over the Gaussian channel.

Consider the use of method (2) on the examples which were previously worked for the Weldon algorithm. Assume a length-15, distance-5 code with the all-zero transmitted sequence and a received sequence of $\mathbf{r} =$ (0 0 0 0 0 0 0 4 4 5 4 0 0 0 0). Method (2) utilizes as test sequences the hard decision sequences with all possible values of the two least reliable symbols (in positions 8 and 9). Thus, the four test sequences are

$$S_1 = (0\ 0\ 0\ 0\ 0\ 0\ 0\ 1\ 1\ 1\ 0\ 0\ 0\ 0)$$

$$S_2 = (0\ 0\ 0\ 0\ 0\ 0\ 0\ 0\ 1\ 1\ 1\ 0\ 0\ 0\ 0)$$

$$S_3 = (0\ 0\ 0\ 0\ 0\ 0\ 0\ 1\ 0\ 1\ 1\ 0\ 0\ 0\ 0)$$

and

$$S_4 = (0\ 0\ 0\ 0\ 0\ 0\ 0\ 0\ 0\ 1\ 1\ 0\ 0\ 0\ 0)$$

The hard decision decoder will obviously decode S_4 into the all-zero code word which is distance 17 from $\mathbf{r}$. Any of the other test sequences which are not decoded into the all-zero code word must be decoded into a code word with weight at least 5. In all cases such a code word will be at least distance 18 from $\mathbf{r}$. Thus, correct decoding is achieved.

In a similar fashion one can decode the received sequence $\mathbf{r} =$ (1 1 1 0 0 0 0 4 4 4 4 0 0 0). Application of method (2) produces the same four test sequences S_1, S_2, S_3, and S_4. The test sequence S_4 again produces the all-zero code word which is distance 19 from r. However, any nonzero code words produced by test sequences S_1, S_2, or S_3 must have at least weight 5, and they must be at least distance 20 from $\mathbf{r}$. Thus, the Chase decoder correctly decodes an error pattern which was incorrectly decoded using the Weldon algorithm. Method (2) typically performs much closer to maximum-likelihood decoding than the Weldon algorithm because of the ability to correctly decode many patterns with weight greater than $(Q - 1)d/2$.

4.4.2. Variations on the Chase Algorithms

Some simple variations on the Chase algorithm have been investigated by R. W. Boyd[30] in an attempt to find a minimum complexity implementation for a soft decision decoder for a (13, 8) shortened Hamming code. The decoder was required as part of a concatenation scheme and it was essential to obtain near optimum performance at relatively high word error rates.

The simplest scheme investigated by Boyd has been called an aided hard decision decoder. Although it is not actually a variation of the Chase algorithm it forms the basis of a simple variation and will be discussed

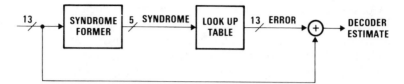

Figure 4-9. Standard hard decision decoder for (13, 8) code.

first. A hard decision decoder for the (13, 8) code is shown in Fig. 4-9. It is basically a syndrome table-lookup decoder and stores the most probable error pattern for each unique syndrome. Of the 31 unique nonzero syndromes, 13 are used to correct single errors and the remainder for two- and three-error patterns. In the case of two- and three-error patterns there are several potential candidates, each of which is an equally good choice when only hard decisions are involved. For soft decisions, however, one of these will obviously be a better choice than the others. Boyd's technique is to use the location of the least reliable symbol to determine which one of the weight-2 or weight-3 error patterns to choose. The choice is made by letting the location of the least reliable symbol be coincident with the location of one of the errors in the weight-2 or -3 error patterns. A circuit which implements this decoder is shown in Fig. 4-10.

This idea can be combined with the Chase algorithm to produce what Boyd has called a multiple-pattern, aided Chase decoder. The simplest case is when the locations of both the least reliable and second least reliable symbol are known. His procedure for this case is as follows:

1. Form the syndrome and use the location of the second least reliable symbol to aid in the table lookup of the error pattern.
2. Repeat step 1 after first inverting the value of the least reliable symbol.
3. Compute the soft decision distance to the actual received sequence

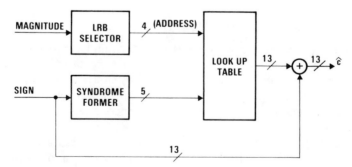

Figure 4-10. Aided hard decision decoder for the (13, 8) code.

and choose the code word which is closest. In case of a tie, always select the word produced in step 2.

The method for resolving the tie in step 3 is based on the observation that the number of hard errors postulated in step 2 will always be greater than or equal to the number of hard errors postulated in step 1. With the usual method of quantizing the channel, this will almost always correspond to a slightly more probable error event. For example, using the transition probabilities given in Section 4.1.3 we see that

$$[Pr(4/0)]^3 = 0.0000186$$

and

$$[Pr(6/0)]^2 = 0.00000441$$

Thus, other things being equal, three soft decision errors each of weight 4 are more probable than two soft decision errors each of weight 6.

Boyd's procedure extends in a very logical way to a three-pattern, aided Chase decoder. In this case he assumes that the locations of the three

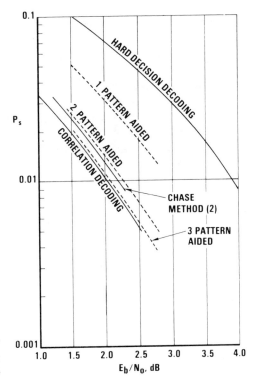

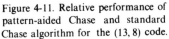
Figure 4-11. Relative performance of pattern-aided Chase and standard Chase algorithm for the (13, 8) code.

least reliable symbols are known and uses the following procedure:

1. Form the syndrome and use the location of the third least reliable symbol to aid in the table lookup of the error pattern.
2. Complement the least reliable symbol, compute the syndrome, and use the second least reliable symbol to aid in the table lookup of the error pattern.
3. Complement the second least reliable symbol, compute the syndrome, and use the third least reliable symbol to aid in the table lookup of the error pattern.
4. Choose the code word which is closest to the received symbol; always resolving ties in favor of higher-weight error patterns.

The relative performance of the three techniques is shown in Fig. 4-11. Also shown is the performance of the standard Chase decoder using method 2. The interesting point is that the two-pattern aided Chase decoder performs almost as well as the standard Chase which uses four patterns, and the three-pattern aided Chase is somewhat better than the standard Chase decoder.

4.5. Information-Set-Decoding Algorithms

The straightforward technique of choosing several different information sets, using each one to generate a code word, and selecting the code word which is closest to the received sequence has already been discussed in Chapter 3. More sophisticated variations on this idea have been investigated by a group at the Jet Propulsion Laboratory (JPL).[27,31] All of these versions of information set decoding require that the received symbols are first ordered according to their reliability. They are all essentially ad hoc schemes with only a modicum of theoretical justification for the selection of the various parameters. Consequently, we have chosen to present these techniques using a case history approach. The algorithms will be discussed in three separate categories; information-set-only algorithms, Omura-type algorithms, and partial syndrome decoding.

4.5.1. Information-Set-Only Algorithms

These algorithms all use the basic information set decoding rule. The only difference is the manner in which the information sets are chosen. In the simplest possible case the k most reliable symbols are chosen to constitute a single information set. This extends in a logical way to dividing the code words into two portions. The j least reliable symbols are always included in the parity set. Several different information sets are then chosen in the remaining $n - j$ positions to cover all combinations of v or fewer errors.

Baumert and McEliece[27] at JPL have used this technique to decode the (48, 24) and the (80, 40) quadratic residue codes. For the (48, 24) code they tried two different values for j, $j = 8$ and $j = 16$. For the $j = 8$ case they chose 130 information sets such that all patterns of four or fewer errors were covered in the remaining 40 symbol positions. For the $j = 16$ case they chose 124 information sets such that all patterns of three or fewer errors were covered in the remaining 32 symbol positions. The second case provided approximately 0.25 dB better performance than the first case and Baumert and McEliece estimated that this was very close to maximum likelihood performance for the code. On the (80, 40) code they reported virtually identical results for two different sets of parameters. One choice was to set $j = 28$ and cover all sets of three or fewer errors in the remaining positions using 130 different information sets. The second choice was to set $j = 36$ and to use 165 different information sets to cover all combinations of two or fewer errors. Both schemes appear to be within 0.25 dB of the maximum likelihood decoder performance.

Baumert and McEliece also attempted to use the same technique on the (128, 64) BCH code and met with only limited success. They then tried dividing the word into four portions. The 16 most reliable symbols were always assumed to be error free and were therefore included in every information set. The 40 next most reliable were assumed to have at most two errors and the next group of 32 to have at most four errors. The 40 least reliable symbols were never included in any information set. Although they do not report on the number of information sets used, one could easily implement the following solution. A total of 64 parity symbols are available to cover the potential error patterns mentioned above. None are used in positions 1 through 16 since these are assumed error free. However, one can use eight parity symbols in positions 17 through 56 to cover all double error patterns in this span. These errors could be covered with 45 covering patterns (generated via the method of Section 3.2.3.2). All four error patterns in positions 57 through 88 can be covered using 16 parity symbols. A set of covering patterns suitable for this case is provided by the nonzero code words of a (32, 5) code (i.e., minimum weight 16). Finally, the remaining 40 parity symbols are always assigned to positions 89–128 which can have any number of errors. Since we generated a total of 31 covering patterns for the 32-symbol group and 45 covering patterns for the 40-symbol group, this solution required a total of 1395 different coverings.

This method resulted in a significant improvement over the earlier technique but did not achieve maximum-likelihood performance. The success of this method led them to postulate an even finer division of the code word. As a result they gathered error statistics as a function of the position of the reliability ordered symbols and then chose information sets at random such that the frequency of zeros was in approximate accordance with the entropy (i.e., $E = -p \log p - q \log q$) of the error statistics. That is, if in a particular symbol position the probability of error was close to 0.5 then the symbol should be included in a parity set most of the time while if the probability of error was very small the symbol is very seldom included in a parity set. The actual weighting function used by Baumert along with several others subsequently tried by Greenberger[31] is shown in Fig. 4-12. The entropy function was scaled such that the first 30 or so symbols were included in every information set. Since the probability that one or more errors will occur in these positions is quite small, this choice has a negligible effect on the overall performance. It was also necessary to deviate slightly from a strictly random selection in order to meet the constraint that each information set contained 64 elements. Baumert found that if information sets were generated using these rules, approximately 1000 were required to obtain performance that was quite close to maximum-likelihood decoding.

Greenberger subsequently investigated several other weighting functions (Fig. 4-12) and found that although Baumert's function was among the best, performance was not particularly sensitive to the exact nature of the weighting function. Greenberger also investigated various ways of generating the information sets and determined that if the sets were arranged so that each one differed from the previous one by a small number

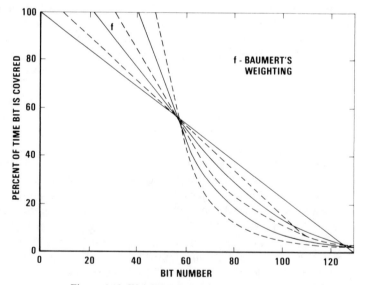

Figure 4-12. Weighting functions for (128, 64) code.

of positions then the amount of computation required could be minimized. Still a relatively large amount of computation is required to generate 1000 different parity check matrices and thus the technique would be limited to rather low transmission rates.

4.5.2. Omura-Type Algorithms

One problem with the information-set-decoding algorithm described in the previous section is that it is essential to have at least one information set that is error free in order to decode correctly. As we have seen, this may require the use of a large number of different covering patterns in order to correct the dominant error patterns. An alternate approach is to use one of the variations of the Omura algorithm discussed in Chapter 3. Recall that in the basic Omura algorithm one starts with a parity check matrix in reduced echelon form. One then computes the syndrome and uses this to define a possible error pattern. The next step is to sequentially add each column of $\mathbf{H}$ in the information set to the syndrome and to determine the weight of the error pattern which corresponds to a postulated error in each of these information set positions. If any of these patterns have lower weight than the original pattern then the new pattern replaces the original. Finally, one interchanges one of the columns in the information set with one in the parity set, determines a new $\mathbf{H}$ matrix, and repeats this operation. One could, of course, extend the initial search for single errors in the information set to include double errors, triple errors, etc. If this procedure is extended to include all patterns of t errors for hard decisions or $d - 1$ errors for soft decisions then it will obviously correct all of the guaranteed correctable errors plus whatever patterns are included by virtue of the parity set. This technique was originally proposed by Omura as an alternative to the random algorithm. Unfortunately, for even moderate values of d, the number of error patterns is too large for this approach to be practical.

If, however, the symbols are first confidence ordered and the initial parity set is placed on the $n - k$ least reliable symbols then one might expect that the actual number of error patterns in the information set would be quite a bit smaller. The group at JPL has pursued this line of thought. They determined for the $(128, 64)$ code that if one could test for all single, double, and approximately 25% of the most probable triple-error patterns then the decoder would perform within about 0.5 dB of maximum-likelihood performance. This requires approximately 10,000 test patterns, which although large might be practical at low data rates. For instance, if one required 50 nsec to test each pattern then it would take approximately 500 μsec to decode a single code word. Since each code word contains 64 information symbols this corresponds to a data rate of 128 kbit/sec.

A logical extension of the above reasoning is to consider the use of more than one information set and to scan a smaller number of test error patterns for each set. Following this same line of thought the JPL group made the following rough determination. In order to achieve a level of performance consistent with using 1000 information sets and not allowing for any errors in these sets one could use 50 information sets and allow for one error, six information sets and allow for two errors, and two information sets and allow for three errors.

Their determination was based on the following experiment. Using the 1000 information sets which had been previously determined, they then generated 10,000 hard decision confidence ordered error vectors. Each error vector was compared with each of the information sets and they noted how many comparisons were required to cover all the errors, all but one of the errors, all but two of the errors, etc. The results of this experiment are shown in Fig. 4-13 for a 2-dB signal-to-noise ratio. The values given above represent the abscissa associated with covering all but 10^{-2} of the errors.

No attempt was made to examine the information sets in any particular order so that the above determination should be quite conservative. One might expect that where a small number of information sets were employed they could be judiciously chosen to obtain somewhat better results. This is entirely consistent with the determination that it was only necessary to examine about 25 % of the triple-error patterns if the parity set was chosen in the $n - k$ least reliable positions.

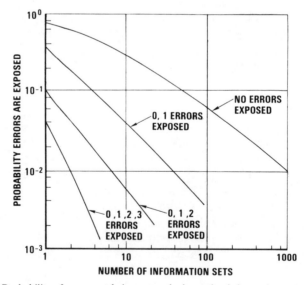

Figure 4-13. Probability of errors not being covered when using information sets of 64 symbols out of 128.

4.5.3. Partial Syndrome Decoding

The most significant defect with the last algorithm is that the total number of multiple error patterns that must be examined in the information set positions can be quite large. This number can be reduced significantly by using the following artifice. Suppose for a particular choice of parity set and information set the parity check matrix **H** has been reordered to be in reduced echelon form with the identity portion to the extreme right. In this case the matrix, error pattern, and syndrome can be partitioned as

$$
\left[\begin{array}{c|c} \mathbf{P}_1 & 0 \\ \hline \mathbf{P}_2 & \mathbf{I} \end{array} \right] \left(\begin{array}{c} \mathbf{e}_1 \\ \hline \mathbf{e}_2 \end{array} \right) = \left(\begin{array}{c} \mathbf{S}_1 \\ \hline \mathbf{S}_2 \end{array} \right) \tag{4-23}
$$

where $\mathbf{P}_1$ is a matrix consisting of r rows and $k + r$ columns, $\mathbf{e}_1$ is a $(k + r)$-component vector, and $\mathbf{S}_1$ is a r-component vector. $\mathbf{S}_1$ is known as the partial syndrome, and the following relationships hold:

$$
\mathbf{P}_1 \mathbf{e}_1 = \mathbf{S}_1 \tag{4-24}
$$

and

$$
\mathbf{e}_2 = \mathbf{P}_2 \mathbf{e}_1 + \mathbf{S}_2 \tag{4-25}
$$

Since $\mathbf{e}_1$ corresponds to the error pattern in the first $k + r$ positions and $\mathbf{e}_2$ to the error pattern in the remaining $n - k - r$ positions one can use (4-24) to postulate various low-weight error patterns in the information set and use (4-25) to determine the remaining errors in the parity set. Thus, if one is willing to restrict the number of errors in the information set plus the first r symbols of the parity set to be some small number ω, then one can use these equations to find all of the potential error vectors that satisfy the parity check equations. Although it is usually convenient to start with the **H** matrix in reduced echelon form, it is actually only necessary that the last $n - k - r$ columns contain the identity matrix as shown.

As an example, assume that one has a code with eight information symbols, and it is determined that $r = 4$ is a satisfactory choice. Assume that when the **H** matrix has been partitioned as in (4-23)

$$
\mathbf{P}_1 = \begin{array}{c} \begin{array}{cccccccccccc} 1 & 2 & 3 & 4 & 5 & 6 & 7 & 8 & 9 & 10 & 11 & 12 \end{array} \\ \left[\begin{array}{cccccccccccc} 1 & 1 & 1 & 0 & 1 & 0 & 1 & 1 & 1 & 0 & 0 & 0 \\ 0 & 0 & 1 & 1 & 1 & 1 & 0 & 1 & 0 & 1 & 0 & 0 \\ 0 & 0 & 0 & 0 & 1 & 1 & 1 & 0 & 0 & 0 & 1 & 0 \\ 1 & 1 & 0 & 1 & 0 & 0 & 1 & 1 & 0 & 0 & 0 & 1 \end{array} \right] \end{array}
$$

Table 4-2. Location of Error Pairs

First component	Possible locations of first error	Second component	Possible locations of second error
1 0 0 0	9	0 1 1 0	6
0 1 0 0	10	1 0 1 0	None
0 0 1 0	11	1 1 0 0	3
0 0 0 1	12	1 1 1 1	None
1 0 0 1	1, 2	0 1 1 1	None
0 0 1 1	None	1 1 0 1	8
0 1 0 1	4	1 0 1 1	7
0 0 0 0	—	1 1 1 0	5

and that the partial syndrome S_1 is

$$S_1^T = (1 \quad 1 \quad 1 \quad 0)$$

Assume that it is desirable to find all double-error patterns in the first 12 symbols that can produce this pattern. By decomposing S_1^T into all possible two vector sums we can construct Table 4-2. From Table 4-2 we see that the following error pairs are possible: (9, 6), (11, 3), (4, 7), and (5). Thus, there are four possible error patterns of weight 2 or less that satisfy the partial syndrome constraint. This is in contrast to 36 possible one- and two-error pairs in the first eight positions if one were not using the partial syndrome technique.

The reduction in the number of error patterns that must be examined does not come without a price. In this case the price is to reduce the effective size of the parity set by r symbols. Thus one might expect that for a given level of performance and choice of r, one might require more parity sets. For a particular application this is best determined experimentally by postulating a group of parity sets and observing the number of errors that cannot be covered when one tests this group with a large number of different error patterns.

4.5.4. Some Performance Predictions

Most of the techniques discussed in this section are fairly current and very little practical data regarding their performance have been amassed. All of these techniques attempt to approximate the performance of a maximum-likelihood decoder, and thus one might expect that, properly implemented, their performance would be quite close to that predicted by a union bound. To give the reader some feeling for what might be possible, the union bound on bit error rate for several block codes of moderate length is shown in Fig. 4-14. With the exception of the (128, 64) code, the union

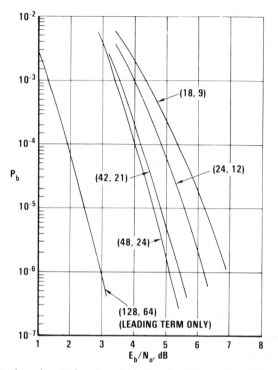

Figure 4-14. Union bound on P_b for selected block codes with maximum-likelihood decoding (infinite quantization and PSK signaling).

bound was calculated using a tabulation of the weight structure of the code. In the case of the (128, 64) code only the number of minimum-weight words is known.[32] Hence, this curve does not represent a true upper bound on performance and is undoubtedly optimistic.

Problems

4-1. Following the treatment in Section 4.1.1, derive the threshold decoding rule (4-1) for estimating e_m. [Hint: Note that $\Pr(e_m = 0) \neq \Pr(e_m = 1)$ and therefore these terms cannot be canceled.]

4-2. Supply the missing steps in the derivation of the Hartmann–Rudolph algorithm.

4-3. Write the decision function specified by the HR algorithm for a (7, 6) single parity check code.

4-4. A code word from a (7, 3) code is received such that the relative confidence of the symbols is given by the vector (1, 4, 2, 3, 7, 5, 6), where the lowest index corresponds to the least reliable symbol. If the parity check matrix for this code is given by

$$H = \begin{bmatrix} 1 & 0 & 1 & 1 & 0 & 0 & 0 \\ 1 & 1 & 0 & 0 & 1 & 0 & 0 \\ 1 & 1 & 0 & 0 & 0 & 1 & 0 \\ 0 & 1 & 1 & 0 & 0 & 0 & 1 \end{bmatrix}$$

then find the new **H** matrix which is appropriate for the Greenberger approximation to the HR algorithm. Write down the decision functions for the most reliable symbol which correspond to the first three iterations of this algorithm.

4-5. The all-zero code word from a $(7, 3)$ distance-4 code is received having the soft decision values $(0, 0, 4, 4, 4, 0, 1)$. Show that the Weldon decoder will decode this pattern correctly.

4-6. Would a Chase decoder which used only the least reliable symbol always decode the error pattern in Problem 4-5 correctly? If not, would it decode correctly using the two least reliable symbols?

4-7. Consider the $(7, 3)$ cyclic code generated by $1 + x + x^3$. Find the proper coset leaders for the correctable two-error patterns assuming that the selection is to be aided by a least reliable symbol in the left-hand-most position.

4-8. Assume that the APP decoding algorithm is used with the $(21, 11)$ code and a three-level channel. Each term in the decision function, f_{d_i}, can be regarded as an independent random variable that assumes the values $+1$, -1, or 0 depending on whether B_i is estimating a 0, 1, or erasure. Let p be the probability of symbol error, q be the probability a symbol is correct, and $1 - p - q$ be the probability a symbol is erased. Recall that each of the estimates B_1 through B_5 contain four independent error terms while the estimate B_0 contains only one. (a) Assuming that the all-zero sequence is transmitted, compute $\Pr(f_{d_i} = 1)$, $\Pr(f_{d_i} = 0)$, and $\Pr(f_{d_i} = -1)$ for $i = 1$–5 and for $i = 0$. (b) Write a general expression or algorithm which uses these probabilities to compute the probability of decoded error.

5

Algebraic Techniques for Multiple Error Correction

Classes of algebraically defined multiple-error-correcting block codes have been known to exist for many years (e.g., the BCH codes). Extensive research has produced a number of practical decoding techniques for these codes, and the theory is quite well developed. It is the algebraic structure imposed on these codes that has allowed computationally efficient decoding algorithms to be developed. We shall present several of these techniques, but the treatment will not be mathematically rigorous. The reader will be referred to the literature for some detailed proofs.

Error correction is accomplished with these codes by solving sets of algebraic equations. As a simple example, consider the double-error-correcting binary BCH code with block length 15. This code has a parity check matrix that can be written as

$$\mathbf{H} = \begin{bmatrix} a_1\, a_2 \cdots a_{15} \\ a_1^3\, a_2^3 \cdots a_{15}^3 \end{bmatrix}$$

The $\{a_i\}, i = 1,\dots, 15$ are distinct nonzero elements of $\mathrm{GF}(2^4)$. If errors occur in positions i and j of the received word $\mathbf{y}$, then the syndrome

$$\mathbf{S} = \mathbf{H}\,\mathbf{y}$$
$$= \mathbf{H}\,\mathbf{e}$$

produces two equations in two unknowns, viz.,

$$a_i + a_j = S_1$$

and

$$a_i^3 + a_j^3 = S_3$$

If one could solve these equations for a_i and a_j, then the error locations i and j would be known. Error correction would then consist of inverting the received symbols in these locations. Unfortunately, the equations are nonlinear, and any method of solving them directly is certainly not obvious. However, we can begin by eliminating one of the variables. Thus, solving the first equation for a_i and substituting into the second equation yields

$$a_j^2 + S_1 a_j + S_1^2 + S_3/S_1 = 0$$

Had one solved the first equation for a_j then the resulting equation would be the same with a_i replacing a_j. Consequently both a_i and a_j are solutions (or roots) of the same polynomial:

$$z^2 + S_1 z + S_1^2 + S_3/S_1 = 0$$

This polynomial is called an *error-locator polynomial*. One method of finding its roots is simple trial and error. Thus, by substituting each of the nonzero elements from $GF(2^4)$ into this equation one is guaranteed to find the location of both errors. This approach represents a significant savings in effort compared with testing all possible error pairs to determine which ones are solutions to the original parity equations.

The procedure which has just been outlined extends in a natural way to both binary and nonbinary t-error-correcting BCH codes. For these codes it is always possible to find an error-locator polynomial directly from the syndrome provided that the number of errors does not exceed the guaranteed error-correcting capability of the code. This error-locator polynomial has exactly the same properties as in the example except that it turns out to be more convenient to let the reciprocal of the roots correspond to the error locations. In this chapter we will discuss two efficient iterative algorithms for determining the error-locator polynomial. Once this polynomial is known its roots may be found directly by an efficient trial-and-error procedure or by an alternate approach which will be presented. For the nonbinary codes one has the additional problem of determining the error values once the error locations are known. This turns out to be easily accommodated in the algorithms to be presented.

There are several ways to introduce the theory of multiple-error-correcting codes. In this chapter we have chosen to use transform techniques. Most engineers are basically familiar with ordinary Fourier transforms. The ones used here are very similar except they are defined using finite field elements. Their use will enable us to derive the basic properties of both BCH and Reed–Solomon codes in a simple way. In addition the use of transforms over finite fields has recently been proposed as a vehicle for reducing decoder complexity.[33-36] In developing this theory we will present the material in a manner similar to Blahut, [37] who has also observed that the transform approach can be quite useful for tutorial purposes.

The interested reader will undoubtedly find Blahut's paper to be useful supplementary reading.

5.1. Finite-Field Transforms

Since the code word $\mathbf{c}$ is an n-tuple, the transform of interest is one with a discrete time variable. Also, since each of the components of the n-tuple are elements of a Galois field, the transform should assume values in a finite field. The properties of this type of transform are described in detail by Pollard.[38] A closely related transform has already been used in Chapter 4.

The discrete, finite-field Fourier transform, $\mathbf{C}$, of the vector, $\mathbf{c}$, is defined in the following fashion:

Definition. Let $\mathbf{c} = (c_0, c_1, \ldots, c_{n-1})$ be an n-tuple with coefficients from $GF(q)$ (where n divides $q^m - 1$ for some m), and let α be an element of $GF(q^m)$ of order n. Then the *finite-field Fourier transform* of $\mathbf{c}$ is the n-tuple $\mathbf{C} = (C_0, C_1, \ldots, C_{n-1})$ defined over $GF(q^m)$ and given by

$$C_j = \sum_{i=0}^{n-1} \alpha^{ij} c_i, \qquad j = 0, 1, \ldots, n-1 \tag{5-1}$$

This definition is analogous to the definition of a discrete Fourier transform of a vector of real numbers in which the Fourier kernel α would be replaced by the nth root of unity in the field of complex numbers, $e^{-j2\pi/n}$. In keeping with engineering notation, the discrete index i will be called "time" while the index j will be called "frequency." Thus, the vectors $\mathbf{c}$ and $\mathbf{C}$ will be called the time domain function and frequency domain function (or spectrum), respectively. These two vectors are a Fourier transform pair, i.e.,

$$\mathbf{c} \iff \mathbf{C}$$

This Fourier transform pair is related by the following theorem, which is given without proof.

Theorem. Over $GF(q)$, a field of characteristic p, a vector and its spectrum are related by

$$C_j = \sum_{i=0}^{n-1} \alpha^{ij} c_i \tag{5-2}$$

$$c_i = \frac{1}{n(\bmod p)} \sum_{j=0}^{n-1} \alpha^{-ij} C_j \tag{5-3}$$

Many of the familiar properties of Fourier transforms (e.g., convolution) are observed with this finite-field transform pair. Several of these properties are extremely useful in demonstrating the properties of certain multiple-error-correcting codes. For example, suppose that the components c_i are the product of components from two other vectors $\mathbf{f}$ and $\mathbf{g}$, i.e.,

$$c_i = f_i g_i, \qquad i = 0, 1, \ldots, n-1$$

Then the Fourier transform of $\mathbf{c}$ may be written

$$
\begin{aligned}
C_j &= \sum_{i=0}^{n-1} \alpha^{ij} f_i g_i \\
&= \sum_{i=0}^{n-1} \alpha^{ij} f_i \left(\frac{1}{n} \sum_{k=0}^{n-1} G_k \alpha^{-ik} \right) \\
&= \frac{1}{n} \sum_{k=0}^{n-1} G_k \left(\sum_{i=0}^{n-1} \alpha^{i(j-k)} f_i \right) \\
&= \frac{1}{n} \sum_{k=0}^{n-1} G_k F_{j-k}
\end{aligned}
\tag{5-4}
$$

where all subscripts are defined modulo n. Thus, multiplication in the time domain is equivalent to convolution in the frequency domain (just as with linear system theory), and the reverse is also true.

The vectors $\mathbf{c}$ and $\mathbf{C}$ can also be represented as polynomials $c(x) = c_{n-1}x^{n-1} + \cdots + c_1 x + c_0$ and $C(z) = C_{n-1}z^{n-1} + \cdots + C_1 z + C_0$. Using this representations, the jth spectral component is

$$
\begin{aligned}
C_j &= \sum_{i=0}^{n-1} c_i \alpha^{ij} \\
&= c(\alpha^j)
\end{aligned}
$$

Thus, $C_j = 0$ if and only if α^j is a root of $c(x)$. Similarly, the ith time domain component is

$$
\begin{aligned}
c_i &= \frac{1}{n} \sum_{j=0}^{n-1} C_j \alpha^{-ij} \\
&= \frac{1}{n} C(\alpha^{-i})
\end{aligned}
$$

Thus, we have the following theorem.

Theorem. The relationship between roots of the polynomial in one domain and the components in the other domain is as follows: (a) The polynomial $c(x)$ has a root α^j if and only if $C_j = 0$. (b) The polynomial $C(z)$ has a root α^{-i} if and only if $c_i = 0$.

We see that defining roots of the polynomial in one domain is equivalent to setting the appropriate components to zero in the other domain. This property is quite useful in understanding the error correction properties of cyclic codes.

Example. Consider the vector

$$\mathbf{c} = (0, 0, 1, 3)$$

over GF (5). The finite-field Fourier transform is also defined over GF (5) with components given by (5-1) to be

$$C_j = \sum_{i=0}^{3} 2^{ij} c_i$$

$$= 1 \cdot 2^{2j} + 3 \cdot 2^{3j}, \qquad j = 0, 1, 2, 3$$

We have selected 2 as the element of order 4 from GF(5). Addition and multiplication tables over this field were given in Table 2-3. Performing the indicated operations provides

$$\mathbf{C} = (4, 3, 3, 0)$$

as the desired Fourier transform. Note that GF(5) has characteristic 5 and that $4^{-1} = 4$. Thus, the inverse transform of $\mathbf{C}$ is defined by (5-3) to be

$$c_i = 4 \sum_{j=0}^{3} 2^{-ij} C_j$$

$$= 4(4 + 3 \cdot 2^{-i} + 3 \cdot 2^{-2i}), \qquad i = 0, 1, 2, 3$$

Again, performing the indicated operations gives the original vector $\mathbf{c}$. Note that the vector $\mathbf{c}$ has two zero components, c_0 and c_1. From the previous theorem this implies that the transform polynomial

$$C(z) = 4 + 3z + 3z^2$$

must have as roots 2^0 and 2^{-1}. Direct substitution of these values in $C(z)$ verifies that they are indeed roots of this polynomial.

5.2. BCH Codes

The BCH codes[39-41] are an important class of multiple-error-correcting codes which have the cyclic property. The remainder of this chapter is devoted to a discussion of the implementation and performance considerations for these codes. These codes were defined in Section 2.3.1 in terms of the roots of the generator polynomial, $g(x)$, of degree $n - k$. Then any code word can be represented as $c(x) = d(x)g(x)$, where $d(x)$ is the data polynomial of degree $k - 1$. This polynomial multiplication is equivalent to a time domain convolution, i.e.,

$$c_i = \sum_{k=0}^{n-1} g_{i-k} d_k \tag{5-5}$$

However, this convolution is also equivalent to multiplication in the frequency domain so that $C_j = G_j D_j$. Since $d(x)$ and hence $D(z)$ are arbitrary, the code can be specified either by specifying the roots $\alpha^{j_1}, \alpha^{j_2}, ..., \alpha^{j_{n-k}}$ of $g(x)$ or by specifying the frequency spectrum $G(z)$ [or $C(z)$] to always be zero in the components $j_1, j_2, ..., j_{n-k}$. We can now give an alternate definition of primitive BCH codes.

Definition. A *primitive t-error-correcting BCH code* over $GF(q)$ of block length $n = q^m - 1$ is the set of all words over $GF(q)$ whose spectrum is zero in the $2t$ consecutive components $m_0, m_0 + 1, ..., m_0 + 2t - 1$.

The fact that $g(x)$ contains $2t$ consecutive powers of α as roots (or equivalently that the spectrum has $2t$ consecutive zero components) is the important property of BCH codes that allows correction of t errors. This is demonstrated in the following theorem.

Theorem (BCH Bound). If any nonzero vector $\mathbf{c}$ has a spectrum $\mathbf{C}$ with $d - 1$ consecutive zero components ($C_j = 0 : j = m_0, m_0 + 1, ..., m_0 + d - 2$), then $\mathbf{c}$ has at least d nonzero components.

Proof. Assume that $\mathbf{c}$ has $v < d$ nonzero components at locations $i_1, i_2, ..., i_v$ and that the transform of $\mathbf{c}$ is identically zero in locations m_0, $m_0 + 1, m_0 + 2, ..., m_0 + d - 2$. Define the locator polynomial

$$\Lambda(z) = \prod_{k=1}^{v} (1 - z\alpha^{i_k})$$

$$= \Lambda_v z^v + \Lambda_{v-1} z^{v-1} + \cdots + \Lambda_0 \tag{5-6}$$

Then the vector $\mathbf{\Lambda}$ is a frequency spectrum that is defined such that its inverse transform $\mathbf{\lambda}$ has zero components ($\lambda_i = 0$) at all i for which $c_i \neq 0$. Thus, $\lambda_i c_i = 0$ in the time domain, and the corresponding convolution in the frequency domain is zero, i.e.,

$$\sum_{k=0}^{n-1} \Lambda_k C_{j-k} = 0 \tag{5-7}$$

However, $\Lambda(z)$ has degree at most $d - 1$ and $\Lambda_0 = 1$ so that (5-7) becomes

$$C_j = -\sum_{k=1}^{d-1} \Lambda_k C_{j-k} \tag{5-8}$$

This is recognized as the equation for the output, C_j, of a linear feedback shift register with feedback polynomial $\Lambda(z)$. Using (5-8), the entire sequence of C_j's can be generated from any block of C_j's of length $d - 1$. However, we assumed that there exists a block of $d - 1$ that are all zero (specifically at indexes $m_0, m_0 + 1, \ldots, m_0 + d - 2$). Using this block as an initial condition, then all $C_j = 0$ and the vector $\mathbf{c}$ must be all zero. Thus, any nonzero vector must have at least d nonzero components. QED

Several aspects of this theorem are deserving of comment. First, if $g(x)$ contains a block of $2t$ consecutive powers of α as roots, then a code minimum distance of at least $d = 2t + 1$ is guaranteed (sometimes referred to as the designed distance). In addition, the $2t$ consecutive zero components of $\mathbf{C}$ are quite important in the error-correction process. These zero components allow calculation of a syndrome by calculating the transform of the received vector. The components of this transform in the positions corresponding to the $2t$ consecutive zero components of $\mathbf{C}$ provide the desired syndrome. As it turns out, the correct error pattern can be computed from this syndrome using algebraic techniques if no more than t errors occurred. Occasionally, a BCH code which has only $2t$ consecutive zero components of $\mathbf{C}$ will have an actual minimum distance of $2t + 3$ or larger. Unfortunately, while the code has the capability of correcting all $(t + 1)$-error patterns, the standard BCH decoding algorithm would require $2t + 2$ consecutive zero components of $\mathbf{C}$ to accomplish the desired correction. Thus, the BCH bound provides not only a lower bound on the actual minimum distance, but it gives the maximum degree of error correction that can be accomplished by the standard BCH decoding algorithms. Finally, an interesting relationship has been demonstrated between nonzero time domain components and a linear feedback shift register defined in the frequency domain. The importance of this relationship will become clear later when decoding techniques for these codes are presented.

BCH codes are a subclass of cyclic codes. Once the code generator polynomial $g(x)$ is defined, the code may be encoded using a feedback shift register as shown in Fig. 2-4. This approach was treated in Chapter 2 and will not be discussed further. It is also possible to encode using transform techniques as discussed in Blahut.[37] Although this approach is interesting, it does not appear to be of practical significance.

An important subclass of BCH codes are the Reed–Solomon codes[42] also introduced in Section 2.3.1. For these codes, both the code words and their transforms have components over $GF(q)$. The generator polynomial is

$$g(x) = (x - \alpha)(x - \alpha^2)\cdots(x - \alpha^{d-1}) \tag{5-9}$$

and the resulting code achieves minimum distance d with $d - 1$ parity symbols. Note that $d = n - k + 1$ is the maximum possible value for any code with the same n and k so that these codes are called *maximum-distance separable*.

5.3. BCH Decoding Techniques

Now consider the problem of decoding BCH codes. The received vector $\mathbf{r}$ is assumed to be the sum of a code word and an error vector $\mathbf{e}$ with addition over $\mathrm{GF}(q)$, i.e.,

$$\mathbf{r} = \mathbf{c} + \mathbf{e} \tag{5-10}$$

The decoder, of course, can only observe $\mathbf{r}$ and must find the minimum Hamming weight $\mathbf{e}$ vector that could produce the observed $\mathbf{r}$ vector. It must select this $\mathbf{e}$ vector from the 2^k possibilities (one for each possible $\mathbf{c}$). A number of decoding techniques have been formulated both in the frequency domain and the time domain. Conventional approaches have operated primarily in the time domain and these will be discussed. However, in recent years frequency domain approaches have been proposed, and they may have computational advantages in certain applications. Since the frequency domain approach is easier to understand, it will be covered first.

5.3.1. Frequency Domain Decoding

In a frequency domain implementation one first computes the Fourier transform, $\mathbf{R}$, of the received vector, $\mathbf{r}$. This transform can be written in terms of the transforms of the code word and the error vector as

$$\mathbf{R} = \mathbf{C} + \mathbf{E} \tag{5-11}$$

From the definition of BCH codes we observe that each transformed code word $\mathbf{C}$ is zero in a block of $2t$ consecutive components. Thus, (5-11) provides a set of $2t$ known values of E_j which are commonly called the *syndromes* and which can be relabeled and reindexed as S_{j-m_0}, i.e.,

$$E_j = R_j$$
$$= S_{j-m_0}: \qquad j = m_0, m_0 + 1, \ldots, m_0 + 2t - 1 \tag{5-12}$$

By using the definition of the transform the reader can verify that this definition is equivalent to the third form of the syndrome discussed in Section 2.2.7. One way of viewing the decoding operation is that starting with the $2t$ known values of E_j one must determine the $n - 2t$ other components that produce an estimate, $\hat{\mathbf{E}}$, of the error transform with the minimum weight inverse, ê. At this point it is not clear that this is possible, but it will become obvious shortly.

Now suppose that there are $v \leq t$ errors. Define the error-locator polynomial, $\Lambda(z)$, as in (5-6) where the indices $i_1, i_2, \ldots, i_v$ correspond to the time indices of the v errors ($e_{i_k} \neq 0; k = 1, 2, \ldots, v$). Then in the time domain, $\lambda_i = 0$ whenever $e_i \neq 0$ so that $\lambda_i e_i = 0$ for all i and

$$\sum_{k=0}^{n-1} \Lambda_k E_{j-k} = 0 \tag{5-13}$$

(note that $\Lambda_k = 0$ for $k > v$ and $\Lambda_0 = 1$). The set of equations indicated by the convolution (5-13) is the "key" to solving the decoding problem and (5-13) is usually referred to as the *key equation*. Since $2t$ consecutive components of E are known, we can use (5-13) to write v equations in the v unknown coefficients Λ_k. These equations are linear and may be solved using ordinary matrix inversion techniques. They can also be solved using one of the iterative techniques to be discussed in Section 5.4. Once $\Lambda(z)$ is found, the unknown components of Ê can be calculated by recursive extension through (5-13). Finally, once Ê is known, an inverse Fourier transform gives the estimated error pattern e, and the decoder output is

$$\hat{e} = r - \hat{e} \tag{5-14}$$

A block diagram of this process is shown in Fig. 5-1. Modifications of this approach are also possible. For example, one might do a complete transform initially and store R in the buffer rather than r. Then E would be found in the same way, but the estimated code word transform would be formed as

$$\hat{C} = R - \hat{E} \tag{5-15}$$

Finally, the decoder output ĉ would be determined by taking the inverse transform of Ĉ.

The initial Fourier transform that produces the syndromes has a

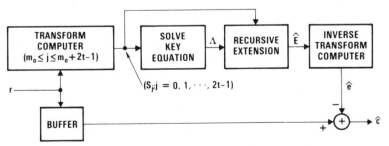

Figure 5-1. Frequency domain decoder for BCH codes.

simple realization. Note that for a code generator with roots $\alpha^{m_0}, \alpha^{m_0+1}, \ldots,$ α^{m_0+2t-1}, the $2t$ syndrome symbols are defined as

$$S_j = E_{j+m_0}$$
$$= \sum_{i=0}^{n-1} e_i \alpha^{i(j+m_0)}$$
$$= e(\alpha^{j+m_0}), \qquad j = 0, 1, \ldots, 2t-1 \tag{5-16}$$

Writing the jth syndrome symbol as

$$S_j = r_{n-1}\alpha^{(n-1)(j+m_0)} + r_{n-2}\alpha^{(n-2)(j+m_0)} + \cdots + r_0 \tag{5-17}$$

it is obvious that the circuit shown in Fig. 5-2 will perform this calculation.

The realization of the block which solves the key equation is a critical portion of the design. The computational effort required using conventional techniques is quite large if more than just a few errors are to be corrected. Fortunately, there exist two iterative techniques which produce the same result with significantly less effort. Although both of these techniques are easy to apply, they require somewhat lengthy explanations and are treated separately in Section 5.4.

The recursive extension to obtain $\hat{E}$ is easily accomplished using the circuit of Fig. 5-3. This is an ordinary feedback shift register with a connection polynomial defined by $\Lambda(z)$.

The final inverse Fourier transform may be computed using circuitry similar to that shown in Fig. 5-2 but with n rather than $2t$ such devices. In general, the complexity of implementing BCH decoders in the frequency domain will not be less than a time domain implementation because of the complexity of the final inverse Fourier transform operation. In some cases the complexity of this operation can be reduced significantly by using fast Fourier transform (FFT) techniques. This requires selecting codes with

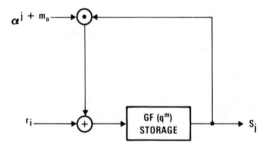

Figure 5-2. Circuit for calculating S_j.

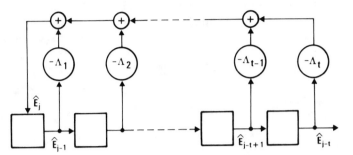

Figure 5-3. Feedback shift register for recursively extending the error transform.

unique properties to allow an FFT to be defined. Recent work in this area has demonstrated computational savings relative to time domain techniques.[43]

5.3.2. Time Domain Decoding

Time domain decoding techniques for BCH codes have seen wider application than frequency domain techniques. However, in many respects the techniques are quite similar. A block diagram of a time domain decoder is shown in Fig. 5-4. In comparing this with Fig. 5-1 the identical elements are the transform computer (to produce the syndromes), the buffer (to store $\mathbf{r}$), and the solution of the key equation for the error-locator polynomial, $\Lambda(z)$. The time domain decoder also requires an error-evaluator

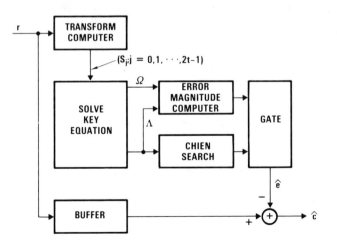

Figure 5-4. Time domain decoder for BCH codes.

polynomial $\Omega(z)$ (for nonbinary codes) which can be generated when solving the key equation. The device labeled "Chien search" finds the error locations given the $\Lambda(z)$ polynomial (by simply finding the inverse roots). For binary codes this is all that is needed to produce **e**. However, for nonbinary codes the error-evaluator polynomial and the error-locator polynomial can be used to find the error magnitude at each error location in a manner to be described shortly.

Define the syndrome components as

$$S_j = \sum_{i=0}^{n-1} e_i \alpha^{i(m_0 + j)}$$

$$= \sum_{k=1}^{v} e_{i_k} \alpha^{i_k(m_0 + j)}$$

where it is assumed that v errors occurred in locations corresponding to time indexes $i_1, i_2, \ldots, i_v$. Hence the error locator for position i_k is defined to be α^{i_k}. Now define the infinite degree syndrome polynomial (though we only know the first $2t$ coefficients) as

$$S(z) = \sum_{j=0}^{\infty} S_j z^j$$

$$= \sum_{j=0}^{\infty} z^j \sum_{k=1}^{v} e_{i_k} \alpha^{i_k(m_0 + j)}$$

$$= \sum_{k=1}^{v} \frac{e_{i_k} \alpha^{i_k m_0}}{1 - z \alpha^{i_k}} \tag{5-18}$$

Defining the error-locator polynomial as in (5-6), the product $\Lambda(z) S(z)$ may be written as

$$\Lambda(z) S(z) = \sum_{k=1}^{v} e_{i_k} \alpha^{i_k m_0} \prod_{l \neq k}^{v} (1 - z \alpha^{i_l})$$

$$\overset{\Delta}{=} \Omega(z) \tag{5-19}$$

The polynomial $\Omega(z)$ is the previously mentioned error-evaluator polynomial. Note that for an error-locator polynomial of degree v, the error-evaluator polynomial has degree less than v.

Actually the polynomial $S(z)$ is known only through the coefficient of z^{2t-1} so that in terms of the known syndrome components we can write (5-19) as

$$\Lambda(z) S(z) \equiv \Omega(z) \bmod z^{2t} \tag{5-20}$$

Equation (5-20) defines exactly the same set of $2t - v$ equations as (5-13) and consequently is also referred to as the key equation. This becomes more apparent if we observe that $\Omega(z)$ is at most of degree $v - 1$ and consequently (5-20) may be used to write

$$[\Lambda(z) S(z)]_v^{2t-1} = 0 \tag{5-21}$$

where $[a(z)]_m^n$ denotes the terms of $a(z)$ from z^m through z^n (i.e., $a_m z^m + a_{m+1} z^{m+1} + \cdots + a_n z^n$). Once $\Lambda(z)$ has been found, then $\Omega(z)$ is given by (5-20).

The next important decoding step is to find the actual error locations $\alpha^{i_1}, \alpha^{i_2}, \dots, \alpha^{i_v}$. Note that $\Lambda(z)$ has roots $\alpha^{-i_1}, \alpha^{-i_2}, \dots, \alpha^{-i_v}$. An easily implemented procedure for finding these roots and hence the error locations was given by Chien.[44] Observe that an error occurs in position i if and only if $\Lambda(\alpha^{-i}) = 0$ or

$$\sum_{k=0}^{v} \Lambda_k \alpha^{-ik} = 0$$

Define

$$\Lambda_k^i = \Lambda_k \alpha^{-ik} \tag{5-22}$$

and as a result

$$\Lambda_k^{i-1} = \Lambda_k^i \alpha^k \tag{5-23}$$

Then

$$\Lambda(\alpha^{-i}) = \sum_{k=0}^{v} \Lambda_k^i \tag{5-24}$$

This suggests the implementation shown in Fig. 5-5. Potential error locations are tested in succession starting with time index $n - 1$. Summing the register

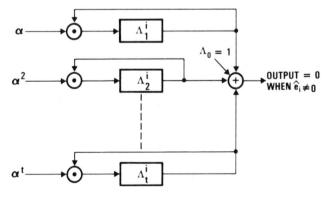

Figure 5-5. Block diagram of Chien search for finding error locations.

outputs at index i tests to see if $\Lambda(\alpha^{-i}) = 0$. Then to test at index $i - 1$ only requires multiplying the kth register contents by α^k for all k as indicated by (5-23) and summing the contents again. This procedure is repeated until index 0 is reached. The initial loading of the kth register is $\Lambda_k \alpha^{-nk}$, which is simply Λ_k if α has order n. After loading the initial conditions, the first multiplication is performed, and the test for $\hat{e}_{n-1}$ is made.

For binary codes the decoding process is now complete (one only needs to confirm the presence or absence of an error to allow correction). However, for nonbinary codes the error value must also be determined. This information can be obtained by utilizing the error-evaluator polynomial. Note that evaluating this polynomial at an actual error location corresponding to time index i_m gives

$$\Omega(\alpha^{-i_m}) = e_{i_m}\alpha^{m_0 i_m}\prod_{l \neq m}(1 - \alpha^{i_l}\alpha^{-i_m}) \tag{5-25}$$

A similar expression is obtained by taking the formal derivative of $\Lambda(z)$ and also evaluating it at $z = \alpha^{-i_m}$. This gives

$$\Lambda'(\alpha^{-i_m}) = -\alpha^{i_m}\prod_{l \neq m}(1 - \alpha^{i_l}\alpha^{-i_m})$$

$$= -\frac{\alpha^{i_m}}{e_{i_m}\alpha^{m_0 i_m}}\Omega(\alpha^{-i_m}) \tag{5-26}$$

Thus, the error value is given by

$$e_{i_m} = -\alpha^{i_m(1-m_0)}\frac{\Omega(\alpha^{-i_m})}{\Lambda'(\alpha^{-i_m})} \tag{5-27}$$

For the commonly encountered case of $m_0 = 1$ this becomes

$$e_{i_m} = -\frac{\Omega(\alpha^{-i_m})}{\Lambda'(\alpha^{-i_m})} \tag{5-28}$$

The calculation indicated by (5-28) is almost identical to the calculation required in the Chien search, and the realization shown in Fig. 5-5 can be used to evaluate each of the polynomials $\Omega(\alpha^{-i_m})$ and $\Lambda'(\alpha^{-i_m})$. The error value can then be calculated by using the circuit indicated in Fig. 5-6. Actually, the function evaluating $\Lambda'(z)$ for certain codes is simpler than the Chien search function shown in Fig. 5-5. With codes defined over a field of characteristic 2, the coefficients of the odd powers of z disappear so $\Lambda'(z)$ is a function of the form

$$\Lambda'(z) = \Lambda_1 + \Lambda_3 z^2 + \Lambda_5 z^4 + \cdots \tag{5-29}$$

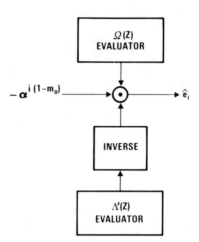

Figure 5-6. Block diagram of error value calculation.

5.4. Solution of the Key Equation

A common step in both time domain and frequency domain decoding is the determination of the minimum-degree error-locator polynomial $\Lambda(z)$ which satisfies the key equation (5-20). In addition, the time domain decoder requires the determination of the error-evaluator polynomial $\Omega(z)$. If $\Lambda(z)$ is thought of as specifying the feedback connections for a linear feedback shift register (as in Fig. 5.3.), then the solution of the key equation is equivalent to finding the minimum-length linear feedback shift register that will produce the first $2t$ terms of the syndrome polynomial. We will find that both viewpoints are useful.

The earliest published procedures for solving the key equation made use of standard matrix inversion techniques, but Berlekamp later devised a simple iterative scheme that today provides the most practical implementation.[2] Recently, it has been shown that Euclid's algorithm can also be used to obtain this solution.[4 5] One nice feature of Euclid's algorithm is that it is considerably easier (than with Berlekamp's algorithm) to understand why it works. In addition, the algorithm is quite closely related to the Berlekamp algorithm. For these reasons, we will present both algorithms.

5.4.1. Euclid's Algorithm

Euclid's algorithm is a recursive technique for finding the greatest common division of two integers or two polynomials. The most basic statement of the algorithm is as follows. Let a and b represent the two integers or polynomials, where $a \geq b$ if they are integers or $\deg(a) \geq \deg(b)$

if they are polynomials. Divide a by b. If the remainder, r, is zero, then $d = b$ is the greatest common divisor. If the remainder is not zero then replace a with b, replace b with r, and repeat. The validity of the algorithm is easily established by considering a simple example.

Suppose that the two integers are $a = 186$ and $b = 66$. Then

$$186 = (66)2 + 54$$
$$66 = (54)1 + 12$$
$$54 = (12)4 + 6$$
$$12 = 6 \cdot 2 + 0$$

Since d divides 186 and 66 it must also divide 54. Since it divides 66 and 54 it must also divide 12. Finally, since it divides 54 and 12, it must also divide 6. Consequently, 6 is the greatest common divisor. In the process of finding the greatest common divisor, the algorithm calculates two numbers, or polynomials, f and g such that

$$fa + gb = d \qquad (5\text{-}30)$$

The useful aspect of this process, however, is not the final answer but the partial results. At each iteration a set of numbers or polynomials f_i, g_i, and r_i are generated such that

$$f_i a + g_i b = r_i \qquad (5\text{-}31)$$

The recursive relationships for f_i, g_i, and r_i become apparent by reexamining the previous example. The first equation in the example is already in the desired form and thus the first iteration of the algorithm produces

$$1(186) - 2(66) = 54$$

The second equation in the example contains the directions for the second iteration. That is, if we write the equation $66 = 66$ and subtract the above equation, we have

$$-(186) + 3(66) = 12$$

Likewise, the third and fourth equation produce

$$5(186) - 14(66) = 6$$

and

$$-11(186) + 31(66) = 0$$

respectively. We note that at each stage the next equation on the list is produced by combining the two previous equations according to the rule

$$\text{Eq}(i) = \text{Eq}(i - 2) - q_i \, \text{Eq}(i - 1)$$

where q_i is just the quotient of the two previous remainders.

The recursive relationships can now be stated formally. Since we are basically interested in polynomials, all of the subsequent discussion will be restricted to this case.

Define $q_i(z)$ to be the quotient polynomial that is produced by dividing $r_{i-2}(z)$ by $r_{i-1}(z)$. That is

$$q_i(z) = \left[\frac{r_{i-2}(z)}{r_{i-1}(z)} \right]_0^\infty \tag{5-32}$$

where $[\quad]_0^\infty$ denotes the nonnegative powers of z. Then

$$r_i(z) = r_{i-2}(z) - q_i(z)r_{i-1}(z)$$
$$f_i(z) = f_{i-2}(z) - q_i(z)f_{i-1}(z) \tag{5-33}$$

and

$$g_i(z) = g_{i-2}(z) - q_i(z)g_{i-1}(z)$$

Initial conditions for the algorithm are

$$f_{-1}(z) = g_0(z) = 1$$
$$f_0(z) = g_{-1}(z) = 0 \tag{5-34}$$
$$r_{-1}(z) = a(z)$$

and

$$r_0(z) = b(z)$$

The nature of the algorithm is such that

$$\deg[r_i(z)] < \deg[r_{i-1}(z)] \tag{5-35}$$

for all i. Thus, the algorithm will always converge to a zero remainder in a finite number of steps. A simple example with $a(z) = z^3 + 1$ and $b(z) = z^2 + 1$ is shown in Table 5-1. The last nonzero remainder $r_i(x) = z + 1$ is the greatest common divisor.

Our principal interest is not in finding a greatest common divisor but

Table 5-1. An Example of Euclid's Algorithm with
$a(z) = z^3 + 1$ **and** $b(z) = z^2 + 1$

i	$f_i(z)$	$g_i(z)$	$r_i(z)$	$q_i(z)$
-1	1	0	$z^3 + 1$	—
0	0	1	$z^2 + 1$	—
1	1	z	$z + 1$	z
2	$z + 1$	$z^2 + z + 1$	0	$z + 1$

in solving the key equation. The applicability of Euclid's algorithm for this problem becomes apparent by noting that (5-31) may be written

$$g_i(z)\, b(z) \equiv r_i(z) \bmod a(z) \tag{5-36}$$

By choosing $a(z) = z^{2t}$, this equation becomes

$$g_i(z)\, b(z) \equiv r_i(z) \bmod z^{2t} \tag{5-37}$$

which is the key equation with $g_i(z) = \Lambda_i(z)$ and $r_i(z) = \Omega_i(z)$. Thus, Euclid's algorithm produces a sequence of solutions to the key equation.

To see that this approach produces the desired solution to the key equation, we need to utilize the property of Euclid's algorithm that states

$$\deg[g_i(z)] + \deg[r_{i-1}(z)] = \deg[a(z)] \tag{5-38}$$

Letting $a(z) = z^{2t}$ then

$$\deg[g_i(z)] + \deg[r_{i-1}(z)] = 2t$$

and

$$\deg[g_i(z)] + \deg[r_i(z)] < 2t \tag{5-39}$$

Remember that if $v \le t$ errors occur, then the solution of interest has $\deg[\Omega(z)] < \deg[\Lambda(z)] \le t$. There exists only one polynomial $\Lambda(z)$ (to within a constant) with degree no greater than t which satisfies the key equation. Also, if $\deg[r_{i-1}(z)] \ge t$ and thus $\deg[g_i(z)] \le t$ and $\deg[r_i(z)] < t$, then $\deg[g_{i+1}(z)] > t$. This means that the partial results at the ith step provide the only solution to the key equation that is of interest. Thus, to solve the key equation simply requires applying Euclid's algorithm until $\deg[r_i(z)] < t$. This technique is summarized in Table 5-2.

The application of the algorithm shall be shown via two examples. Arithmetic operations will be performed over $GF(2^4)$ using Table 5-3.

Example. Consider the triple-error-correcting binary BCH code of length 15. Take $m_0 = 1$ so that the generator polynomial has $\alpha, \alpha^2, \ldots, \alpha^6$ as roots, where α is a primitive element of $GF(2^4)$. Suppose the received word is

Table 5-2. Solution of Key Equation for t-Error-Correcting BCH Codes

---

1. Apply Euclid's algorithm to $a(z) = z^{2t}$ and $b(z) = S(z)$.
2. Use the initial conditions of (5-34).
3. Stop when $\deg[r_n(z)] < t$.
4. Set $\Lambda(z) = g_n(z)$ and $\Omega(z) = r_n(z)$.

Table 5-3. The Ordinary and Logarithmic Representation of the Nonzero Elements of the Field GF(2^4)
with $p(x) = x^4 + x + 1$

i	α^i	i	α^i
0	0 0 0 1	8	0 1 0 1
1	0 0 1 0	9	1 0 1 0
2	0 1 0 0	10	0 1 1 1
3	1 0 0 0	11	1 1 1 0
4	0 0 1 1	12	1 1 1 1
5	0 1 1 0	13	1 1 0 1
6	1 1 0 0	14	1 0 0 1
7	1 0 1 1		

$$r(x) = x^3 + x^{10} \tag{5-40}$$

and time domain decoding is used. Then the syndrome coefficients of interest
may be calculated from (5-17) as

$$S_j = (\alpha^{j+1})^3 + (\alpha^{j+1})^{10}: \qquad j = 0, 1, ..., 5$$

This gives the syndrome polynomial

$$S(z) = \alpha^{14}z^5 + \alpha^{10}z^4 + \alpha^3 z^3 + \alpha^7 z^2 + \alpha^9 z + \alpha^{12} \tag{5-41}$$

Starting with $a(z) = z^6$ and $b(z) = S(z)$, Euclid's algorithm finds the solution
to the key equation and produces the error locator polynomial as shown in
Table 5-4. Note that the solution is reached at the i for which $\deg[\Omega_i(z)] <$
$t = 3$. The resulting error-locator polynomial

$$\Lambda(z) = \alpha^{12}z^2 + \alpha^{11}z + \alpha^{14}$$

Table 5-4. Solution of the Key Equation for $S(z)$ Given by (5-41)

i	$\Lambda_i(z)$	$\Omega_i(z)$	$q_i(z)$
-1	0	z^0	—
0	1	$S(z)$	—
1	$\alpha z + \alpha^{12}$	$\alpha^3 z^4 + \alpha^2 z^3 + \alpha^2 z^2 + z + \alpha^9$	$\alpha z + \alpha^{12}$
2	$\alpha^{12}z^2 + \alpha^{11}z + \alpha^{14}$	α^{11}	$\alpha^{11}z + \alpha^6$

has roots α^{-3} and α^{-10} which can be found either by exhaustive substitution or by a Chien search. Thus, the double-error pattern $\hat{e}(x) = x^3 + x^{10}$, is hypothesized. The resulting decoder output is simply $c(x) = r(x) - \hat{e}(x) = 0$.

A frequency domain decoder would utilize the error-locator polynomial in a different fashion. The polynomial $\Lambda(z)$ specifies the taps of a feedback shift register that can generate the transform of the error vector through (5-13) if a block of v components are given. Thus, for this example, (5-13) can be written

$$\hat{E}_j = \alpha^{12}\hat{E}_{j-1} + a^{13}\hat{E}_{j-2}$$

The known values are $\hat{E}_j = S_{j-1}$ ($j = 1, 2, \ldots, 6$). By recursive extension the error transform is found to be

$$\hat{\mathbf{E}} = (0, \alpha^{12}, \alpha^9, \alpha^7, \alpha^3, \alpha^{10}, \alpha^{14}, \alpha^7, \alpha^6, \alpha^{11}, \alpha^5, \alpha^{11}, \alpha^{13}, \alpha^{13}, \alpha^{14})$$

Then taking the inverse transform the hypothesized error vector is

$$\hat{\mathbf{e}} = (0, 0, 0, 1, 0, 0, 0, 0, 0, 0, 1, 0, 0, 0, 0)$$

Performing the indicated correction again produces the all-zero code word.

Example. As a second example consider the triple-error-correcting Reed–Solomon code of length 15 defined over $GF(2^4)$. The roots of the generator polynomial are $\alpha, \alpha^2, \ldots, \alpha^6$. Suppose the received word is

$$r(x) = \alpha^7 x^3 + \alpha^{11} x^{10} \tag{5-42}$$

and time domain decoding is used. The syndrome coefficients may be calculated as

$$S_j = \alpha^7 (\alpha^{j+1})^3 + \alpha^{11}(\alpha^{j+1})^{10}, \qquad j = 0, 1, \ldots, 5$$

The resulting syndrome polynomial is

$$S(z) = \alpha^{14} z^5 + \alpha^{14} z^4 + \alpha^{12} z^3 + \alpha^6 z^2 + \alpha^{12} z + \alpha^7 \tag{5-43}$$

Application of Euclid's algorithm to find the error-locator and error-evaluator polynomials is shown in Table 5-5. The resulting error-locator polynomial,

$$\Lambda(z) = \alpha^9 z^2 + \alpha^8 z + \alpha^{11}$$

has roots at α^{-3} and α^{-10}. The error-evaluator polynomial,

$$\Omega(z) = \alpha^2 z + \alpha^3$$

and the derivative of $\Lambda(z)$,

$$\Lambda'(z) = \alpha^8$$

Table 5-5. Solution of the Key Equation for $S(z)$ Given by (5-43)

i	$\Lambda_i(z)$	$\Omega_i(z)$	$q_i(z)$
-1	0	z^6	—
0	1	$S(z)$	—
1	$\alpha z + \alpha$	$\alpha^6 z^4 + \alpha^5 z^3 + \alpha^5 z^2 + \alpha^3 z + \alpha^8$	$\alpha z + \alpha$
2	$\alpha^9 z^2 + \alpha^8 z + \alpha^{11}$	$\alpha^2 z + \alpha^3$	$\alpha^8 z + \alpha^{11}$

may be used to find the error values at α^{-3} and α^{-10} through (5-28). This gives $\hat{e}(x) = \alpha^7 x^3 + \alpha^{11} x^{10}$, and the resulting decoder output is simply $c(x) = r(x) - \hat{e}(x) = 0$.

In summary, application of Euclid's algorithm to $S(z)$ produces a sequence of solutions to the key equation increasing in degree for $\Lambda_i(z)$ and decreasing in degree for $\Omega_i(z)$. However, it was shown that for any pattern of t or fewer errors with BCH codes the $\deg[\Lambda(z)] > \deg[\Omega(z)]$. Thus, the only solution that guarantees the desired degree of error correction is obtained at the first i for which $\deg[\Omega_i(z)] < t$.

5.4.2. Berlekamp's Algorithm

The solution of the key equation when $S(z)$ is given involves finding the polynomial $\Lambda(z)$ of minimum degree v for which

$$[\Lambda(z)\,S(z)]_v^{2t-1} = 0 \qquad (5\text{-}44)$$

As was pointed out earlier, the problem of finding the minimum-degree solution to the key equation is precisely the same problem as finding the minimum-length shift register which generates the first $2t$ terms of $S(z)$. Berlekamp's algorithm is probably best understood when approached from this viewpoint. The essential features of the Berlekamp algorithm are as follows. First, find the shortest feedback shift register which will predict S_1 from S_0. Next, test this feedback shift register to see if it will also predict S_2. If it does, it is still the best solution and one should continue to test successive syndrome symbols until the test fails. At this point, modify the register in such a manner that

1. it predicts the next symbol correctly,
2. it does not change any of the previous predictions, and
3. the length of the register is increased by the least possible amount.

Continue in this fashion until the first $2t$ symbols of $S(z)$ can be generated. The details will be illustrated by the following example.

Example. Consider the problem of finding the minimum-length shift register which will produce the sequence 1 1 1 0 1 1 0 0 0 0 1 0 1. The simplest feedback shift register which will generate the first two symbols is shown in the first row of Fig. 5-7. This register also generates the next symbol but fails on the fourth symbol. At this point it becomes apparent that the length of the register must be increased to at least 3 since one cannot combine two ones to produce a one (third symbol), and then on the next shift combine

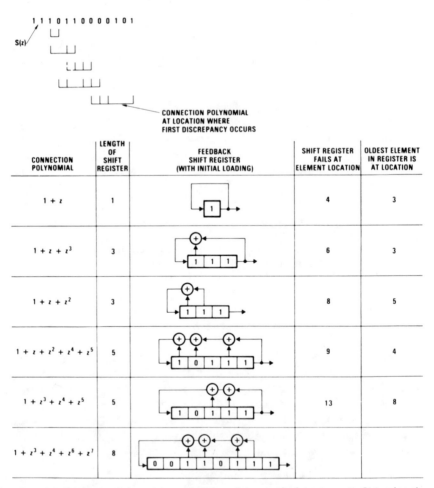

Figure 5-7. Synthesis of the minimum-length register required to generate $S(z)$ using the Berlekamp algorithm.

two ones to produce a zero (fourth symbol). A length-3 feedback shift register which generates the fourth symbol of $S(z)$ correctly is shown in the second row of Fig. 5-7. This feedback shift register is not unique. Any length-3 register with an even number of feedback connections would work equally well. This register may be used to generate the fifth symbol in $S(z)$, but it fails on the sixth. One way to correct this situation is to find some linear combination of the symbols of $S(z)$ which, when added to the current prediction for the sixth symbol, will change it from a zero to a one. This correction term must have the property that all earlier shifts of this linear combination of symbols sum to zero. Finally, one would like to pick this linear combination of symbols so that it is displaced from the sixth symbol by the least possible amount. The connection polynomial for the first generator provides the answer. One may observe that since this register failed at location 4, the sum of the third and fourth symbols is one while the sums of the second and third symbols and the first and second symbols are both zero. Consequently, if we were to add the term $z^2(1 + z)$ to the current connection polynomial, it would change the prediction of the sixth symbol to be a one (by adding the third and fourth symbols), but it would not change either of the previous predictions. Hence, it satisfies the first two requirements of the Berlekamp algorithm. It also satisfies the third requirement since the length of the shift register did not increase. We further note that although the new connection polynomial is now of degree 2, the length of the feedback shift register cannot be reduced to less than 3 since this would violate the earlier restrictions.

If one now continues to use this new feedback shift register, the seventh symbol is predicted correctly, but the eighth is not. At this point we observe that it is again possible to remove the discrepancy at symbol 8 by using the first connection polynomial (which adds symbols 3 and 4 to the current prediction) or by using the second connection polynomial (which adds symbols 3, 5, and 6 to the current prediction). In either case the length of the new register must increase to 5 since the oldest symbol associated with either of these choices occurs in location 3. For reasons which will become clear later, the Berlekamp algorithm always makes the first choice. Thus, the current connection polynomial is modified by adding $(1 + z)z^4$. The reader may now verify that no register of length less than 5 can produce the desired result.

First, a three-stage register would not work. Note that when the register is loaded with the pattern 1 1 0, the feedback term would have to be one at the fifth symbol and zero at the eighth symbol. Similarly, a four-stage register would have to predict a one for the sixth symbol, a zero for the seventh symbol, and a zero for the eighth symbol from registers loaded with the patterns 1 1 0 1, 1 0 1 1, and 0 1 1 0, respectively. This is clearly impossible with a linear device since the modulo-2 sum of 1 1 0 1 and 1 0 1 1 is 0 1 1 0.

If the feedback connections were selected to predict the correct values for the sixth and seventh symbols, then a one will always be predicted incorrectly for the eighth symbol because of the linearity requirement.

Continuing with the example we see that the new five-stage register fails at the ninth symbol. Following the same procedure there are now three possible choices to remove the discrepancy. The register in row 3 is clearly the best choice since the location of the oldest symbol in this register is 5 at the point where the register fails while the oldest symbol in the other two registers is in location 3 when they fail. Hence, we add the correction term $z(1 + z + z^2)$, and the new connection polynomial is $1 + z^3 + z^4 + z^5$. The shift register associated with this polynomial now predicts correctly until it reaches symbol 13. At this point the best choice for determining a correction polynomial is again register 3, and the new connection polynomial becomes $1 + z^3 + z^4 + z^6 + z^7$. At first one might think that a seven-stage register would be sufficient. However, this is not the case. If one were to attempt to use the newly formed connection polynomial to define the symbol in location 8 (as it must with a seven-stage register), then the contribution due to the third shift register (i.e., the correction term) would have to predict the symbol in location 3 from the values in locations 1 and 2. This is impossible and violates the conditions under which register 3 was selected. Consequently, an eight-stage register must be used as shown in Fig. 5-7. The new register length could, of course, have been computed by subtracting the location of the oldest symbol in the correction register from the location of the most recent discrepancy. This will always be the correct register length whenever this value exceeds the degree of the new connection polynomial. Otherwise, the degree of the connection polynomial can be taken as the register length.

From this example it is clear that a simple iterative procedure can be defined for constructing the shortest register that will produce a given sequence. We note that at each iteration this algorithm must retain both a connection polynomial and a possible correction term. For each new symbol of $S(z)$, the algorithm tests to determine whether the current connection polynomial will predict this symbol correctly. If so, the connection polynomial is left unchanged and the correction term is modified by multiplying by z. If the current connection polynomial fails, then it is modified by adding the correction term. One then checks to determine whether the new register length has increased. If it has not, then the correction polynomial that is currently being maintained is at least as good a choice as any other. If the length of the new register increases, then the previous connection polynomial becomes a better choice since the oldest symbol associated with this register must have a higher index than the oldest symbol associated with the current correction term. A slight modification of the algorithm is required when $S(z)$ is defined over an arbitrary field rather than $GF(2)$. In this case the discrepancies associated with both the connection polynomial and the correction term can take on values other than one. This is accounted for by normalizing the correction polynomial to produce a discrepancy of one each time a new correction polynomial is selected. This normalized polynomial is then multiplied by the value of the current discrepancy when a correction is made.

The Berlekamp algorithm may now be stated formally as follows. Let $\Lambda_n(z)$ be the connection polynomial at the nth iteration, let v_n be the degree of the connection polynomial, and L_n be the length of the associated feedback shift register. Let $D_n(z)$ be the correction term, k_n be the location of the oldest symbol in the feedback shift register at the point where the register fails, and define the discrepancy, d_n, as

$$d_n = \sum_{i=0}^{v_n} \Lambda_{n,i} S_{n-i} \qquad (5\text{-}45)$$

Then the new connection polynomial,

$$\Lambda_{n+1}(z) = \Lambda_n(z) - d_n D_n(z) \qquad (5\text{-}46)$$

has length

$$L_{n+1} = \begin{cases} L_n, & d_n = 0 \\ \mathrm{Max}(L_n, n - k_n), & d_n \neq 0 \end{cases}$$

When $L_{n+1} = L_n$ the correction polynomial is

$$D_{n+1}(z) = z D_n(z)$$

with

$$k_{n+1} = k_n$$

or when $L_{n+1} > L_n$ the correction polynomial is

$$D_{n+1}(z) = z \frac{\Lambda_n(z)}{d_n}$$

with

$$k_{n+1} = n - L_n$$

Initial conditions are chosen as $n = 0$, $k_0 = -1$, $L_0 = 0$, $\Lambda_0(z) = 1$, and $D_0(z) = z$. Although we have not offered a rigorous proof, the Berlekamp algorithm will always find the shortest feedback shift register that generates the first $2t$ terms of $S(z)$. The interested reader is referred to the discussions by Berlekamp,[2] Massey,[46] or Gallager[3] for the details of the proof.

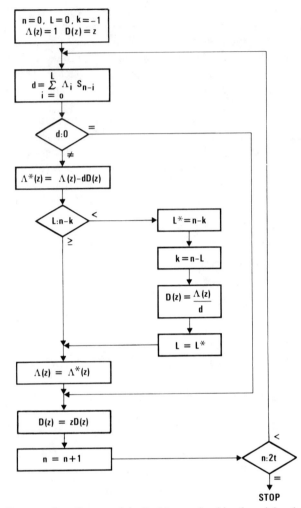

Figure 5-8. Computer flow diagram of the Berlekamp algorithm for solving the key equation.

A computer flow chart for the algorithm is shown in Fig. 5-8. In this figure one should interpret the equal sign in the usual computer language sense as meaning to replace. The subscripts have been dropped to avoid confusion. When solving the key equation in the time domain it is also necessary to find the error-evaluator polynomial $\Omega(z)$. Since $\Omega(z)$ can be found from $\Lambda(z)$ by multiplying by $S(z)$, it follows that one can define a sequence of polynomials $\Omega_n(z)$ which obeys the same recursion relationship as $\Lambda_n(z)$. To avoid confusion let the correction term associated with finding $\Omega(z)$ be

$A_n(z)$. Then

$$\Omega_{n+1}(z) = \Omega_n(z) - d_n A_n(z) \tag{5-47}$$

and

$$A_{n+1}(z) = \begin{cases} z A_n(z), & L_{n+1} = L_n \\ z \dfrac{\Omega_n(z)}{d_n}, & L_{n+1} > L_n \end{cases}$$

By choosing $\Omega_0(z) = 0$ and $A_0(z) = -1$, the reader may verify that the key equation is satisfied at each stage in the iteration.

The two examples which follow illustrate the use of the Berlekamp algorithm to solve the key equation. The first example is for a binary BCH code, and we find only the error-locator polynomial. The second example is for a Reed–Solomon code, and we find both $\Lambda(z)$ and $\Omega(z)$.

Example. Consider the triple-error-correcting BCH code of length 15 with received word and syndrome polynomial given by (5-40) and (5-41). The steps leading to the solution of the key equation using Berlekamp's algorithm are shown in Table 5-6. Note that only one new syndrome polynomial coefficient is utilized on each step in contrast to Euclid's algorithm, which utilizes the entire polynomial. However, the final solution for $\Lambda(z)$ is identical (after normalization) to that found in Table 5-4 using Euclid's algorithm.

Suppose now that we rework the same example, but with an additional error in position 1, i.e.,

$$r(x) = x + x^3 + x^{10} \tag{5-48}$$

The syndrome polynomial is easily calculated to be

$$S(z) = \alpha^8 z^5 + z^4 + \alpha^7 z^3 + \alpha^4 z^2 + \alpha^{11} z + \alpha^{13} \tag{5-49}$$

Table 5-6. Solution of the Key Equation for $S(z)$ Given by (5-41) Using Berlekamp's Algorithm

n	S_n	$\Lambda_n(z)$	d_n	L	$D_n(z)$
0	α^{12}	1	α^{12}	0	z
1	α^9	$1 + \alpha^{12}z$	0	1	$\alpha^3 z$
2	α^7	$1 + \alpha^{12}z$	α^{10}	1	$\alpha^3 z^2$
3	α^3	$1 + \alpha^{12}z + \alpha^{13}z^2$	0	2	$\alpha^5 z + \alpha^2 z^2$
4	α^{10}	$1 + \alpha^{12}z + \alpha^{13}z^2$	0	2	$\alpha^5 z^2 + \alpha^2 z^3$
5	α^{14}	$1 + \alpha^{12}z + \alpha^{13}z^2$	0	2	$\alpha^5 z^3 + \alpha^2 z^4$
6		$1 + \alpha^{12}z + \alpha^{13}z^2$			

Table 5-7. Solution of the Key Equation for $S(z)$ Given by (5-49) Using Berlekamp's Algorithm

n	S_n	$\Lambda_n(z)$	d_n	L	$D_n(z)$
0	α^{13}	1	α^{13}	0	z
1	α^{11}	$1 + \alpha^{13}z$	0	1	$\alpha^2 z$
2	α^4	$1 + \alpha^{13}z$	α^{14}	1	$\alpha^2 z^2$
3	α^7	$1 + \alpha^{13}z + \alpha z^2$	0	2	$\alpha z + \alpha^{14}z^2$
4	α^0	$1 + \alpha^{13}z + \alpha z^2$	α^0	2	$\alpha z^2 + \alpha^{14}z^3$
5	α^8	$1 + \alpha^{13}z + \alpha^{14}z^3$	0	3	$z + \alpha^{13}z^2 + \alpha z^3$
6		$1 + \alpha^{13}z + \alpha^{14}z^3$			

Application of Berlekamp's algorithm is shown in Table 5-7. It is easily verified that the solution for $\Lambda(z)$ has roots α^{-1}, α^{-3}, and α^{-10}, which would again allow the decoder to recover the all-zero code word.

Tables 5-6 and 5-7 illustrate an important property of the Berlekamp algorithm for binary BCH codes. For these codes, $d_n = 0$ for all odd values of n. This means, of course, that calculation of d_n at odd values of n is unnecessary so all the odd steps can be skipped. One needs only to multiply $D_n(z)$ by z^2 to account for merging two steps into one.

Example. Consider the block length 15, triple-error-correcting Reed–Solomon code with $r(z)$ and $S(z)$ given by (5-42) and (5-43). Application of Berlekamp's algorithm to solve the key equation for this $S(z)$ is shown in Table 5-8. Calculation of $\Omega(z)$ is also shown. The solutions $\Lambda(z)$ and $\Omega(z)$ are identical (within a constant factor) to those found previously using Euclid's algorithm.

Table 5-8. Solution of the Key Equation for $S(z)$ Given by (5-43)

n	S_n	$\Lambda_n(z)$	d_n	L	$D_n(z)$	$\Omega_n(z)$	$A_n(z)$
0	α^7	1	α^7	0	z	0	-1
1	α^{12}	$1 + \alpha^7 z$	α^5	1	$\alpha^8 z$	α^7	0
2	α^6	$1 + \alpha^5 z$	α^3	1	$\alpha^8 z^2$	α^7	0
3	α^{12}	$1 + \alpha^5 z + \alpha^{11}z^2$	α^2	2	$\alpha^{12}z + \alpha^2 z^2$	α^7	$\alpha^4 z$
4	α^{14}	$1 + \alpha^{12}z + \alpha^{13}z^2$	0	2	$\alpha^{13}z + \alpha^3 z^2 + \alpha^9 z^3$	$\alpha^7 + \alpha^6 z$	$\alpha^4 z^2$
5	α^{14}	$1 + \alpha^{12}z + \alpha^{13}z^2$	0	2	$\alpha^{13}z^2 + \alpha^3 z^3 + \alpha^9 z^4$	$\alpha^7 + \alpha^6 z$	$\alpha^4 z^3$
6		$1 + \alpha^{12}z + \alpha^{13}z^2$					

5.5. Implementation Considerations

BCH decoders are typically implemented using the time domain techniques. The top-level block diagram is usually of the form of Fig. 5-4. However, the details of implementing the various functions shown in this block diagram may change significantly in different implementations depending on the code and the required data rate. The best decoder architecture to use for any given problem is not nearly as obvious with BCH decoders as it is for certain other techniques such as table look-up decoders, threshold decoders, and Viterbi decoders for convolutional codes.

Decoding a received BCH code word requires execution of three successive computational processes with all computations performed over the field $GF(q^m)$. These processes are syndrome calculation, solution of the key equation, and Chien search (with error magnitude calculation for nonbinary codes). In a low-data-rate implementation the decoder architecture might be most efficiently organized as a special-purpose processor with the following functions:

1. $GF(q^m)$ central processing unit (CPU),
2. random access memory (RAM) for buffering (two code words) and temporary storage of intermediate results,
3. read-only memory (ROM) for storage of computational subroutines and control of the decoding process,
4. interfaces with the rest of the system (demodulator and data sink).

The CPU will probably require one or more fast $GF(q^m)$ multipliers (to be discussed later) depending on required data rate. To achieve a very efficient design the architecture of the CPU must be carefully matched to the hardware requirements of the computational subroutines. This is not an easy problem, but there is at least one such efficient design that is commercially available.

An increase in data rate for a given code and decoder architecture can be obtained by taking advantage of certain situations which are easy to decode. For example, when the syndrome polynomial is zero, no error correction need be done. Thus, all of the operations associated with the Berlekamp algorithm and Chien search may be skipped. In addition, the single-error case is easily accommodated through special calculations. This case is detected if there exists a value of k for which

$$\frac{S_j}{S_{j+1}} = \alpha^{-k}$$

for all j in the range $0 \leq j \leq 2t - 2$. The value k is the error location of the single error, and the error value is easily calculated as

$$e_k = S_0 \alpha^{-k}$$

Thus, in certain applications the zero or one-error case may occur a high percentage of the time, and very little computation is required for these code words. This can substantially reduce the average computational requirement. In order to take advantage of this property one must provide a code word buffer to hold incoming code words whenever one must perform full Berlekamp algorithm/Chien search decoding. The required buffer size is dependent on the code, the decoder architecture, the data rate, and the channel error rate.

As the data rate requirements increase, the architecture must be modified. One approach is to utilize special-purpose peripheral computational devices to ease the computational burden on the $GF(q^m)$ CPU. For example, a syndrome computer "front-end" built with devices of the form shown in Fig. 5-2 will remove the requirement for calculating syndromes by the CPU. At still higher data rates one may have to perform the Chien search in a peripheral unit leaving only the solution of the key equation in the CPU. The Berlekamp algorithm for solving the key equation provides a technique that can be implemented in the central processor at data rates in the several Mb/s range. Note that the special-purpose circuits shown in Figs. 5-2 and 5-5 for implementing the syndrome calculation and the Chien search are basically shift registers with $GF(q^m)$ multipliers and adders attached. One can utilize similar structures for implementing the Berlekamp algorithm at very high data rates (tens of Mb/s). This structure will be discussed later in this section.

The manner in which $GF(q^m)$ multipliers are implemented is a very important part of the design optimization process. There are several ways in which the multiplication can be accomplished. As an example, consider the ordinary and logarithmic representation of $GF(2^4)$ shown in Table 5-3. This field will be used in all subsequent examples in this chapter. One can multiply by adding the logarithms (in the first column) modulo 15. However, the representation in the second column is most convenient for addition. Thus, to implement multiplication via logarithms one needs a modulo-15 adder and two look-up tables for conversion between ordinary and logarithmic representations (easily implemented in a ROM). One may also use the Zech logarithm discussed in Chapter 2 to eliminate one of the look-up table accesses. This approach is particularly attractive when the characteristic of the field is large.

An alternate approach which can achieve the greatest speeds is to calculate the product directly using a logic function. Consider taking the product of two field elements a and b represented as

$$a = a_3 x^3 + a_2 x^2 + a_1 x + a_0$$

and

$$b = b_3 x^3 + b_2 x^2 + b_1 x + b_0$$

By performing the indicated multiplication ($c' = a \cdot b$), we obtain a degree-6 polynomial for c' with coefficients

$$
\begin{aligned}
c'_0 &= a_0 b_0 \\
c'_1 &= a_0 b_1 + a_1 b_0 \\
c'_2 &= a_0 b_2 + a_1 b_1 + a_2 b_0 \\
c'_3 &= a_0 b_3 + a_1 b_2 + a_2 b_1 + a_3 b_0 \qquad\qquad (5\text{-}50) \\
c'_4 &= a_1 b_3 + a_2 b_2 + a_3 b_1 \\
c'_5 &= a_2 b_3 + a_3 b_2 \\
c'_6 &= a_3 b_3
\end{aligned}
$$

The polynomial c' can be reduced modulo $p(x)$ to a degree-3 polynomial $c(x)$ with coefficients

$$
\begin{aligned}
c_0 &= c'_0 + c'_4 \\
c_1 &= c'_1 + c'_4 + c'_5 \\
c_2 &= c'_2 + c'_5 + c'_6 \qquad\qquad (5\text{-}51) \\
c_3 &= c'_3 + c'_6
\end{aligned}
$$

Thus, the product may be accomplished entirely with two-input AND gates and modulo-2 adders.

For small fields [say $GF(2^6)$ and smaller] the multiplier may be more conveniently implemented in a ROM. A two-input multiplier for $GF(2^m)$ requires $2m$ input lines and m output lines. This can be provided by a $2^{2m} \times m$ ROM.

A circuit such as shown in Fig. 5-2 only requires multiplication by a constant. In this case the logic function indicated by (5-50) and (5-51) is significantly simplified. For example, let $a = x^2$ ($a_0 = a_1 = a_3 = 0$ and $a_2 = 1$). Then $c = x^2 \cdot b$ has coefficients

$$
\begin{aligned}
c_0 &= b_2 \\
c_1 &= b_2 + b_3 \\
c_2 &= b_0 + b_3 \qquad\qquad (5\text{-}52) \\
c_3 &= b_1
\end{aligned}
$$

Thus, no AND gates and only two modulo-2 adders are required.

One can also implement the multiplication using a "shift-and-add" technique. A feedback shift register of the form in Fig. 2-6 with connection polynomial $p(x)$ is used as an accumulator. The multiplier and multiplicand are held in two separate registers. The multiplier bits are successively tested starting with the most significant bit and for each bit that is 1 the multiplicand is added into the accumulator register, which is then shifted once. On each shift the accumulator feedback reduces the partial result modulo $p(x)$. After m shifts the accumulator contains the desired result. Obviously, this technique will be slower than direct calculation, but it can lead to substantial hardware savings for large m (say $m \geq 9$).

The multiplier function is a prime candidate for implementation as a custom integrated circuit. In fact, the function shown in Fig. 5-2 provides a natural building block for BCH decoders. If the multiplication indicated in this figure was done using a general $GF(q^m)$ multiplier (for some selected q and m), then a custom integrated circuit performing this function could be used in many places in the decoder. Several such circuits would be used in the syndrome calculation, the Chien search, and the CPU. Such a circuit would substantially reduce implementation complexity of very-high-speed decoders.

The syndrome computation can be simplified in certain cases. Assume a BCH code defined over $GF(q)$, where q is prime. Let a and b be defined over $GF(q^m)$. If $q = 2$, then note that

$$(a + b)^2 = a^2 + 2ab + b^2$$
$$= a^2 + b^2$$

Similarly, for any prime q,

$$(a + b)^q = a^q + b^q$$

Then S_j^q may be written

$$S_j^q = \left[\sum_{k=1}^{v} e_{i_k} \alpha^{i_k (m_0 + j)} \right]^q$$
$$= \sum_{k=1}^{v} e_{i_k}^q \alpha^{i_k q(m_0 + j)} \tag{5-53}$$

Noting that $a^q = a$ for q prime, (5-53) may be written

$$S_j^q = S_{q(j+m_0) - m_0} \tag{5-54}$$

Thus, not all S_j need to be calculated directly. Some may be calculated as powers of other S_j. For the most common case of $q = 2$ and $m_0 = 1$, (5-54) becomes

$$S_{2j+1} = S_j^2 \qquad (5\text{-}55)$$

In this case only the S_j for j even need be calculated directly.

The Berlekamp algorithm could be implemented at moderate data rates in a special-purpose processor with an associated $GF(q^m)$ arithmetic unit and memory as suggested previously. The flow diagram of Fig. 5-8 or the versions presented in Berlekamp[2] or Massey [46] could be used for control of the processor.

A highly parallel realization suggested by Massey[46] has been used to achieve very high data rates (40 Mb/s).[47] A functional block diagram of the processor that executes the Berlekamp algorithm is shown in Fig. 5-9. Storage of polynomials $S(z), \Lambda_n(z), D_n(z), \Omega_n(z)$, and $A_n(z)$ is in fast shift registers R1 through R5. The flow diagram used for control is shown in Fig. 5-10. The most important difference between this diagram and the previous flow diagram is the manner in which one determines that the register length increases. It turns out that the length always increases when $n \geq 2L$, and so this test is made. In addition, the new length of $\Lambda_n(z)$ is given by $L = n - L + 1$. The parameter d^* is the value d_n^{-1} used in $D_n(z) = d_n^{-1} \Lambda_n(z)$. Note that it is not necessary to keep the zero-order coefficients in R2 and R3.

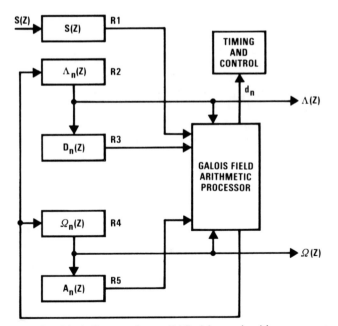

Figure 5-9. Block diagram of a parallel Berlekamp algorithm processor.

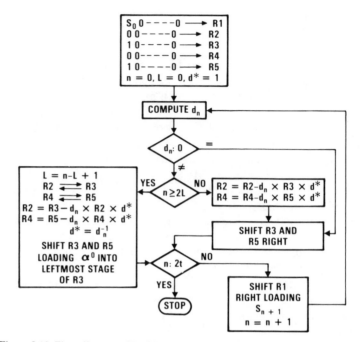

Figure 5-10. Flow diagram of Berlekamp algorithm for the realization of Fig. 5-9.

5.6. Correction of Errors and Erasures

The BCH decoding algorithm does not generalize to utilize demodulator soft decisions in a direct fashion. The only approaches that have been used require multiple application of the hard decision decoding algorithm. Algorithms that operate in this fashion were discussed in the previous chapter, e.g., the Weldon algorithm and the Chase algorithm. However, in the special case of demodulator erasures, an algebraic decoding approach is possible. That is, if the code is defined over the field $GF(q)$, then the demodulator is assumed to have the normal q outputs plus a $(q + 1)$th output corresponding to an erasure. The erasure output is assumed to contain absolutely no information about which of the q possible symbols was actually transmitted. Unfortunately, this decoding algorithm does not provide significant additional gain on the Gaussian channel. It is of benefit, though, on certain interference channels where the interference may be detected and treated as erasures.

If a code has minimum distance d and a received word contains e erasures, then the e "erased" positions can be ignored in the decoding process. However, even ignoring these positions, all code words will differ in at least $d - e$ of the remaining positions allowing correction of up to

$$t_m = \left\lfloor \frac{d - e - 1}{2} \right\rfloor \tag{5-56}$$

channel errors in addition to the erasures. Thus, all combinations of v channel errors and e erasures are correctable providing that

$$2v + e < d \tag{5-57}$$

Erasure and error correction with binary BCH codes is accomplished with the following algorithm. Assume that a BCH code is designed to correct t errors, and that the received word contains e erasures. The decoding algorithm is very simple. First, replace the e erased positions with zeros and decode the resulting word using the standard BCH decoding algorithm. Next, replace the e erased positions with ones and decode in the same fashion. Finally, of the two code words obtained, select the one corresponding to the smallest number of errors corrected outside the e erased positions. It is easily shown that this algorithm always results in correct decoding when (5-57) is satisfied. Assume that when the e erasures are replaced by zeros, $e^* \le e/2$ channel errors are introduced in those e positions making a total of $v + e^* \le t$ errors which are guaranteed correctable when $2v + e \le 2t$. On the other hand, if $e^* > e/2$ then only $e - e^* < e/2$ channel errors are introduced when replacing the e erasures by ones. In this case the resulting pattern of $v + (e - e^*) < t$ errors is also correctable. Note that this algorithm is actually closely related to the Chase algorithm and is not strictly algebraic in nature.

Correction of erasures is more difficult with nonbinary BCH codes. As in (5-6) denote the erasure-locator polynomial by

$$\Gamma(z) = \prod_{k=1}^{e} (1 - z\alpha^{j_k}) \tag{5-58}$$

Knowing the erased positions one can compute $\Gamma(z)$ and substitute arbitrary received symbols in these positions to allow computation of $S(z)$. The substitution of e arbitrary received symbols into the erased positions can introduce up to e additional errors. The resulting key equation is

$$S(z) \Lambda(z) \Gamma(z) \equiv \Omega(z) \bmod z^{2t} \tag{5-59}$$

The decoding problem is to find the solution of (5-59) which results in the minimum $\deg[\Lambda(z)]$. Assuming v channel errors and e additional errors added in the erased positions, then $\deg[\Omega(z)] < v + e$. This means that

$$[S(z) \Lambda(z) \Gamma(z)]_{v+e}^{2t-1} = 0 \tag{5-60}$$

provides a set of $(2t - v - e)$ linear equations in v unknown coefficients of $\Lambda(z)$. This set of equations can always be solved if $2v + e \leq 2t$. Once the $\Lambda(z)$ polynomial is found by solving the key equation, then the polynomial $\Lambda(z)\Gamma(z) = \Psi(z)$ represents the combined erasure and error locator polynomial, and $\Omega(z)$ is the corresponding error-evaluator polynomial. Then the errors may be located via the Chien search technique discussed in Section 5.3, and in a similar fashion the erasure and error values may be calculated from

$$e_{i_m} = -\alpha^{i_m(1-m_0)} \frac{\Omega(\alpha^{-i_m})}{\Psi'(\alpha^{-i_m})} \tag{5-61}$$

Note also that if an erasure occurs in position i_m and the substituted symbol does not contain an error, then both $\Gamma(z)$ and $\Omega(z)$ contain $(1 - z\alpha^{i_m})$ as a factor. Error evaluation via (5-61) will give a zero value for this symbol.

Solution of the key equation may be obtained using either Euclid's or Berlekamp's algorithm. In applying Euclid's algorithm with a t-error-correcting BCH code one should use Forney's "modified" syndrome polynomial[48]

$$T(z) \equiv S(z)\Gamma(z) \bmod z^{2t} \tag{5-62}$$

The technique is summarized in Table 5-9.

It is easily verified that the set of initial conditions specified in step 2 produces a sequence of solutions to (5-59) [and each $g_i(z)$ contains $\Gamma(z)$ as a factor]. Thus,

$$g_i(z) = \Gamma(z)\Lambda_i(z) \tag{5-63}$$

In addition, the algorithm has properties similar to (5-39) in that

$$\deg[\Lambda_i(z)] + \deg[\Omega_{i-1}(z)] = 2t$$

and   (5-64)

$$\deg[\Lambda_i(z)] + \deg[\Omega_i(z)] < 2t$$

The desired solution is the polynomial $\Lambda_i(z)$ which has degree less than

Table 5-9. Solution of Key Equation with Errors and Erasures for T-Error-Correcting BCH Codes

---

1. Apply Euclid's algorithm to $a(z) = z^{2t}$ and $b(z) = T(z)$.
2. Use the initial conditions $g_{-1}(z) = 0$, $g_0(z) = \Gamma(z)$, $r_{-1}(z) = z^{2t}$, and $r_0(z) = T(z)$.
3. Stop when $\deg[r_n(z)] < t + e/2$ (e even) or when $\deg[r_n(z)] < t + [(e-1)/2]$ (e odd) where $e = \deg[\Gamma(z)]$.
4. Set $\Psi(z) = \Gamma(z)\Lambda(z) = g_n(z)$ and $\Omega(z) = r_n(z)$.

---

t_m where

$$t_m = \begin{cases} t - e/2, & e \text{ even,} \\ t - \dfrac{e+1}{2}, & e \text{ odd,} \end{cases} \qquad (5\text{-}65)$$

is the maximum number of correctable errors given the occurrence of e erasures. That the stopping rule of step 3 in Table 5-9 produces the desired solution can be shown in a straightforward manner.

Consider the case of e even. Then if $\deg[\Omega_{i-1}(z)] \geq t + e/2$ and thus $\deg[\Lambda_i(z)] \leq t - e/2 = t_m$ and $\deg[\Omega_i(z)] < t + e/2$, then $\deg[\Lambda_{i+1}(z)] > t - e/2 = t_m$. Thus, the solution at the ith step is the desired solution to the key equation. The case for e odd follows in a similar manner. In this case it turns out that one may use a stopping threshold of either $t + (e - 1)/2$ or $t + (e + 1)/2$ since the desired $\Omega_i(z)$ will have decreased in degree by at least 2 from the preceding polynomial.

Example. Again consider the block length 15, triple-error-correcting RS code with $r(x)$ given by

$$r(x) = \alpha^0 x + \alpha^1 x^2 + \alpha^7 x^3 + \alpha^{11} x^{10} \qquad (5\text{-}66)$$

where the positions corresponding to x and x^2 are identified as erasures. Note that $r(x)$ is identical to the previous $r(x)$ given in (5-42), but with the erasures added. Then

$$\Gamma(z) = \alpha^3 z^2 + \alpha^5 z + 1 \qquad (5\text{-}67)$$

$$S(z) = \alpha^3 z^5 + \alpha^0 z^4 + \alpha^5 z^3 + \alpha^{12} z^2 + \alpha^{13} z + \alpha^0 \qquad (5\text{-}68)$$

and

$$T(z) = \alpha^7 z^5 + \alpha^{10} z^4 + \alpha^{12} z^2 + \alpha^7 z + \alpha^0 \qquad (5\text{-}69)$$

Since $e = 2$, Euclid's algorithm should terminate when $\deg[\Omega_i(z)] < 4$. Application of Euclid's algorithm to solve the key equation (5-59) is shown in Table 5-10. The desired solution for the error-locator polynomial, $\Lambda(z)$,

Table 5-10. Solution of the Key Equation for $T(z)$ Given by (5-69)

i	$\Gamma(z)\,\Lambda_i(z)$	$\Omega_i(z)$	$q_i(z)$
-1	0	z^6	—
0	$\Gamma(z)$	$T(z)$	—
1	$\Gamma(z)[\alpha^8 z + \alpha^{11}]$	$\alpha^6 z^4 + \alpha^5 z^3 + \alpha^2 z^2 + \alpha^{13} z + \alpha^{11}$	$\alpha^8 z + \alpha^{11}$
2	$\Gamma(z)[\alpha^9 z^2 + \alpha^8 z + \alpha^{11}]$	$\alpha^2 z^3 + \alpha^{11} z^2 + \alpha^{13} z + \alpha^{11}$	$\alpha z + \alpha$

Table 5-11. Solution of Key Equation with Erasures and Errors via Berlekamp's Algorithm for t-Error-correcting BCH Codes

1. Compute $\Gamma(z)$ and $e = \deg[\Gamma(z)]$.
2. Use flow diagram of Fig. 5-8 replacing $\Lambda_n(z)$ by $\Psi_n(z)$.
3. Set initial conditions $n = L = e$, $\Psi_e(z) = \Gamma(z)$, and $D_e(z) = z\Gamma(z)$.
4. Proceed as before to calculate $\Psi(z)$.

is identical to that obtained previously. Thus, $\Lambda(z)$ has roots at α^{-3} and α^{-10} while the erasure-locator polynomial has roots at α^{-1} and α^{-2}. The error-evaluator polynomial is

$$\Omega(z) = \alpha^2 z^3 + \alpha^{11} z^2 + \alpha^{13} z + \alpha^{11}$$

while the derivative of $\Psi(z) = \Gamma(z)\Lambda(z)$ is

$$\Psi'(z) = \alpha^{10} z^2 + \alpha^{10}$$

Then using (5-61) the estimated error polynomial is $\hat{e}(x) = \alpha^0 x + \alpha^1 x^2 + \alpha^7 x^3 + \alpha^{11} x^{10}$, and the resulting decoder output is $c(x) = 0$.

The erasures-and-errors decoding problem can also be solved using Berlekamp's algorithm. One way involves utilizing the modified syndrome polynomial $T(z)$ as suggested by Berlekamp.[2] However, one can solve the problem without computing the polynomial $T(z)$.[37] Using this method, the only change to the Berlekamp algorithm of Fig. 5-8 is a change in initial conditions. The procedure for utilizing Berlekamp's algorithm is shown in Table 5-11. Since the solution $\Psi(z)$ must contain $\Gamma(z)$ as a factor, this polynomial is used to initialize the algorithm at step $n = e = \deg[\Gamma(z)]$. The algorithm then proceeds as before.

However, this process does not yield the error-evaluator polynomial $\Omega(z)$. Instead, this polynomial must be computed from $S(z)$ and $\Psi(z)$ using (5-59). Then the error values may be calculated from (5-61).

Table 5-12. Solution for the Key Equation for $\Gamma(z)$ and $S(z)$ Given by (5-67) and (5-68) via Berlekamp's Algorithm

n	S_n	$\Psi_n(z)$	d_n	L	$D_n(z)$
0	α^0				
1	α^{13}				
2	α^{12}	$1 + \alpha^5 z + \alpha^3 z^2$	α^{12}	2	$z + \alpha^5 z^2 + \alpha^3 z^3$
3	α^5	$1 + \alpha^{14} z + \alpha^6 z^2 + z^3$	α^9	3	$\alpha^3 z + \alpha^8 z^2 + \alpha^6 z^3$
4	α^0	$1 + \alpha^5 z + \alpha^3 z^2$	α^{10}	3	$\alpha^3 z^2 + \alpha^8 z^3 + \alpha^6 z^4$
5	α^3	$1 + \alpha^5 z + \alpha^8 z^2 + \alpha^3 z^3 + \alpha z^4$	α^7	4	$\alpha^5 z + \alpha^{10} z^2 + \alpha^8 z^3$
6		$1 + \alpha^{14} z + z^2 + \alpha^{14} z^3 + \alpha z^4$			

Example. Consider again the RS code example with $\Gamma(z)$ and $S(z)$ given by (5-67) and (5-68). Solution of the key equation via Berlekamp's algorithm is shown in Table 5-12. The erasure and error locator polynomial found has the same roots as that found using Euclid's algorithm. In addition, the reader can verify that the solutions for the error values at these locations are identical to those found previously.

5.7. Performance Results

Performance of BCH codes may be estimated in a simple manner using the performance estimation techniques of Section 1.3.1. Since code weight structure is known for only a small subset of BCH codes, we will assume no knowledge of the structure.

The standard BCH decoding algorithm is a bounded-distance algorithm. That is, no error patterns of more than t errors can be corrected. This property simplifies performance calculations. For a t-error-correcting BCH code of length n the sequence error probability is given by the probability that more than t errors occur in a code word, i.e.,

$$P_s = \sum_{i=t+1}^{n} \binom{n}{i} p^i (1-p)^{n-i} \tag{5-70}$$

where p is the symbol error rate on the channel of interest. The quantity P_s includes the contributions due both to undetected errors and to a failure to decode.

To upper bound bit error probability for binary BCH codes, we make the pessimistic assumption that a pattern of i channel errors $(i > t)$ will cause the decoded word to differ from the correct word in $i + t$ positions and thus a fraction $(i + t)/n$ of the k information symbols to be decoded erroneously. Thus,

$$P_b \leq \sum_{i=t+1}^{n} \frac{i+t}{n} \binom{n}{i} p^i (1-p)^{n-i} \tag{5-71}$$

Figure 5-11 shows this bound on P_b for a number of interesting BCH codes with $R \approx 1/2$ when utilizing binary PSK modulation $p = Q[(2RE_b/N_0)^{1/2}]$. Substantial coding gains can be obtained with a practical decoding algorithm. Coding gains (at $P_b = 10^{-5}$) of about 4 dB are available with block length 511. Longer codes providing slightly more gain can also be effectively utilized. In addition, there is significant flexibility available in terms of code rate. It has been shown that a relatively broad maximum of coding gain versus code rate for fixed n occurs roughly between $1/3 \leq R \leq 3/4$ for BCH codes.[6] Performance on a Gaussian

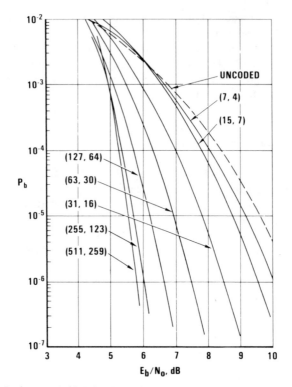

Figure 5-11. Performance of BCH codes with BPSK modulation (hard decision decoding).

channel degrades substantially at very high or very low rates. Thus, the curves of Fig. 5-11 are indicative of the performance that can be achieved at code rates near 1/2 at the block lengths shown.

Similar curves are shown in Fig. 5-12 for orthogonal signaling and noncoherent detection $[p = \frac{1}{2}\exp(-RE_b/2N_0)]$. One has inherently less gain available in a noncoherent system because of the difference in slope of the uncoded performance curve. As a result, only slightly more than 3 dB of gain is obtained at $P_b = 10^{-5}$. The codes shown in this figure are $R \approx 3/4$ since they are slightly more efficient than $R \approx 1/2$ codes in non-coherent systems.

Performance of the erasures-and-errors decoding algorithm presented in the previous section is also easily estimated. The algorithms presented in Tables 5-9 and 5-11 guarantee correction of all combinations of v errors and e erasures for which $2v + e \le 2t$. Any patterns which do not satisfy this constraint will result in either decoding failures or errors. Then assuming a channel error rate p and an erasure rate of s, the sequence error probability

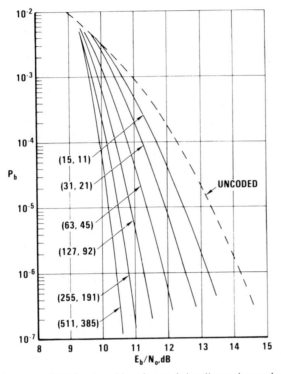

Figure 5-12. Performance of BCH codes with orthogonal signaling and noncoherent detection (hard decision decoding).

is

$$P_s = \sum_{i=0}^{n} \sum_{\substack{j = 2(t-i)+1 \\ (j \geq 0)}}^{n-i} \binom{n}{i,j} p^i s^j (1 - p - s)^{n-i-j} \tag{5-72}$$

where

$$\binom{n}{i,j} = \frac{n!}{i!\,j!\,(n - i - j)!} \tag{5-73}$$

Unfortunately, these decoding algorithms are not very effective on a Gaussian channel. For PSK modulation with optimum three-level quantization of the demodulator output, the performance improvement amounts to only about 0.2 dB relative to hard decision decoding. This is in contrast to an expected performance improvement of about 1 dB which is predicted by ensemble coding results.

An optimum algorithm for decoding erasures and errors would decode the received word into the code word that differs in the fewest number of positions outside the erased positions. This can be accomplished by replacing the e erased positions by all possible values in those positions and then decoding each of these cases (using the standard "errors only" algorithm). A total of q^e code words must be decoded. Then the resulting code word in this set that differs from the received word in the fewest positions outside the erased positions should be selected. This approach is a variant of the Chase algorithm. Note that the first algorithm discussed in Section 5.5 for binary codes tried only two decoding operations (replacing the erasures by either all zeros or all ones). This simpler algorithm will not correct all of the error and erasure patterns (with $2v + e > 2t$) that would be correctable using the Chase version of the algorithm. However, this algorithm is slightly better (by about 0.1 dB) than the algorithms of Tables 5-9 and 5-11 since it will correct some of the patterns with $2v + e > 2t$.

Thus, the purely algebraic algorithm for correcting errors and erasures appears to be of marginal utility on a Gaussian channel. However, there are other situations in which it could prove quite useful. One such situation is a channel with occasional strong interference which can be detected and treated as an erasure. In these cases the algorithms of Tables 5-9 and 5-11 would provide most, though not all, of the benefits of erasures and errors decoding.

All of the codes presented in the performance curves of Figs. 5-11 and 5-12 have primitive block lengths of the form $n_m = 2^m - 1$. Many system applications may impose a requirement on block length. Such a requirement can typically be satisfied by shortening a longer code. An (n_m, k_m) code is shortened by s positions by always setting the first s information digits to 0 and not transmitting them. This results in an (n, k) code with parameters $(n, k) = (n_m - s, k_m - s)$. The parameter s may take on any value in the range $1 \leq s \leq k_m - 1$. However, only a limited amount of code shortening can be done *efficiently*. The reason is that the number of parity digits in a t-error-correcting code with maximum length n_m is proportional to mt. This code can be efficiently shortened to $n = (n_m + 1)/2$, but at lengths less than this it is more efficient to shorten the t-error-correcting code with maximum length n_{m-1} since the latter code requires only about $(m - 1)t$ parity digits. Thus, the parameter m should be selected as the smallest integer such that $n < 2^m$ to maximize code efficiency.

5.8. Remarks

This chapter has been devoted principally to a discussion of the implementation and performance of BCH codes, the most widely known class of multiple-error-correcting block codes. Finite field Fourier transforms were utilized in the tutorial discussion because of the ease with which this approach can demonstrate the key properties of BCH codes. In addition, there are situations in which transform techniques provide a practical method for implementing the decoding operations.

The techniques discussed have simple implementations, even at very high data rates. This is due in large part to the remarkable ease with which the key equation can be solved via the Berlekamp algorithm. Solution of this equation using Euclid's algorithm was also presented because it is easier to understand and is closely related to the Berlekamp algorithm.

Substantial amounts of coding gain (4dB and greater) on a hard decision PSK channel are available with codes of moderate length. These gains are available at code rates in the range $1/3 \le R \le 3/4$. At very high or very low rates, the coding gain degrades substantially.

Unfortunately, the algebraic decoding techniques do not generalize easily to utilize soft decisions. The generalization of Section 5.5 to include the correction of erasures has marginal utility on a Gaussian channel, but it may be quite useful in other applications such as interference environments. It appears that, at present, soft decisions can only be effectively utilized via other techniques (e.g., the Chase algorithm) in combination with the standard hard decision BCH decoding algorithm. However, an excellent alternate approach is to use *concatenated codes*. Thus, an inner code that utilizes soft decision decoding can be combined with a BCH code (such as a Reed–Solomon code) that uses a hard decision algorithm. This approach is treated in detail in Chapter 8.

There are several problem areas where additional research is warranted. One, of course, is the soft decision decoding problem just discussed. Another is the problem of decoding beyond the BCH bound. The techniques presented in this chapter are bounded distance decoding algorithms which allow correction of all patterns of t or fewer errors for a BCH code with designed distance $2t + 1$. Algorithms which can correct error patterns with weight greater than t have been formulated but have been found to be impractical because of their complexity. Finally, the BCH codes are asymptotically poor in that if one holds R fixed, the ratio $t/n \to 0$ as $n \to \infty$. Recently, with the discovery of Goppa codes, it has been shown that there exist long Goppa codes whose actual minimum distance lies very close to the Gilbert bound and, hence, are much better than BCH codes.[49,50] Unfortunately, a constructive procedure for finding the good, long Goppa codes has not yet been found. These codes can be decoded out to the designed distance $2t + 1$ using essentially the BCH decoding algorithm.[45] However, the appeal of these codes is that their actual minimum distance is much larger than the designed distance. So far, no practical decoding algorithm has yet been found which takes advantage of this additional error correction capability.

Problems

5-1. Consider the double-error-correcting binary BCH code with block length 15 which has α, α^2, α^3, and α^4 as roots of its generator polynomial. Assume arithmetic operations over $GF(2^4)$ are defined by the primitive polynomial $\alpha^4 = \alpha + 1$. Draw block diagrams (to the level of individual modulo-2 adders) for syndrome computation registers to compute $S_0 = r(\alpha)$ and $S_2 = r(\alpha^3)$.

5-2. Continuing the previous example, draw the block diagram for the Chien search unit.

5-3. Let α be a primitive element of $GF(2^{11})$ of order 23. The minimum function of β, $m(x)$, can be taken as the generator polynomial of the (23, 12) Golay code which has $d = 7$. What is the error-correcting capability of this code as predicted by the BCH bound, and hence how many errors could be corrected using the standard BCH decoding algorithm.

5-4. Redraw the flow diagram of Fig. 5-8 for the special case of binary BCH codes to minimize decoding operations.

5-5. Consider a triple-error-correcting binary BCH code of length 15 with the received word

$$r(x) = x + x^5 + x^8$$

Find the error-locator polynomial, $\Lambda(z)$, using Euclid's algorithm. Complete the decoding process by finding the inverse roots of $\Lambda(z)$. Use $GF(2^4)$ as defined in Table 5-3.

5-6. Using the same code and a received word

$$r(x) = x^5 + x^{10} + x^{12}$$

find $\Lambda(z)$ using Berlekamp's algorithm. Find the error transform through recursive extension, and complete the decoding process via an inverse transform.

5-7. Perform erasures and errors decoding for the same binary BCH code. Suppose

$$r(x) = x$$

and the positions corresponding to x^4, x^6, x^{10}, and x^{11} are identified as erasures.

5-8. Find the generator polynomial of the double-error-correcting RS code of block length 15. Operations over $GF(2^4)$ are defined in Table 5-3.

5-9. For the same RS code and

$$r(x) = \alpha^{10}x^2 + \alpha x^{10}$$

find $\Lambda(x)$, $\Omega(z)$, and $\hat{e}(x)$.

5-10. For the triple-error-correcting RS code of length 15 we have

$$r(x) = \alpha^5 x^6$$

and the positions corresponding to x, x^5, x^9, and x^{13} are identified as erasures. Perform erasures and errors decoding using Euclid's algorithm to solve the key equation.

5-11. For the same RS code

$$r(x) = x^{10} + \alpha^{11}x^{12}$$

and the positions corresponding to x^2 and x^5 are identified as erasures. Perform erasures and errors decoding using the Berlekamp algorithm to solve the key equation.

5-12. Using the properties of finite-field Fourier transforms, (a) express the inverse transform of $zD(z)$ as a function of $d(x)$, the inverse transform of $D(z)$; (b) express the mth discrepancy

$$d_m = \sum_{i=0}^{L} \Lambda_i S_{m-i}$$

in terms of the received polynomial, $r(x)$, and the inverse transform of the error-locator polynomial, $\lambda(x)$, at the mth step of the Berlekamp algorithm.

Then using these results draw a flow diagram of the Berlekamp algorithm as in Fig. 5-8 to operate entirely in the time domain and to produce the inverse transform of the error-locator polynomial, $\lambda(x)$. How may the error locations be identified from this polynomial? How should the flow diagram be augmented to produce the inverse transform, $\omega(x)$, of the error-evaluator polynomial? Finally, how may e_{i_m} as given by (5-28) be evaluated using $\lambda(x)$ and $\omega(x)$? (The procedure outlined in this problem was suggested by R. E. Blahut in a talk "Transform Decoding Without Transforms" at the 1980 IEEE Communication Theory Workshop.)

6

Convolutional Code Structure and Viterbi Decoding

The codes and decoding techniques presented in the previous chapters apply only to block codes. With these codes the information sequence is segmented into blocks which are encoded independently. Thus, the coded sequence becomes a sequence of fixed-length independent code words. This is not done with convolutional codes. Instead, redundant symbols are generated as a function of a span of preceding information symbols. The transmitted sequence is actually a single "semi-infinite" code word.

The development of the convolutional coding theory has been quite different from that for block codes. With block codes algebraic properties have been very important in constructing good classes of codes and in developing decoding algorithms. This has not been the case with convolutional codes. Most good convolutional codes have been found by computerized searches of large numbers of codes to locate those with good distance properties. Also, there is no convolutional decoding algorithm similar to the algebraic techniques for multiple-error correction presented in the previous chapter. The maximum-likelihood decoding algorithm for convolutional codes (the Viterbi algorithm) is presented in this chapter, and the other important convolutional decoding techniques are treated in Chapter 7.

In recent years convolutional coding with Viterbi decoding has become one of the most widely used forward-error-correction techniques. This is due to both the simplicity of implementation and the relatively large coding gains that can be achieved. The achievement of such coding gains results principally from the ease with which this algorithm can utilize demodulator soft decisions and thereby provide approximately 2 dB more gain than the corresponding hard decision decoder. The Viterbi algorithm is itself suggested by the sequential-state machine nature of a convolutional encoder, and a detailed study of the algorithm provides significant insight into how the code properties influence performance. This crucial connection between the codes and the decoding algorithm will be established in this chapter. Using this as a base, some useful performance estimation techniques will be presented, and these in turn will be used to explore the practical problem of optimizing the hardware without undue sacrifice in performance. Finally, a set of performance curves will be provided for several situations of practical interest.

6.1. Binary Rate-1/2 Convolutional Codes

A constraint length k convolutional encoder consists of a k-stage shift register with the outputs of selected stages being added modulo-2 to form the encoded symbols. A simple example is the $R = 1/2$ convolutional encoder shown in Fig. 6-1. Information symbols are shifted in at the left, and for each information symbol the outputs of the modulo-2 adders provide two channel symbols. The connections between the shift register stages and the modulo-2 adders are conveniently described by generator polynomials. The polynomials $g_1(x) = 1 + x^2$ and $g_2(x) = 1 + x + x^2$ represent the upper and lower connections, respectively (the lowest-order coefficients represent the connections to the leftmost stages). The input information sequence can also be represented as a power series $I(x) = i_0 + i_1 x + i_2 x^2 + \cdots$, where i_j is the information symbol (0 or 1) at the jth symbol time. With this representation the outputs of the convolutional encoder can be described as a polynomial multiplication of the input sequence, $I(x)$, and the code generators. For example, the output of the upper encoded channel sequence of Fig. 6-1 can then be expressed

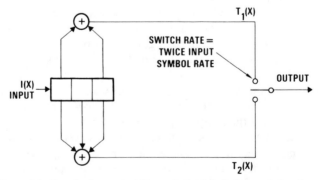

Figure 6-1. Encoder for a rate-1/2, constraint-length-3 convolutional code.

as $T_1(x) = I(x) g_1(x)$, where the polynomial multiplication is carried out in $GF(2)$.

The code used in this example is a *nonsystematic* code because neither $T_1(x)$ nor $T_2(x)$ is identically equal to the information sequence $I(x)$. Conversely, a *systematic* code would have either $g_1(x) = 1$ or $g_2(x) = 1$ so that the information sequence would appear directly in the output sequence. For reasons which will subsequently become clear, nonsystematic codes are usually preferred in systems which utilize Viterbi decoding.

As one might suspect from the name, the output of the encoder can also be thought of as the convolution of the impulse response of the encoder with the input sequence. For example, if a single one followed by all zeros $[I(x) = 1]$ is shifted into the encoder of Fig. 6-1, the resultant output sequence is $1\,1\,01\,1\,1\,0\,0\,0\,0\ldots$. The reader may verify that the response to any arbitrary input sequence may be produced by adding modulo-2 appropriately shifted versions of this sequence. This suggests that a generator matrix, $\mathbf{G}$, can be constructed as

$$\mathbf{G} = \begin{bmatrix} 11\ 01\ 11\ 00\ 00\ 00\ \ldots \\ 00\ 11\ 01\ 11\ 00\ 00\ 00\ \ldots \\ 00\ 00\ 11\ 01\ 11\ 00\ 00\ 00\ldots \\ \vdots \end{bmatrix} \qquad (6\text{-}1)$$

In contrast with block codes, convolutional codes do not need to have the information sequence in blocked form, i.e., the code words may have infinite length. Thus, the generator matrix as shown in (6-1) is sometimes said to be a *semi-infinite* matrix. The code word corresponding any input sequence $\mathbf{X}$ may be found by multiplying the input vector by $\mathbf{G}$, i.e.,

$$\mathbf{Y} = \mathbf{X}^T \mathbf{G} \qquad (6\text{-}2)$$

Thus, the output sequence corresponding to $\mathbf{X}^T = 1\,1\,1\,0\,0\,0\ldots$ is obtained by adding rows 1, 2, and 3 of $\mathbf{G}$ to give $\mathbf{Y}^T = 11\ 10\ 01\ 10\ 11\ 00\ 00\ldots$.

Another convenient method of describing the relationship between input and output sequences is a *code tree* such as shown in Fig. 6-2. Each branch of the tree represents a single input symbol. The convention used is that an input 0 corresponds to the upper branch and an input 1 corresponds to the lower branch. Thus, any input sequence traces out a particular path through the tree. Specifically, a 10110 input sequence causes a 11 01 00 10 10 output sequence. This path is indicated in Fig. 6-2. Of course, as the input sequence grows in length, the number of possible paths grow exponentially, thus limiting the role of such a tree diagram to a conceptual one.

Fortunately, the exponential growth of the code tree can be "tamed"

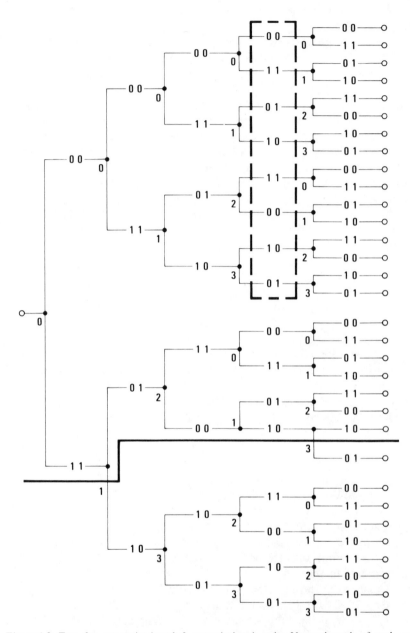

Figure 6-2. Tree for constraint-length-3 convolutional code. Upper branch of each pair corresponds to an input polarity of zero, lower branch corresponds to an input polarity of one.

somewhat by observing that the convolutional code structure is repetitive. In Fig. 6-2 each node of the tree is labeled with a number from 0 to 3 corresponding to the contents of the two left-most positions of the encoder register at that point in the tree (the lsb is the left-most position). This number is called the *state* of the encoder. Note that the branches of the tree emanating from any two nodes in the same state are identical. For example, at level 3 in the tree the upper half tree and the lower half tree are identical from that point on and may be *merged*. Similarly, at level 4 there are four sets of four nodes that may be merged. This ability to merge paths that enter the same state provides the key to the Viterbi algorithm.

6.2. Viterbi Decoding Algorithm

Figure 6-3 represents an alternate way of viewing the code tree of Fig. 6-2 and has been called a *trellis structure* by Forney.[51] As in Fig. 6-2, the convention is that an input 0 corresponds to the selection of the upper branch and an input 1 to the selection of the lower branch. As before, each possible input sequence corresponds to a particular path through the trellis. For instance, an input sequence of 1 0 1 1 0 can be seen to provide an output sequence of 1 1 0 1 0 0 1 0 1 0, which is identical to the result that was obtained from Fig. 6-2.

The problem of decoding a convolutional code can be thought of as attempting to find a path through this trellis diagram by making use of some decoding rule. Ideally, one would like to choose a path which minimizes the actual number of symbol errors that would be made if one were to compare the encoder input with the decoder output. Although it is possible to conceive of a decoding algorithm based on this approach, attempts to do so have not resulted in hardware that can be easily

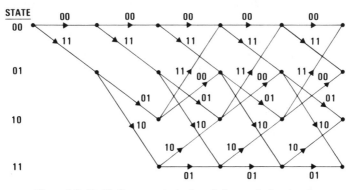

Figure 6-3. Trellis for a constraint-length-3 convolutional code.

implemented. As with block codes, it turns out to be more practical to attempt to choose the path which matches the received sequence as closely as possible; that is, to attempt to minimize the sequence error probability. While this procedure does not guarantee that the actual bit error rate will be minimized, one is guaranteed that making the sequence error rate small will also make the bit error rate small for all but pathologically bad codes. Any performance difficulties that result from this nonoptimality are of no practical importance.

The significance of the trellis viewpoint is that the number of nodes in the trellis does not continue to grow as the number of input symbols increases, but remains at 2^{k-1}, where k is the constraint length of the code (the number of shift register stages necessary in the coder). This, of course, is because the redundant portions of the code tree have been merged. A consequence of this merging is that if at some point an incorrect path is chosen, it is possible (and for good codes highly probable) that the correct path will merge with it at a later time.

Since the number of paths grows exponentially with sequence length, it would appear at first to be a hopelessly impossible task to actually implement a true maximum-likelihood sequence estimator for a convolutional code. Surprisingly, however, it is not only possible but is also quite simple. One readily discovers this process by simply taking the brute force approach and attempting to compute a metric for every possible path through the trellis. At first the number of paths does indeed grow exponentially with sequence length. Quickly, however, it becomes possible to discard a number of paths at every node that exactly balances the number of new paths that are created. Thus, it is possible to maintain a relatively small list of paths that are always guaranteed to contain the maximum-likelihood choice. This simple iterative process is known as the Viterbi algorithm. Although it was discovered independently and first applied to convolutional codes by Viterbi[52] in 1967, it was at that time a well-known technique in operations research[53] and is also a special case of forward dynamic programming.[54]

The Viterbi algorithm is best explained using the trellis diagram for the $R = 1/2$, $k = 3$ code as a reference. First note that at level 3 for any node in the trellis there are two paths entering that node. Since those two paths will be identical from that point on, a maximum-likelihood decoder may make a decision at that point with no loss in performance. Once this is done, the same process may be repeated at level 4, etc. This is exactly what the Viterbi algorithm does. It compares the two paths entering each node, and only the path with the best metric is retained. The other path is discarded since its likelihood can never exceed that of the path which is retained no matter what data are subsequently received. The retained paths are called *survivors*. For this $k = 3$ code no more than four survivor paths will ever be retained.

Example. As an example of how the Viterbi algorithm corrects a specific error pattern, consider the following situation. First, assume that the demodulator provides only hard decisions. In this case the Hamming distance between the received word and the code words can be used as a metric. Suppose that the all-zero sequence is transmitted and the received word is 1 0 00 1 0 00 00.... Since the code in our example has a minimum distance of 5, all double-error patterns are correctable. A series of incomplete trellis diagrams appears in Fig. 6-4 that illustrate the status of the decoder after each received branch is processed. The numbers attached to each node indicate the Hamming distance accumulated by the surviving path at that node.

At level 2 in the trellis only one of the two channel errors has occurred and the surviving paths with the respective metrics are as shown in Fig. 6-4a. At level 3 (shown in Fig. 6-4b) each of the paths at level 2 is extended in two directions (making a total of eight paths—two at each state). The metrics for the two paths entering each node are then compared, and only the best path is retained (again giving a total of four paths). This process is continued each time a new branch is received. Note that at level 5 (shown in Fig. 6-4c) the all-zero path has a metric that is better than that of any of the other paths. Since no more errors will occur, it is now obvious that the correct path will eventually be selected. From this example it is also obvious that the surviving paths can disagree over long spans. However, at level 10 (shown in Fig. 6-4e) the surviving paths merge into a single path over the first eight branches. In this way the Viterbi algorithm inherently makes symbol decisions, i.e., all survivor paths tend to lead back to a unique node (and hence to a unique information symbol).

The depth at which this merging takes place cannot be guaranteed but is a random variable which depends upon the severity of the error pattern. Hence, in a practical implementation one does not rely on this mechanism but rather establishes a fixed decoding depth. Each time a new branch is processed the oldest symbol in one of the survivor sequences is released from the decoder. The optimum technique is to select the sequence with the best path metric. Other techniques are sometimes used, however, to reduce complexity.

Suppose now that an uncorrectable error pattern occurs, e.g., 1 1 01 00 00 .... Several incomplete trellis diagrams for this case are shown in Fig. 6-5. Note that by level 3 the correct path has been eliminated so an error will definitely occur. In addition, all surviving paths agree on the first branch of the path. Then at level 5 all surviving paths agree on the first two branches. Finally, at level 11 all surviving paths agree in the first 9 branches. The interesting point is that although an error has occurred, the erroneous path selected disagrees with the correct path over a rather short span (only three branches). Actually the information sequence corresponding to this path is 1 0 0 0 0 0 ... so only a single symbol error is made.

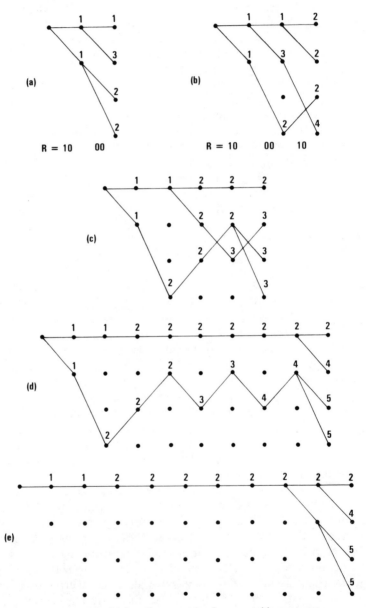

Figure 6-4. Viterbi decoding example of a correctable error pattern.

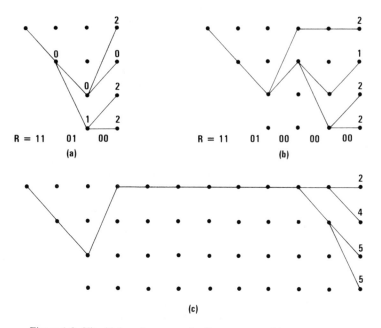

Figure 6-5. Viterbi decoding example of an uncorrectable error pattern.

This demonstrates a characteristic of typical error sequences of convolutional codes when decoded by a Viterbi decoder. That is, error events typically cover a rather short span (less than several constraint lengths) and result in a short burst of symbol errors.

Soft decision decoding is a trivial modification of the procedure discussed above. The Hamming metric is simply replaced by a soft decision metric (see Section 1.3.4). All other decoding operations remain the same. Thus, as one might expect, the implementation complexity of a soft decision decoder is not significantly different from that of a hard decision decoder. This is a major advantage of the Viterbi decoding algorithm.

The most significant implementation problem with Viterbi decoding is that the decoder complexity increases exponentially with code constraint length k, i.e., the number of states is 2^{k-1}. This limits the values of k to be relatively short, say $k \leq 10$, and makes it necessary to optimize the codes which are used.

6.3. Rate-*m/n* Convolutional Codes

Generalizations to code rates other than $1/2$ are very straightforward. The $R = 1/n$ case is the simplest since the same encoder structure is retained

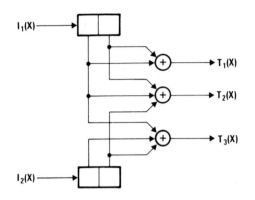

Figure 6-6. Encoder for a rate-2/3, $k =$ 4 code.

but the two modulo-2 adders are now replaced by n modulo-2 adders to form n channel symbols for each information symbol shifted into the encoder. The only change to the trellis structure is that each branch now has n channel symbols rather than 2. The changes in the Viterbi algorithm are rather trivial. Now the branch metrics must simply account for the n symbols per branch. The other operations remain unchanged.

The situation is more cumbersome for $R = m/n$ codes with $m > 1$. A typical case is shown in Fig. 6-6 for a $R = 2/3$, $k = 4$ code. For each branch two symbols are shifted into the encoder [one on $I_1(x)$ and one on $I_2(x)$], and three channel symbols are calculated by the modulo-2 adders [one each on $T_1(x)$, $T_2(x)$, and $T_3(x)$]. The code is described by six generator polynomials which describe the six sets of connections between the two registers and the three modulo-2 adders. The input–output relationships may be described in matrix form as

$$[T_1(x) \; T_2(x) \; T_3(x)] = [I_1(x) \; I_2(x)] \begin{bmatrix} g_{11}(x) & g_{12}(x) & g_{13}(x) \\ g_{21}(x) & g_{22}(x) & g_{23}(x) \end{bmatrix} \quad (6\text{-}3)$$

For this example the generator polynomial matrix is

$$\mathbf{G}(x) = \begin{bmatrix} 1 + x & 1 + x & 1 \\ 0 & x & 1 + x \end{bmatrix}$$

The trellis structure for this code is shown in Fig. 6-7. Note that there are still four states as there were in the $R = 1/2$, $k = 3$ case, but now there are four paths entering each state rather than two. The operation of the Viterbi algorithm is the same except that at each node the single best path out of the four possibilities must be selected, i.e., a 4-ary rather than a 2-ary comparison must be made. In general, for a $R = m/n$ code a 2^m-ary comparison must be made. It turns out that this is a fairly serious imple-

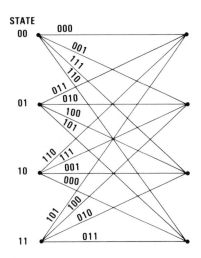

Figure 6-7. Trellis structure for a rate-2/3, $k = 4$ code.

mentation difficulty, particularly at high data rates (several Mb/s). Fortunately, one can select certain codes that eliminate this difficulty.

The key to simplifying the algorithm is to take a $R = 1/n$ code and *puncture* or delete certain channel symbols[55] thereby producing a $R = m/n$ code. Now consider the $R = 1/2, k = 3$ code. Suppose every fourth encoder output is deleted. The new code produces three channel symbols for every two information symbols, i.e., it is a $R = 2/3$ code. In fact, if the symbol from the generator $1 + x^2$ is deleted from every other branch, the resulting code is identical to the $R = 2/3$ code in the previous example. This code has the trellis shown in Fig. 6-8 where X indicates the deleted symbols. Note that the transitions between states and the resulting transmitted

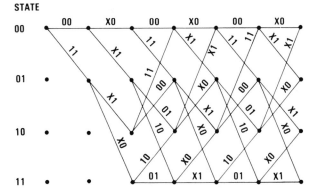

Figure 6-8. Trellis diagram of a rate-2/3, $k = 4$ code produced by periodically deleting symbols from a rate-1/2, $k = 3$ code.

symbols are identical in Figs. 6-7 and 6-8, but in Fig. 6-8 the transition is through a set of intermediate states since only one symbol at a time is shifted into the encoder rather than two. Obviously, the same code has been generated in a different manner. Punctured codes have been shown to offer nearly equivalent performance compared to the best previously known codes and thus are to be preferred in most cases because of their simplicity.[55] For example, in this case the direct method requires four 4-ary comparisons (each equivalent to three 2-ary comparisons) for each three channel symbols while the punctured code approach uses two levels of four 2-ary comparisons. Hence, there is a 3:2 advantage with the punctured code approach.

Note that the $R = 1/2, k = 3$ and $R = 2/3, k = 4$ codes have the same number of states. Often another designation for code constraint length is used. The constraint length v is defined to be the base-2 logarithm of the number of states. Thus, the number of states will always be 2^v. For rate-$1/n$ codes, v is always one less than the number of storage elements. For rate-m/n codes it may be even less depending on how the encoder is implemented. The two examples that have been used ($R = 1/2$ and $2/3$) then both have $v = 2$. This designation will be used subsequently to facilitate comparisons among codes with different rates.

6.4. Finite-State Machine Description and Distance Properties of Convolutional Codes

A convolutional encoder is a finite-state machine, and the preceding discussion demonstrated that the finite number of encoder states plays a key role in the Viterbi algorithm. The state of a convolutional encoder is determined by the most recent v information symbols shifted into the encoder shift register. The state is assigned a number from 0 to $2^v - 1$, which is usually obtained from the binary representation of the most recent v information symbols (with the most recent symbol being the least significant bit). The current state and output of the encoder are always uniquely determined by the previous state and current input. A complete description of the encoder as far as input and resulting output are concerned can be obtained from a state diagram as shown in Fig. 6-9 for the $R = 1/2$, $v = 2$ ($k = 3$) code. The symbol along the top of the transition line indicates the information symbol at the encoder input that causes the transition and the symbols below the transition line show the resulting channel symbols at the encoder output. Any sequence of information symbols dictates a path through the state diagram, and the channel symbols encountered along this path constitute the resulting encoded channel sequence.

Because of the topological equivalence between the state diagram

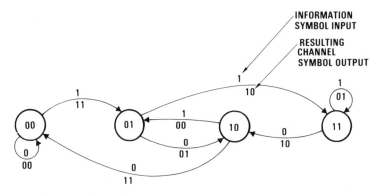

Figure 6-9. State diagram for the $R = 1/2$, $v = 2$ convolutional code.

of Fig. 6-9 and a signal flow graph (Mason and Zimmerman[56]), the properties and theory of signal flow graphs have been applied to the study of convolutional code structure and performance.[57, 58] Any quantity that can be accumulated linearly on a per branch basis can be computed using a signal flow graph by writing that quantity as the exponent of some indeterminate and letting the result be the gain of the branch. The gain of the overall graph between two states of interest will then provide a generating function for the quantity under consideration. Thus, generating functions for such quantities as weight, length, transitions. etc., can be computed using the signal flow graph. Assuming the all-zero transmitted sequence, the paths of interest are those which start and end in the zero state and do not return to the zero state anywhere in between. For each of these paths it is often desirable to determine the length of the input sequence, the weight of the input sequence, and the weight of the output sequence. Such a determination can be made using signal flow graphs as illustrated in the following example.

Let L be the indeterminate associated with the length of the input sequence, N be the indeterminate associated with the weight of the input sequence, and D be the indeterminate associated with the weight of the output sequence. Each branch in the flow graph will have a gain, g, given by $LN^{w_i}D^{w_o}$, where w_i is the weight of the input required to drive the coder between the two nodes connected by the branch and w_o is the weight of the output associated with the branch. Now consider the $R = 1/2$, $v = 2$ code whose state diagram is shown in Fig. 6-9. Since the only paths of interest are those paths that start and end in state zero and do not go through zero in between, the self loop on the zero state is removed and the state is split into an input and an output. The resulting flow graph for this code is shown in Fig. 6-10. When the gain of this graph is computed using

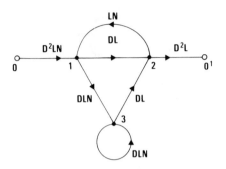

Figure 6-10. Signal flow graph for the $R = 1/2$, $v = 2$ convolutional code.

standard methods (Mason and Zimmerman[56]), the generating function

$$T(D, L, N) = \frac{D^5 L^3 N}{1 - DL(1 + L)N} \tag{6-4}$$

is obtained. When expanded by long division,

$$
\begin{aligned}
T(D, L, N) = D^5 L^3 N + D^6 L^4 (1 + L) N^2 \\
+ D^7 L^5 (1 + L)^2 N^3 + \cdots \\
+ D^{5+k} L^{3+k} (1 + L)^k N^{1+k} + \cdots
\end{aligned}
$$

Thus, there is one path of input weight 1, output weight 5, and length 3. There are two paths of input weight 2, output weight 6, and lengths 4 and 5. There are four paths of input weight 3 and output weight 7. One path has length 5, two have length 6, and one has length 7. This information is the weight structure of the code, which in turn determines the code performance in exactly the same fashion as the weight structure of a block code. The weight of the minimum-weight path which begins and ends in state zero is called the *free distance*, d_f (e.g., $d_f = 5$ for this $R = 1/2$, $v = 2$ code).

There is an alternate technique for evaluating $T(D, L, N)$ that is sometimes useful in performance estimation. This involves writing the state equations, i.e., for this example,

$$x_1 = D^2 L N + L N x_2$$

$$x_2 = D L x_1 + D L x_3$$

$$x_3 = D L N x_1 + D L N x_3$$

$$x_4 = D^2 L x_2$$

where x_i represents the value of the accumulated path gain from state 0 to state i as influenced by all other states. As in Fig. 6-10, state 0 is separated into an originating state and a terminating state with the terminating state, $0'$, represented as 2^v. These equations can be written in matrix form as

$$x = Ax + x^0 \qquad (6\text{-}5)$$

where

$$x^T = (x_1, x_2, x_3, x_4)$$

and

$$x^{0T} = (D^2LN, 0, 0, 0)$$

The state transition matrix for this example is given by

$$A = \begin{bmatrix} 0 & LN & 0 & 0 \\ DL & 0 & DL & 0 \\ DLN & 0 & DLN & 0 \\ 0 & D^2L & 0 & 0 \end{bmatrix} \qquad (6\text{-}6)$$

In general, this matrix is a $2^v \times 2^v$ matrix.

The formal solution to these equations can be written as

$$x = (I - A)^{-1} x_0 \qquad (6\text{-}7)$$

or as the matrix power series

$$x = [I + A + A^2 + A^3 + \cdots] x_0$$

Since the paths in state $0'$ are represented by x_{2^v}, $T(D, L, N)$ is equal to x_{2^v}. The power series solution is intuitively satisfying if we think of the state diagram as being drawn as a semi-infinite trellis and interpret each successive multiplication by A as the change in the state vector that is incurred by progressing one level deeper in the trellis. It also provides an algorithm for the rapid computation of $T(D, L, N)$ when the number of states is large. One should note that multiplication by A is quite simple since most of the entries in A will be zero. For a small number of states one may compute the matrix inversion in (6-7) in the standard way. For example, if A is given by (6-6), $(I - A)$ can be inverted to yield the $T(D, L, N)$ as given in (6-4). It can be shown[59] that this procedure will always yield the correct results except for codes that permit catastrophic error propagation. The conditions which exclude these codes are discussed later in this chapter. A sufficient test, however, is that the determinant $|I - A|$ not be zero.

Now, consider the function obtained by setting $L = 1$ and $N = 1$ in $T(D, L, N)$. The resulting function $T(D)$ is a generating function such that the coefficient, n_j, of D^j is equal to the number of code paths of weight j, i.e.,

$$T(D) = \sum_{j=0}^{\infty} n_j D^j \tag{6-8}$$

For the $R = 1/2$, $v = 2$ code

$$T(D) = \sum_{j=5}^{\infty} 2^{j-5} D^j$$

so that there is one path of weight 5, two of weight 6, four of weight 7, etc. This information can be used to estimate the probability of event error in a manner to be discussed in the next section. Another interesting quantity is the total information weight, w_j, of all paths of weight j. This can be found as the coefficient of D^j in the function

$$\left.\frac{\partial T(D, N)}{\partial N}\right|_{N=1} = \sum_{j=0}^{\infty} w_j D^j \tag{6-9}$$

For the $R = 1/2$, $v = 2$ code this becomes

$$\left.\frac{\partial T(D, N)}{\partial N}\right|_{N=1} = \sum_{j=5}^{\infty} (j - 4) 2^{j-5} D^j$$

Thus, the total information weight of weight 5 paths is 1, of weight 6 paths is 4, of weight 7 paths is 12, etc. This information is useful in estimating the probability of bit error in a manner that will also be discussed in the next section.

The state diagram generating function approach is not the only technique for determining the weight structure of a code. One can perform a search through the code trellis to tabulate the n_j and w_j. There are several acceptable techniques for accomplishing this, and some of the other decoding algorithms for convolutional codes can be adapted to this purpose.

6.5. Performance of Convolutional Codes with Viterbi Decoding

The most useful techniques for estimating the performance of convolutional coding are union bounds and computer simulation. The usefulness of computer simulation is limited by the long computation times that are required to get a good statistical sample (it may take several hours to get a single point). Consequently, the greatest usefulness of this technique is in special coding applications where union bounds cannot be used to provide a good performance estimate. The union bound for block codes was introduced in Chapter 1 as a technique that is computationally efficient. The union bound approach for convolutional codes is virtually identical, and it provides performance estimates accurate to within a small fraction of a decibel for all signal-to-noise ratios large enough to give an error rate of 10^{-3} or less. In addition, this approach may also be used to predict performance for some specific hardware configurations (used in implementing the decoder). The performance measures considered are first-event error probability, P_e, and the more widely used bit error probability, P_b.

6.5.1. Union Bounds

Before proceeding with the union bound, consider the case of comparing the correct path with another path that differs from it in j positions. Denote the probability of error in this comparison by P_j. For the binary symmetric channel (BSC) with channel symbol error rate p, the probability of error in this comparison is simply the probability that more than $j/2$ errors occur plus $1/2$ the probability that $j/2$ errors occur, as given by $\Pr(B_j'' | A_0)$ in (1-30).

Since the codes considered are linear codes, the all-zero path may be assumed to be the transmitted path without loss of generality. A first-event error is made at the ith received branch if the all-zero path is eliminated at this point by another path merging with it. The probability, P_j, that the all-zero path is eliminated by a path of weight j merging with it is dependent only on the weight of that path. The number of possible paths of weight j merging with the all-zero path is n_j, and the total information weight of these paths is w_j. (Note that the weight distribution of paths entering state zero is identical to that for paths leaving state zero.)

A union bound on the probability of a first-event error, P_e, at branch i may be obtained by summing the error probabilities for all possible paths which merge with the all-zero path at this point. This overbound is then simply

$$P_e < \sum_{j=0}^{\infty} n_j P_j \qquad (6\text{-}10)$$

A union bound on the probability of bit error, P_b, may be obtained from (6-10) by weighting each term by the corresponding number of bit errors (the information weight for each path). However, for a $R = m/n$ code there are m symbols decoded on each branch. Thus, P_b is bounded by

$$P_b < \frac{1}{m} \sum_{j=0}^{\infty} w_j P_j \qquad (6\text{-}11)$$

The validity of the second equation, (6-11), is not immediately obvious. However, this can be established by considering the example of Fig. 6-11. Whenever an error event occurs, then one or more symbol errors will occur. Suppose an event error occurs at step k, and the correct path is eliminated by path $\mathbf{x}'$. Now the comparison at the next step will not be between $\mathbf{x}''$ and the correct path but between $\mathbf{x}''$ and $\mathbf{x}'$. At this comparison the probability that $\mathbf{x}''$ is the survivor is upper-bounded by the probability of first-event error because the correct path has been replaced by a path with a larger metric. To compute an upper bound on the expected number of symbol errors in N steps one can sum the probabilities of error of all paths multiplied by the number of symbol errors per path and then summed for each of the N steps. But note that N information symbols will be decoded so the average bit error rate is the sum of N identical terms divided by N, which gives (6-11).

The evaluation of (6-10) and (6-11) requires a knowledge of the path weight distribution of the code, which can be obtained using the generator functions discussed in Section 6.4. One can also perform various manipulations with the generator functions to provide approximate closed-form expressions for bounds on P_e and P_b. We first observe that (6-8) and (6-10) would be identical if the term P_j could be written in the form a^j. This turns out to be possible for a few channels of practical interest. For example, it can be shown [58] that for the BSC, P_j is over-bounded by

$$P_j < \{2[p(1-p)]^{1/2}\}^j \qquad (6\text{-}12)$$

Thus, from (6-8), (6-10), and (6-12) one may write the bound on first event

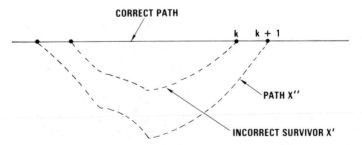

Figure 6-11. Typical error event with Viterbi decoding.

error as

$$P_e < T(D, N)|_{N = 1, D = 2[p(1 - p)]^{1/2}} \tag{6-13}$$

which for the $R = 1/2$, $v = 2$ code is

$$P_e < \frac{\{2[p(1 - p)]^{1/2}\}^5}{1 - 4[p(1 - p)]^{1/2}}$$

This technique is quite useful when using the matrix approach for finding $T(D, L, N)$. Note that (6-13) can be evaluated by performing the matrix inversion indicated by (6-7). This inversion is straightforward since after substituting $N = 1$ and $D = 2[p(1 - p)]^{1/2}$, it becomes a simple numerical matrix inversion. Alternately, one can approximate this inversion by calculating the first few terms of the series $\mathbf{I} + \mathbf{A} + \mathbf{A}^2 + \cdots$. It turns out that this is a rapidly convergent series (less than 100 terms are needed for almost any code of interest). In a similar fashion, one can write the bit error probability as

$$P_b < \frac{1}{m} \frac{\partial T(D, N)}{\partial N}\bigg|_{N = 1, D = 2[p(1 - p)]^{1/2}} \tag{6-14}$$

which for the $R = 1/2$, $v = 2$ code is

$$P_b < \frac{\{2[p(1 - p)]^{1/2}\}^5}{\{1 - 4[p(1 - p)]^{1/2}\}^2}$$

Again the matrix inversion technique can be used in calculating (6-14). The derivative at $N = 1$ can be computed numerically by using the normalized first difference, i.e.,

$$\frac{\partial T(D, N)}{\partial N}\bigg|_{N = 1, D = 2[p(1 - p)]^{1/2}}$$

$$\approx \frac{T(2[p(1 - p)]^{1/2}, 1 + \delta) - T(2[p(1 - p)]^{1/2}, 1)}{\delta} \qquad \delta \ll 1$$

Thus, one can evaluate the union bound without first knowing the code weight structure since the matrix inversion effectively accomplishes this computation. A union bound which is valid for demodulator soft decisions may be evaluated in a similar fashion (See Problem 6-5). In specific cases one may also improve the accuracy of the calculation by choosing

$$P_j \approx (ap^{1/2})^j$$

and letting a be adjusted so that this accurately approximates P_j for the range of j that is important for the particular code.

6.5.2. Optimum Codes

For most channels of practical interest the values of P_j in (6-11) obey the relationship $P_{j+1} < P_j$. Further, at high signal-to-noise ratios P_{j+1}/P_j tends to zero so that (6-11) is dominated by the first nonzero term. Thus, good codes are ones which have as large a free distance as possible and for which the w_j corresponding to the minimum distance term is as small as possible. At high signal-to-noise ratios these codes are optimum. At other signal-to-noise ratios the performance depends upon several of the terms in (6-11) so that this optimality is not necessarily preserved. For situations of practical interest the differences are slight.

A considerable amount of effort has been devoted to finding and tabulating optimum codes using exhaustive searches. For the shorter constraint lengths both the free distance and the multiplier, w_j, are usually used as criteria. For longer constraint-length codes, the free distance alone is used in order to reduce the amount of computation that is required.

Nonsystematic codes are preferred over systematic codes because a larger free distance can be achieved for a given value of v. This translates directly into a performance advantage which increases with code rate. Since systematic codes are just as difficult to decode as nonsystematic codes, they are rarely, if ever, used in Viterbi decoders.

In selecting good codes, the so-called *catastrophic error-propagating* codes should be avoided. In cases where catastrophic error propagation occurs there exist certain infinite Hamming weight input sequences that produce code sequences which have a finite Hamming weight. Thus, a small number of errors occurring in the communication channel can transform a particular code word into a second code word for which the corresponding input sequence differs in an infinite number of positions. This effect is discussed in considerable detail by Massey and Sain.[60] For rate $1/n$ codes one must avoid generators which have a common factor. For rate m/n codes one avoids generators such that the determinants of the $\binom{n}{m}$ distinct m by m submatrices of the m by n generator matrix, $G(x)$, do not have a common factor. As an example, consider the $R = 1/2$, $k = 3$ code with $g_1(x) = 1 + x^2$ and $g_2(x) = 1 + x$. The code generators have the common factor $1 + x$. Note that an all-zero input sequence gives the output sequence 00 00 00 ... while an all-ones input sequence gives the output sequence 11 10 00 00 00 ... . Then two channel errors in the first three positions will result in an infinite number of information symbol errors, obviously a catastrophic failure.

Tables of the best code generators for short-constraint-length codes are given in Appendix B. Tables B-1 and B-2 provide the optimum generators and the significant terms of the code weight structure for $R = 1/3$ and $1/2$. Similarly, punctured code generators for $R = 2/3$ and $3/4$ are given in Tables B-3, B-4, and B-5. The weight structures given allow accurate computation of performance.

One might wonder how efficient these short-constraint-length codes are. Note from Tables B-1 through B-5 that some rather large minimum free distances are achieved with these codes. Plotkin-type upper bounds on free distance have also been derived for these codes.[61] In almost all cases the free distances achieved by these codes are equal to the upper bounds, indicating that the codes are indeed very efficient.

6.5.3. Performance with Coherent PSK

If coherent PSK is used on a Gaussian channel with infinite quantization of the demodulator output, then the probability of error in comparing two sequences that differ in j positions is given by (1-39). This value of P_j has been used to compute the performance of the optimum codes given in Appendix B. In performing this computation the first five nonzero values of w_j were computed for each of the codes and then used in a truncated version of (6-11). The results of this calculation are shown in Figs. 6-12 through 6-15. These performance curves are for code rates of $1/3$, $1/2$, $2/3$, and $3/4$ and include the range of constraint lengths which can be practically implemented. When eight-level quantization is used rather than infinite quantization, the curves degrade by about 0.25 dB. For binary quantization (hard decisions) the degradation is approximately 2.2 dB. Note that the achievable coding gains are quite significant. These predicted gains are very real and can be achieved by a practical decoder to within a few tenths of a decibel (which is usually sacrificed to obtain a more efficient design). Note also that, as expected, the coding gain for a specific code increases as the desired P_b decreases.

Increasing the code rate at a constant value of v will require slightly more E_b/N_0. These curves indicate that, for these code rates, the additional E_b/N_0 required is rather consistent and is roughly 0.4 dB to change to the next higher code rate (maintaining constant v). Decoders with the same v for each of these code rates have roughly the same complexity. Finally, as v is increased the coding gain increases by 0.3 to 0.4 dB for each increase of 1 in v at moderate values of v. However, at high data rates (requiring a parallel implementation), this increase is typically accompanied by significant increases in decoder complexity.

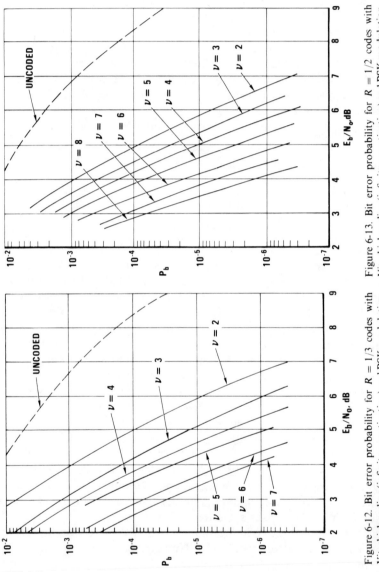

Figure 6-13. Bit error probability for $R = 1/2$ codes with Viterbi decoding (infinite quantization) and PSK modulation.

Figure 6-12. Bit error probability for $R = 1/3$ codes with Viterbi decoding (infinite quantization) and PSK modulation.

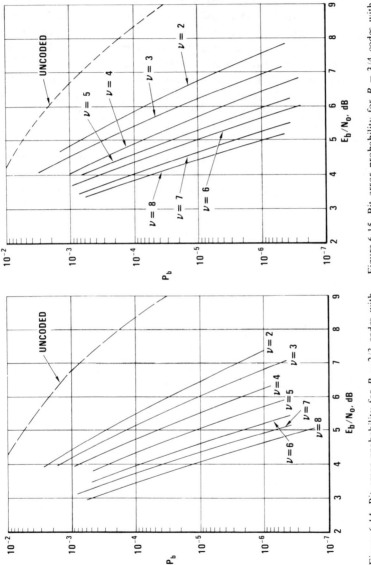

Figure 6-15. Bit error probability for $R = 3/4$ codes with Viterbi decoding (infinite quantization) and PSK modulation.

Figure 6-14. Bit error probability for $R = 2/3$ codes with Viterbi decoding (infinite quantization) and PSK modulation.

6.5.4. Performance with Orthogonal Signaling and Noncoherent Detection

Another interesting case is that of orthogonal signaling with non-coherent detection. Performance with convolutional coding may also be calculated using the union bound technique. However, the calculation is more tedious than with coherent PSK. The difference is in computing the error term, P_j, for a weight j path. Assuming noncoherent detection with square-law detectors, the probability of error in comparing a weight-zero path with a weight-j path is equal to the probability of error for noncoherent combining of j transmissions [given by (1-42) and (1-43)]. The decoded P_b for a $R = m/n$ code is then obtained via the standard union bound by substituting (1-42) into (6-11).

Performance curves for a variety of code constraint lengths with $R = 1/2$ are shown in Fig. 6-16. There are several significant differences in comparing these curves with the corresponding curves for coherent PSK. Of course, performance is obviously worse because of the signal design

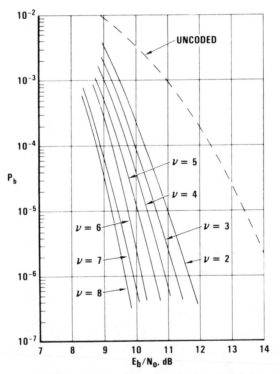

Figure 6-16. Bit error probability for $R = 1/2$ codes with Viterbi decoding (infinite quantization) and orthogonal signaling with noncoherent detection.

and noncoherent detection. However, it is also obvious that coding does not provide as much gain relative to the uncoded system. Whereas, with coherent PSK coding gains of greater than 6.0 dB were observed at the longest constraint lengths $(R = 1/3$ and $1/2$ at $P_b = 10^{-5})$, with orthogonal signaling the coding gains provided in the same situation are 2 dB less. In addition, the incremental coding gain achieved by increasing v by 1 is about 0.1 dB less than that observed with coherent PSK. Finally, in contrast to PSK, there is an optimum code rate in the neighborhood of $R = 1/2$. Similar curves for $R = 1/3, 2/3$, and $3/4$ are uniformly worse than those for the $R = 1/2$ codes. For a fixed value of v the codes with rates $1/3, 2/3$, and $3/4$ typically degrade by about 0.25, 0.15, and 0.3 dB, respectively, relative to the $R = 1/2$ curves. This behavior is significantly different from that observed with coherent PSK because of the noncoherent combining loss which increases with the code free distance. A coherent demodulator would have been able to more effectively utilize the additional code free distance, and these characteristics would not have been observed.

The results presented in Fig. 6-16 may also be applied to differentially coherent PSK modulation (DPSK). As discussed in Chapter 1, DPSK is 3 dB more efficient than binary orthogonal signaling. Thus, performance curves for DPSK are identical to Fig. 6-16, but shifted 3 dB to the left. Note that these results require independent errors at the demodulator output. Since DPSK produces paired errors, one would need to provide some interleaving to make the errors independent over a decoding span.

The problem of degraded performance due to noncoherent combining loss can be alleviated by using M-ary orthogonal signaling with non-coherent detection. The use of either 4-ary or 8-ary signaling can offer substantial performance improvements because of the increased coherence time per symbol. However, with the change in modulation format, coding must be applied carefully to ensure maximum benefit.

One method is to use a rate-1 code as shown in Fig. 6-17. (This implies that for each information symbol shifted in, one M-ary symbol is transmitted.) Note that the use of this approach means that codes should no longer be selected on the basis of maximizing the code free distance. Now the important criterion must be to maximize the number of nonzero

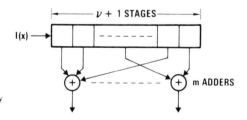

Figure 6-17. Rate-1 encoder for M-ary orthogonal signaling $(M = 2^m)$.

branches in any unmerged span. With orthogonal signaling all differing branches are the same distance (in signal space) from each other independently of how many binary symbols differ. However, after replacing the Hamming weight criterion by the number of nonzero branches, one can search for good codes and determine code weight structures exactly as before. Some good code generators are given in Tables B-6 and B-7 of Appendix B.

Performance estimates are obtained exactly as with the case of binary signaling. The expressions for P_j and ρ are still given by (1-42) and (1-43). In this case E_s refers to the signal energy per M-ary symbol (which is equal to E_b for rate-1 codes). Some typical performance results are given in Fig. 6-18. The use of 8-ary signaling results in a performance improvement of roughly 2.5 dB relative to binary signaling at $\nu = 6$. The cost of this improvement is increased demodulator complexity.

Other coding approaches for M-ary modulation are possible. For example, the dual-m codes are nonbinary codes as shown in Fig. 6-19 for

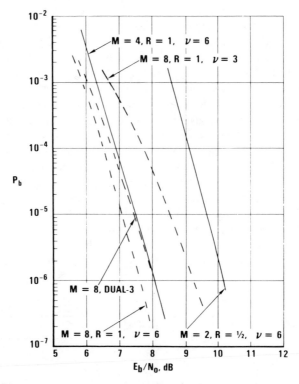

Figure 6-18. Bit error probability with Viterbi decoding for M-ary orthogonal signaling and noncoherent detection (infinite quantization).

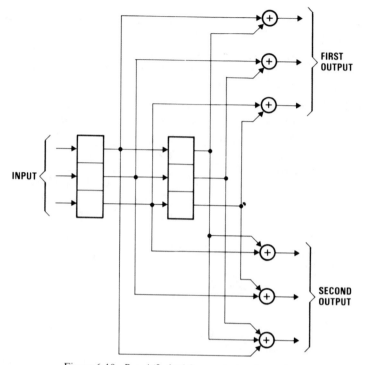

Figure 6-19. $R = 1/2$, dual-3 convolutional encoder.

a $R = 1/2$, dual-3 code. Here the binary information symbols are blocked into three-bit symbols and shifted into the encoder. With each three-bit symbol shifted in, two three-bit channel symbols are transmitted. This code has a free distance of 4. The weight structure for this code is given in Table B-6, and a performance curve is shown in Fig. 6-18. The implementation of this code is quite simple because the decoder has only eight states with eight paths entering and leaving each state.

6.6. Implementation Considerations

The Viterbi algorithm is basically simple and an acceptable hardware version can undoubtedly be produced by a good design engineer using nothing more than the information that has been provided up to this point. There are a few areas, however, where significant savings in the number of logic elements can be achieved through innovative designs and by making approximations which result in relatively minor performance degradations. The overall design problem will be outlined in this section and those areas will be discussed where major savings in parts are possible.

A functional block diagram for a general Viterbi decoder is shown in Fig. 6-20. It accepts at the input a continuous sequence of n-bit binary numbers which represent the successive outputs of the demodulator quantized to one of 2^n levels and delivers the decoded information stream at the output. The machine itself consists of

1. a synchronizer which both provides branch timing information to the decoder and determines the correct bit polarity if this is required;
2. a branch metric computer which determines the appropriate branch metrics for each received code branch;
3. a path metric updating, comparison, and storage device in which the branch metrics are added to the previously stored path metrics, the appropriate comparisons made, and the new path metrics stored;
4. a device for updating and storing the hypothesized information sequences; and
5. an output decision device which determines the decoder output and also provides information to the synchronizer.

In general, a particular block in this diagram can be implemented using parallel or serial logic depending on the operating speed required and the logic line that is being used. With fully parallel TTL logic that is available in the 1979–80 time frame a practical upper limit on the input information rate is on the order of 10 Mbit/s. Much higher information rates can be handled using emitter-coupled logic; however, the number of integrated circuit packages required is much larger than the TTL machine because of the lack of a variety of medium scale integration packages at this time. This situation is changing rapidly with the introduction of new logic lines, and thus it is undoubtedly true that the practical upper frequency limit will be increased.

In addition, advances in the state of the art in LSI and VLSI circuitry are making it possible to build entire decoders on a single chip. Even in

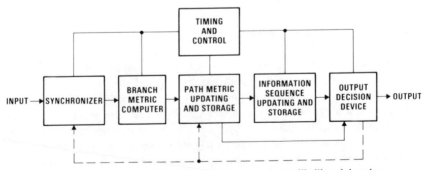

Figure 6-20. Functional block diagram for a maximum-likelihood decoder.

this case the circuit designer will still have to make the same basic types of design trades in order to maximize the performance that can be achieved for a given level of technology.

6.6.1. Synchronizer

The synchronizer is a device for determining the beginning of a branch in the received symbol stream [i.e., to differentiate between $T_1(x)$ and $T_2(x)$ in Fig. 6-1] and for resolving symbol polarity when required. In general, convolutional codes have the property that branch synchronization can be resolved without information being added to the transmitted symbol stream because the received data appear to have an excessive error rate when not properly synchronized. The operation of the synchronizer is straightforward and requires an indication that the input sequence contains a much larger number of errors than normal. Several techniques are possible and almost any way that one can get an error-detecting indication will work. One method for which there are many possible variations is to monitor the rate at which the path metrics are increasing. A second method is to use the merging properties of the trellis. That is, when the decoder is *out of sync*, the surviving paths in the trellis merge much more slowly than when the decoder has correct sync. This condition is easily detected.

When PSK data modulation is used, the potential exists for an ambiguity in the polarity of the received channel symbols. This problem can be solved in one of two ways. First, one could use a so-called *transparent* code. If the generator polynomials all have odd weight, then the code is said to be transparent to phase inversions. This means that all ones at the encoder input produces all ones at the output (after the first v branches). In this case an inversion in the polarity of the decoder input simply produces an inversion in the decoder output (after the initial transient during the first v branches). When this type of code is used, differential encoding/decoding is used outside of the convolutional encoder/decoder to provide the proper polarity for decoded data as shown in Fig. 6-21. The differential encoder maps the actual data $I(x)$ into a sequence $I'(x)$ which changes polarity

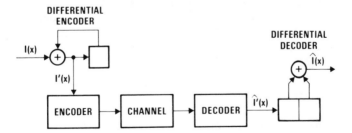

Figure 6-21. Transparent convolutional code with differential encoding/decoding.

for each data one in $I(x)$ and remains unchanged for each data zero. Then if the channel does not change the polarity of the transmitted sequence, the recovered sequence $\hat{I}'(x)$ is identical to $I'(x)$ (except for decoding errors), and of course $\hat{I}(x)$ is then identical to $I(x)$. Now in the event that the channel changes the polarity of the transmitted data, then the use of a transparent code guarantees that the sequence $\hat{I}'(x)$ is the complement of $I'(x)$. However, since the actual data have been encoded differentially, the sequence $\hat{I}(x)$ is identical to $I(x)$. Thus, when a transparent code is used, the synchronizer itself does not accomplish the resolution of symbol polarity. This is done automatically by the differential decoder. Note from the form of this decoder that an error multiplication effect occurs (a single decoding error produces two errors at the differential decoder output). This results in a loss of about 0.1 to 0.2 dB.

When *nontransparent codes* are used, an all-ones input no longer produces an all-ones output. Since the all-ones word is not a code word, a change in polarity at the decoder input results in a word that is not a legitimate code word. As a result, the received data appear to have an excessive error rate. Again, this condition is easily detected. In this case, under typical operating conditions ($P_b \approx 10^{-5}$), there is no performance loss as there is when differential encoding is used.

6.6.2. Branch Metric Computer

Each time a new branch is received the branch metric computer determines a new set of metric values for each of the different branches which appear in the code trellis. For $R = 1/2$ codes there will be four different values or, in general, 2^n values for codes having n symbols per branch. Ideally, the branch metric is proportional to the logarithm of the probability that the specified branch was transmitted given a particular set of received soft decisions from the demodulator. In the most straightforward case the branch metric is simply the sum of the individual metrics associated with each received symbol. For example, assume three-bit quantization with $0 = (0\,0\,0)_2$ representing a very reliable transmitted zero and the quantization values increasing monotonically to $7 = (1\,1\,1)_2$ for a very reliable 1. Further assume that the linear metric assignment scheme discussed in Chapter 1 is being used. If the particular branch in question is represented by 110 and the received soft decisions are, respectively, levels 7, 5 and 3, then the branch metric is $0 + 2 + 3 = 5$. Note that with this convention a small branch metric represents a highly probable event while larger metrics represent less likely events. If this method is used for an $R = 1/2$ code, then the entire set of metric values are represented by the branch metric table shown in Table 6-1. In this table the coordinates of each point are the "error" values associated with each of the received

Table 6-1. Uniform Branch Metric
Assignments

e_1 \ e_2	0	1	2	3	4	5	6	7
0	0	1	2	3	4	5	6	7
1	1	2	3	4	5	6	7	8
2	2	3	4	5	6	7	8	9
3	3	4	5	6	7	8	9	10
4	4	5	6	7	8	9	10	11
5	5	6	7	8	9	10	11	12
6	6	7	8	9	10	11	12	13
7	7	8	9	10	11	12	13	14

symbols when compared with a very reliable one or zero. As an example of how the table is used, suppose that the received symbols fall into quantization levels 2 and 6, respectively. The resulting error pairs associated with the branches 0 0, 0 1, 1 0, and 1 1 are (2,6), (2,1), (5,6), and (5,1), and the metric values are 8, 3, 11, and 6.

Using the linear metric assignment method discussed above, the metric values occupy the range 0 to 14 and consequently four bits are required to represent these values in the decoder. When these branch metric values are later used to compute path metrics, it is clear that the number of bits that will be required to represent each path will be even larger than 4. It turns out that the complexity of the decoder is a rather strong function of the number of bits that are required to represent each of the path metric values. Thus, in many situations, it is desirable to keep this number as small as possible. There are basically two mechanisms that the designer can exploit in order to accomplish this task. First, he can attempt to assign metric values directly to the branch metric table without requiring that a particular entry be the sum of the individual bit metrics. Second, he can attempt to exploit the fact that each time a branch is received, only four values (for the $R = 1/2$ code) are taken from the table. One may add or subtract any number from each of these four values without changing the performance of the decoder. Utilizing both of these properties,

Table 6-2. Nonoptimum Branch Metric
Assignments Which Produce Branch
Metrics Representable with Three Bits

e_1 \ e_2	0	1	2	3	4	5	6	7
0	0	0	0	0	0	1	2	3
1	0	0	0	0	0	1	2	3
2	0	0	0	0	0	1	2	3
3	0	0	0	0	0	1	2	3
4	0	0	0	0	1	2	3	4
5	1	1	1	1	2	3	4	5
6	2	2	2	2	3	4	5	6
7	3	3	3	3	4	5	6	7

one might be led to create the metric table shown in Table 6-2. This table only requires three bits to represent a particular value and has the added benefit that the best metric associated with a given set of four values never produces a change in the path metric to which it is added. This has the effect of minimizing any unnecessary growth in the path metrics and hence minimizing the necessity to rescale frequently. Table 6-2 was given only as an example and not as the optimum choice of metric assignments. Although the degradation associated with Table 6-2 is quite small, one may find even better assignments through a trial-and-error evaluation of each potential candidate by numerically computing the probability of error for some typical code sequences. This can be done using precisely the same technique as was used to generate the curves in Fig. 1-9.

6.6.3. Path Metric Storage and Updating

The trellis diagram for any $R = 1/n$ code can be subdivided into basic modules of the form shown in Fig. 6-22. This suggests that the portion of the decoder concerned with the storage and updating of the path metrics can be implemented by repeated use of the computational element shown

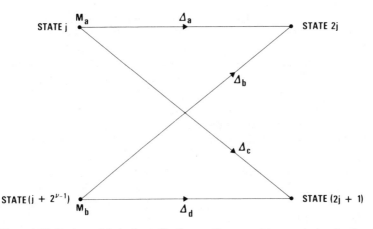

Figure 6-22. Basic module in the trellis diagram for a rate-1/n convolutional code.

in Fig. 6-23. This element consists of two adders, followed by a comparator, followed by a storage element. For obvious reasons, this is called the add–compare–select (ACS) function. For a constraint length v code, this circuit must be duplicated 2^v times. At low speeds all but the memory can be time shared while at high speeds it is necessary to replicate the entire circuit. We see that the complexity of this circuit is linearly proportional to the number of bits that are used to represent the path metric. In a fully parallel machine any savings realized here can be especially significant since it is multiplied by 2^v.

We first show that for any $R = 1/m$ code the maximum metric range need not exceed vm_b, where m_b is the maximum metric value required for a branch. This follows immediately if we assume that the best state has

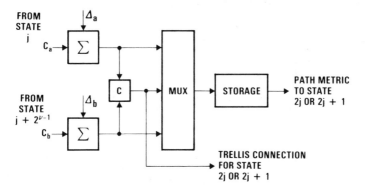

Figure 6-23. Functional block diagram of an add-compare-select circuit for a rate-1/n code including storage for path metric.

been scaled to zero and note that any other path can be made to merge with this path in no more than v steps. On any one step the maximum differential can be at most m_b. Thus, the total range can be no more than vm_b. Of course, one must always rescale the best path metric in order to successfully utilize this limited range. This implies that as long as the metric range is greater than or equal to vm_b and values are scaled to stay in that range, then the only performance degradation will be that associated with the initial branch metric assignment.

Actually one can reduce the total metric range to be significantly less than this value with only minor degradation in performance. Although this seems surprising at first, it becomes more reasonable if we note that paths with large metric values are very unlikely to be the correct path, since they must have encountered several very improbable errors in order to have achieved this large value. Thus, truncating the metric range really only affects the paths that were very unlikely to be the correct one.

Figure 6-24 shows performance with several path metric ranges

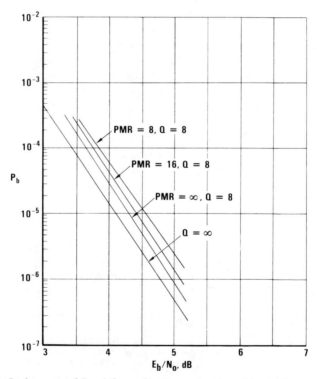

Figure 6-24. Performance of $R = 1/2$, $v = 6$ code as a function of demodulator quantization (Q) and path metric range (PMR) (curves supplied courtesy of R. C. Davis).

(PMR = 8, 16, and ∞) at $Q = 8$ compared with $Q = \infty$ (supplied courtesy of R. C. Davis, private correspondence). The union bound technique was used to compute performance utilizing a computational technique that accounts for the finite values of demodulator quantization and path metric range. The branch metric tables were also optimized using this same procedure. For four-bit path metric (PMR = 16), note that only 0.1 dB degradation relative to $Q = 8$ and PMR = ∞ is observed. One could design a custom LSI chip to accomplish the ACS with the path metric range of only four bits. Note also that metric compression to three bits (PMR = 8) loses only about 0.25 dB relative to PMR = ∞. This is quite remarkable. With the metric compressed to such a small range one can now use a read-only memory (ROM) to perform the ACS function. Both of these approaches have been used in commercially available units.

It should also be observed that an ACS circuit for a $R = (n - 1)/n$ code would normally contain multiple elements of the form of Fig. 6-23. However, if one uses the "punctured code" approach, then the structure of Fig. 6-23 is all that is needed. The metric compression techniques also work just as well. Thus, the use of punctured codes for higher rate implementations results in a significantly simpler ACS circuit.

6.6.4. Storage and Updating of Hypothesized Information Sequences

There are basically two different techniques for handling this portion of the machine. The first method, called *register exchange*, calls for a complete hypothesized information sequence to be stored for each of the 2^v states. The registers in which these sequences are stored are interconnected in precisely the same fashion as the add–compare–select circuits discussed in Section 6.6.3. Each time a new branch is processed by the computer, the registers are interchanged corresponding to which sequences survived the comparison, a new symbol is added at one end of each register, and the oldest symbol in each register is delivered to the output decision device. At high speeds this technique requires all of the shifting to be done in parallel and means that each individual symbol storage device must be equipped with the necessary gates to accept inputs from one of two different locations. For even moderately long decoding spans, this necessitates an enormous amount of hardware and is normally not practical. At very low speeds one may usually design the decoder such that the sequence can be interchanged in a serial fashion through a single routing device.

Fortunately, there is a second method which can be used for high-speed decoders. This is referred to as the *trace-back method*. Using this technique one does not store the actual information sequences but instead stores the results of each comparison (the trellis connections). After several branches have been processed, the trellis connections are recalled in the reverse order in which they were stored and a path is traced back through the trellis diagram. Instead of decoding a single branch of information symbols, this technique requires that one decode several branches at a time at high speeds. One drawback to this technique is that the decoding delay is 2 to 3 times longer than with register exchange since one cannot initiate the trace back until several branches past the minimum decoding depth have been processed. Typically, one might trace back with a 2:1 speed advantage. Thus if L branches is the minimum required decoding depth, one must provide storage for $3L$ sets of branches in order to provide the necessary buffering. For high-speed operation it is also desirable to organize the memory so that one can trace back through two or more levels on a single memory access. This can be done with only a small increase in complexity and will alleviate the requirement for a larger speed advantage.

Since storage must be provided for a number of symbols equal to the number of decoder states times the lengths of the paths stored, it is desirable to minimize the stored path length as well as use an efficient technique for updating these paths.

The length of path that must be stored is strongly dependent on the merging characteristics of the code. These characteristics for the optimum $R = 1/2$, $v = 6$ code and the $R = 3/4$ punctured code derived from it were determined through an exhaustive search of the code weight structure and are shown graphically in Fig. 6-25. This figure shows the number of merged and unmerged paths as a function of decoding depth (in information symbols) for paths that have weight equal to d (the code minimum free distance) as well as $d - 1$ and $d + 1$. Note that at the short decoding depths there are many low-weight unmerged paths which will significantly degrade performance because they have a reasonably high probability of being selected by the output decision device. The interesting point is the difference between the merging characteristics of the $R = 1/2$ and $R = 3/4$ codes with the same constraint length. Satisfactory performance is achieved only at decoding depths at which the large number of low-weight unmerged paths disappear. For the $R = 1/2$ code this occurs by depth 30 and is consistent with the "rule of thumb" that the decoding depth needs to be about $5v$. However, for the $R = 3/4$ code at a decoding depth of 30 there are some weight $d - 1$ unmerged paths. Thus, if this decoding depth were used, the code would perform as if it had a free distance of $d - 1$, rather than d. In addition, the numbers of paths at weight d and $d + 1$ are significantly larger than the final values reached as decoding depth increases. At depth 40, there are no longer any paths of weight $d - 1$. At depth 50, the only paths left at weight d have merged, and the number of paths of weight $d + 1$ has almost reached its final value. By depth 60, all paths at weight $d + 1$ have merged. Thus, it takes the paths approximately twice as long to merge for a $R = 3/4$ code, indicating that the required decoding depth is

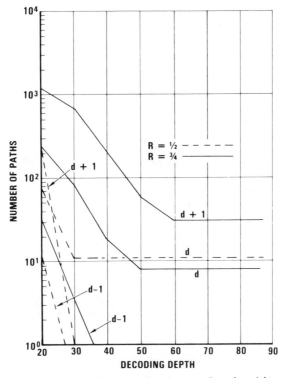

Figure 6-25. Number of merged and unmerged paths at each path weight as a function of decoding depth for $v = 6$ codes.

approximately $10v$. Similar results can be obtained for $R = 2/3$ codes, which indicates a required decoding depth of about $8v$.

6.6.5. Output Device

Selection of the output device involves a trade between two conflicting parameters. From a strict performance viewpoint the optimum procedure is to decode the oldest symbol (or symbols) in the sequence having the minimum path metric. For a given decoding depth this will always result in the lowest probability of error. The procedure for locating this path requires a minimum of $2^v - 1$ binary comparisons. At low rates it is possible to do this sequentially, but at high speed these comparisons must be performed in parallel. It is sometimes easier to select a path whose metric is close to the best by selecting any path whose metric lies below a threshold. Since this path will usually merge with the correct path within a few branches, it normally produces the correct decision. Any differences in

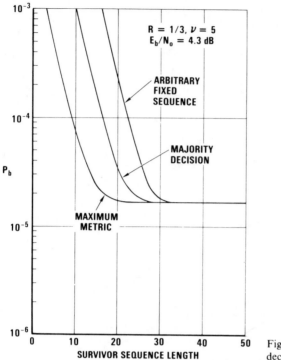

SURVIVOR SEQUENCE LENGTH

Figure 6-26. Comparison of bit decision techniques.

performance can be made up by increasing the decoding depth. This idea may be extended to selecting any arbitrary sequence (or a fixed sequence) if we observe that for sufficiently long decoding spans all paths tend to merge with the correct one. A related technique that can be used when the actual information sequences have been stored is to make a majority decision on the oldest symbol in each sequence.

In a particular design situation it is often useful to rely on computer simulations to select the best combination of decoding length and decision technique. The results of one such set of simulations are shown for the $R = 1/3$, $v = 5$ code in Fig. 6-26.

6.6.6. Demodulator Quantization and AGC

One final design parameter that is worth considering is the degree of demodulator quantization and the accuracy of the AGC (automatic gain control) that is required. The quantization of the demodulator output to Q levels affects both the performance of the Viterbi decoder and its complexity. The performance curves presented in Section 6.5 assume $Q = \infty$.

If $Q = 8$ is used (as is done normally), then the performance degrades only by about 0.25 dB. For coarser quantization the performance degrades rapidly. At $Q = 4$ and $Q = 2$ the resulting degradations are slightly over 0.9 and 2.2 dB, respectively. While reducing Q might be an acceptable trade if it resulted in significant decreases in hardware complexity, in this case any reductions are insignificant. This would only result in slightly reduced metric range, and there are more effective techniques for achieving this. As a result, $Q = 8$ is almost always used except in rather unusual circumstances.

Another interesting question is how accurately the quantizer threshold must be set (or equivalently the AGC). Fortunately, it is not very critical. The optimum quantizer threshold spacing with evenly spaced thresholds is about 0.5σ (σ being the standard deviation of the noise).[62] The spacing can vary by ± 3 dB and cause less than 0.1 dB degradation. In addition, a ± 6-dB variation results in less than 0.5 dB of degradation. Thus, Viterbi decoders do not need an extremely accurate AGC in the demodulator. Performance is quite insensitive to wide variations in AGC gain. Often the thresholds are spaced at some fraction of the signal mean. The spacing is optimized at the nominal operating point and the threshold is held at the same relationship to the signal mean at all signal-to-noise ratios. At the signal-to-noise ratios typically of interest, an acceptable setting for highest threshold with $Q = 8$ is slightly less than the signal mean.

6.7. Remarks

The structural properties of convolutional codes have been presented in this chapter. This forms the basis for all succeeding discussions on convolutional codes and is complemented by a discussion of the Viterbi algorithm. Viterbi decoding has been widely applied in communication systems and presently is the most practical forward-error-correcting technique for providing large coding gains on Gaussian channels. Although the technique is limited to small values of code constraint length because of the exponential increase in decoder complexity with constraint length, excellent performance is obtained by using optimum codes. The widely used $R = 1/2$, $\nu = 6$ code provides 5 dB of coding gain at $P_b = 10^{-5}$. In addition, the technique is easily applied over a broad range of code rates.

Other decoding techniques preceded the Viterbi algorithm historically, and several are quite useful in certain applications. Techniques such as threshold decoding and sequential decoding typically use much longer code constraint lengths than the codes presented in this chapter. These techniques are the subject of the next chapter. The special codes typically used with each of these techniques are also given.

Problems

6-1. For the $R = 1/2, v = 1$ systematic code with $g_1(x) = 1$ and $g_2(x) = 1 + x$ give the code tree, trellis, and state diagram.

6-2. For the $R = 1/3, v = 2$ nonsystematic code with $g_1(x) = 1 + x + x^2, g_2(x) = 1 + x + x^2$, and $g_3(x) = 1 + x^2$ give the code tree, trellis, and state diagram.

6-3. Give the signal flow graph of the code of Problem 6-1 and find the generating function, $T(D, L, N)$, from this graph.

6-4. Give the signal flow graph of the code of Problem 6-2 and find the generating function, $T(D, L, N)$, from this graph.

6-5. Observing that $Q[(x + y)^{1/2}] \le Q(x^{1/2}) e^{-y/2}$, find a closed-form expression in terms of $T(D, L, N)$ for upper bounds on first-event error probability, P_e, and bit error probability, P_b, for coherent PSK on the infinitely quantized Gaussian channel (Viterbi[58]).

6-6. Using the results of Problem 6-5, give closed form expressions for upper bounds on P_e and P_b for the $R = 1/2, v = 2$ code of Fig. 6-1 and the codes used in Problems 6-3 and 6-4.

6-7. Determine which of the following $R = 1/2$ codes are catastrophic:
      (a) $g_1(x) = x^2, g_2(x) = 1 + x + x^3$
      (b) $g_1(x) = 1 + x^2, g_2(x) = 1 + x^3$
      (c) $g_1(x) = 1 + x^2 + x^4, g_2(x) = 1 + x + x^3 + x^4$
      (d) $g_1(x) = 1 + x + x^2 + x^4, g_2(x) = 1 + x^3 + x^4$
      (e) $g_1(x) = 1 + x^4 + x^5 + x^6, g_2(x) = 1 + x + x^3 + x^5$

6-8. The $R = 1/2, v = 2$ code with trellis shown in Fig. 6-3 is used on a BSC. The received sequence is 00 00 11 11 11 00 01 00 00. Trace the decisions (on a trellis diagram) made by a Viterbi decoder using a Hamming metric through the first 10 nodes, and label each node with the survivor metric. Resolve all ties in favor of the upper path.

6-9. Repeat Problem 6-8 using the $R = 1/3, v = 2$ code of Problem 6-2 and a received sequence 000 000 100 110 000 000 000 000 000 000.

6-10. Repeat Problem 6-8 for a binary input, ternary output channel with output alphabet 0, 1, and E (erasure). Assume that an output E gives 0 metric contribution and the received sequence is 00 1E 0E 00 00 00 00 00 00 00.

6-11. By expanding the generating function $T(D, L, N)$ for the $R = 1/2, v = 2$ code [Eq. (6-4)] and for the $R = 1/3, v = 2$ code of Problem 6-4 find the decoding depth as a function of k which will ensure that all paths of weight $d_f + k$ have merged.

6-12. Considering the structure of the generator matrix for a $R = 1/n$ convolutional code, derive an expression for a Plotkin-type upper bound on code free distance, d_f, as a function of n and v. Compare with the values of d_f for the codes in Tables B-1 and B-2.

7

Other Convolutional Decoding Techniques

Convolutional coding techniques have been studied for many years. The technique was first introduced by Elias[63] in 1955. Shortly thereafter, Wozencraft[64] devised a decoding technique which he called "sequential decoding." This technique has been the subject of much research interest for over 20 years. It is a trial-and-error search decoding technique that provides performance that can meet or exceed that of Viterbi decoders and is preferable in some applications. One of the most significant properties of sequential decoding is that the time required to decode a given information symbol is a random variable. In this respect the algorithm is quite similar to the random search techniques for block codes discussed in Chapter 3.

During the first few years of convolutional coding research, it was recognized that although sequential decoding offered very obvious performance advantages, it would at that time have required a very costly implementation. Thus, a number of researchers attempted to find suboptimum, easily implementable algorithms for decoding convolutional codes. Unlike sequential decoding, these techniques are deterministic so that one code branch is completely decoded during each computation cycle. Typically, with these techniques a syndrome sequence which is dependent only on the channel error symbols is calculated. Just as with block codes, the syndrome provides linear equations that can be solved for the minimum-weight channel error sequence. Correction symbols for the received information symbols are formed through an appropriate mapping of a span of syndrome symbols. Hence, we will classify these techniques as syndrome decoding techniques. One widely used syndrome decoding technique is *table look-up feedback decoding*. As the name implies, the mapping from syndrome symbols to correction symbols is accomplished through a look-up table (probably implemented in a read-only memory). This technique is very similar to the Meggitt decoder for block codes introduced in Chapter 3. The other widely used syndrome decoding technique is threshold decoding. The algorithms used for hard and soft decision threshold decoding are virtually identical to those used for block codes in Chapters 3 and 4, respectively. An extensive class of convolutional codes suitable for threshold decoding has been found.

7.1. Syndrome Decoding Techniques

The two most widely used syndrome decoding techniques, table look-up decoding and threshold decoding, map a span of syndrome symbols into the necessary correction symbol. The principal difference between the techniques is that the table look-up feedback decoder typically uses an optimal mapping which allows good performance even when a short span of syndrome symbols is examined. However, since the span must necessarily be short (due to implementation considerations), it is important to optimize the code which is used. Threshold decoders use a special type of mapping which allows much longer spans of syndrome symbols, but only codes with unique properties can be used. Unfortunately, the allowable codes are rather poor so only moderate values of coding gain are possible. Systematic codes are used in the discussion that follows. It will become obvious later that nonsystematic codes offer little advantage, and they are rarely, if ever, used in syndrome decoders.

7.1.1. Basic Concepts

Before discussing table look-up feedback decoders and threshold decoders in detail, it is instructive to examine certain basic properties that apply to both of these approaches. These properties include structural properties such as minimum distance, relationship between syndrome sequence and channel error sequence, and decoder structure. Performance analysis techniques will also be discussed. Finally, the problem of unlimited error propagation in feedback decoders will be examined. An $R = 1/2$, $v = 1$ code is used in this discussion to illustrate important points.

7.1.1.1. Structural Properties. Using the polynomial representation of Chapter 6, the output sequence of a $R = 1/2$ systematic encoder is represented as an information sequence $I(x)$ and a parity sequence $P(x)$, i.e.,

$$T_1(x) = I(x) \tag{7-1}$$

and

$$T_2(x) = P(x)$$

$$= I(x)\,g(x) \tag{7-2}$$

with the code generator polynomial represented by $g(x)$. Denoting the channel error sequences for the information and parity portions as $E^i(x)$ and $E^p(x)$, respectively, the received sequences are denoted as

$$R^i(x) = I(x) + E^i(x) \tag{7-3}$$

and

$$R^p(x) = P(x) + E^p(x) \tag{7-4}$$

The decoding techniques under consideration decode a given information symbol, say i_n, by utilizing prior received symbols and the symbols for L succeeding branches $r_n^i, r_{n+1}^i, \ldots, r_{n+L-1}^i$ and $r_n^p, r_{n+1}^p, \ldots, r_{n+L-1}^p$ including the branch in question. There are basically two methods by which the prior received symbols can be used. The *feedback decoding* technique utilizes the sequence of actual decoding decisions prior to i_n to define a decoder starting state (or code tree starting node) for the next decision. Thus, decoding decisions at any given time will affect decoding decisions in the future. The feedback process can cause an error propagation phenomenon which will be discussed subsequently. In contrast, the *definite decoding* technique does not utilize the preceding decoding decisions, but instead uses the preceding v received information symbols $r_{n-v}^i, r_{n-v+1}^i, \ldots,$ r_{n-1}^i and the succeeding received branches in making a decision on the information symbol i_n. Thus, each decoding decision is determined only by a certain span of received symbols and is not influenced directly by previous decoding decisions. As a result, the aforementioned error propagation phenomenon does not occur with definite decoding.

To develop more insight into the feedback decoding process, consider the $R = 1/2$, $v = 1$ systematic code with $g(x) = 1 + x$ and the code tree shown in Fig. 7-1. A feedback decoder can be constructed that utilizes a span of $L = 2$ branches to make each decoding decision. Assume the following decoding algorithm. The starting node in the code tree is determined by previous decoding decisions. Then, given a starting node, the decoder "looks ahead" for $L = 2$ branches and compares the received sequence with the four-symbol span on each of the four possible code sequences. The code sequence closest to the received sequence in metric (e.g., Hamming distance) is selected, and the initial information symbol determines the decoding decision for i_n. This decoding decision and the previous starting node then define the new starting node for the next decoding decision. This is essentially the algorithm that is employed in a table look-up feed-

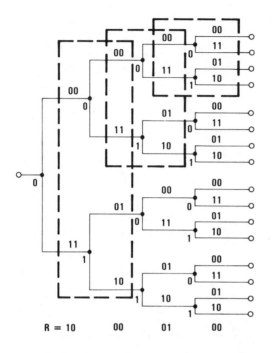

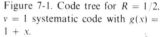

Figure 7-1. Code tree for $R = 1/2$, $v = 1$ systematic code with $g(x) = 1 + x$.

back decoder. For this example, consider the received sequence **R** which has errors in the first and third branches as shown (assuming an all-zero transmitted sequence). Starting in state 0 and looking ahead two branches, we note that the corresponding span of the received sequence 1 0 0 0 is to be compared with each of the four possible code sequences 00 00, 00 11, 11 01, and 11 10. Obviously, the coded sequence closest in Hamming distance is 00 00 so the proper decoding decision for the first informa-tion symbol is $\hat{i}_0 = 0$. This decision means that the new starting node on the next symbol time is still state 0. From this node, the decoder then ob-serves the span of received symbols 00 01 and compares it with the four possible code sequences. Again the decision $\hat{i}_1 = 0$ is made, and again the new starting node will be in state 0. Continuing in this fasion all subsequent decoding decisions will be zeros, and, all channel errors are corrected.

One can decode in a similar fashion using the received syndrome to produce an estimated sequence of error symbols $\hat{E}^i(x)$ to be used in correct-ing the received sequence $R^i(x)$. The syndrome is formed by regenerating parity symbols from the received information symbols and adding the received parity symbols (thus "checking" parity). This gives the syndrome sequence $S(x)$, i.e.,

$$S(x) = R^i(x) g(x) + R^p(x) \tag{7-5}$$

Then using (7-2) through (7-4) the syndrome can be written as a function only of the channel error sequences $E^i(x)$ and $E^p(x)$, i.e.,

$$S(x) = E^i(x)g(x) + E^p(x) \qquad (7\text{-}6)$$

The information sequence estimated by the decoder is

$$\hat{I}(x) = R^i(x) - \hat{E}^i(x)$$
$$= I(x) + [E^i(x) - \hat{E}^i(x)] \qquad (7\text{-}7)$$

Correct decoding is accomplished as long as the estimated error sequence equals the actual error sequence.

The syndrome symbols are related to the channel error symbols through (7-6). For a given span of L syndrome symbols one can obtain a set of L linear equations in the channel error symbols. Typically, this set of equations is solved for the minimum Hamming-weight length $2L$ error sequence to produce an estimate of a desired channel error symbol.

For the $R = 1/2$, $v = 1$ example with $g(x) = 1 + x$ the syndrome is

$$S(x) = E^i(x)(1 + x) + E^p(x)$$

Then the syndrome symbol corresponding to the nth branch is

$$s_n = e^i_{n-1} + e^i_n + e^p_n$$

The error-correction strategy is specified by determining the mapping of a span of syndrome symbols into a correction symbol, $\hat{e}^i_n$, for the nth branch that corresponds to a minimum Hamming-weight error sequence. Note that for a single channel error in the nth information symbol $[E^i(x) = x^n]$ the resulting syndrome is

$$S(x) = x^n + x^{n+1}$$

while for a single error in the nth parity symbol the syndrome is

$$S(x) = x^n$$

Since these two single-error patterns result in different syndromes, all single errors are correctable. The correction algorithm is simply to set $\hat{e}^i_n = 1$ when $s_n = s_{n+1} = 1$ and set $\hat{e}^i_n = 0$ otherwise.

Figure 7-2 shows a block diagram of the encoder, digital channel, and feedback decoder for this code. Note that feedback is used to remove the effect of previous error symbols from the syndrome sequence. When the $(n - 1)$th correction symbol is generated, the syndrome register contains

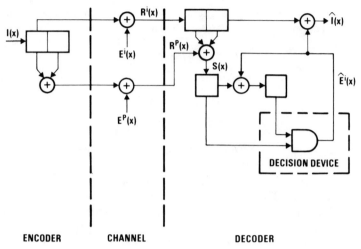

Figure 7-2. Encoder and feedback decoder for $R = 1/2$, $v = 1$ code.

s_n and s_{n-1}. On the next shift, s_n is modified and becomes $s_n^{(1)}$, where

$$s_n^{(1)} = s_n + \hat{e}_{n-1}^i$$
$$= e_{n-1}^i + \hat{e}_{n-1}^i + e_n^i + e_n^p$$
$$= e_n^i + e_n^p \tag{7-8}$$

if $e_{n-1}^i = \hat{e}_{n-1}^i$. This removal of the effect of preceding channel errors is equivalent to specifying the starting node in the code tree of Fig. 7-1. In many cases this results in a performance improvement. The decision device in this example is an AND function which sets

$$\hat{e}_n^i = s_n^{(1)} \cdot s_{n+1}$$
$$= (e_{n-1}^i + \hat{e}_{n-1}^i + e_n^i + e_n^p) \cdot (e_n^i + e_{n+1}^i + e_{n+1}^p) \tag{7-9}$$

and is equivalent to the decoding rule specified in the preceding paragraph. This decoding rule allows correction of all single-error patterns. Codes with greater error-correction capability require utilizing a larger span of syndrome symbols and hence require a much more complex decision device. Such codes will be discussed shortly.

It happens that for this particular code one can use a definite decoder and still correct all single-error patterns. The decoder is identical to that of Fig. 7-2 except there is no feedback. Removing the feedback of $\hat{E}^i(x)$ into the register holding $S(x)$ means that the output of the decision device is then

$$\hat{e}_n^i = s_n \cdot s_{n+1}$$

$$= (e_{n-1}^i + e_n^i + e_n^p) \cdot (e_n^i + e_{n+1}^i + e_{n+1}^p) \tag{7-10}$$

By inspection it is obvious when there is only a single error in the branches at the $(n-1)$th, nth, and $(n+1)$th symbol times, that $\hat{e}_n^i = 1$ if $e_n^i = 1$, and $\hat{e}_n^i = 0$ otherwise. Thus, a single error in i_n is corrected provided that the preceding and following branches are received correctly.

The correction terms for feedback and definite decoding given in (7-9) and (7-10) illustrate an important difference between the two approaches. Assuming no previous errors occurred with a feedback decoder, the correction symbol is dependent only on error symbols that occur over a span of L branches or $2L$ channel symbols. Thus, the *feedback decoding constraint length* is $2L$. In contrast, the correction symbol for definite decoding is dependent on these same symbols plus the v channel error symbols corresponding to the preceding v information symbols. This means that the *definite decoding constraint length* is $2L + v$. The differing correction approaches sometimes result in different minimum distances for feedback and definite decoder implementations of the same code.

Definition. *The feedback decoding minimum distance, $d_{fd}(L)$, for a binary $R = 1/2$ convolutional code is defined as the minimum number of positions in which two encoded sequences with differing values of i_n differ from time n to time $n + L - 1$ assuming the effect of previous information symbols has been removed. This is equivalent to finding the minimum-weight sequence in the lower half code tree (starting from state 0) over the first L branches.*

Note that the minimum distance is defined as a function of L. For a fixed code, performance can be improved by increasing L. In addition, as $L \to \infty$, $d_{fd}(L)$ becomes equal to d_f, the free distance.

Definition. *The definite decoding minimum distance, $d_{dd}(L)$, is defined to be the minimum number of positions in which two encoded sequences with differing values of i_n differ over a definite decoding constraint length.*

In finding the feedback decoding minimum distance a set of sequences that agree in the v preceding information symbols is compared to find the minimum Hamming distance. Then to find the definite decoding minimum distance this set of sequences plus sequences that do not agree in the v preceding information symbols are compared to find the minimum

Hamming distance. Therefore,

$$d_{\text{fd}}(L) \geq d_{\text{dd}}(L) \tag{7-11}$$

for the same code. The resulting error-correcting capabilities for the two approaches are

$$t_f = \left\lfloor \frac{d_{\text{fd}}(L) - 1}{2} \right\rfloor \tag{7-12}$$

and

$$t_d = \left\lfloor \frac{d_{\text{dd}}(L) - 1}{2} \right\rfloor \tag{7-13}$$

7.1.1.2. Performance Analysis. Performance calculations are made by summing the probabilities of all error-causing events. This is readily illustrated by using the $R = 1/2$, $v = 1$ example. The bit error rate for definite decoding is given by the probability that the correction symbol is not equal to the actual error symbol, i.e.,

$$P_{\text{dd}} = \Pr[\hat{e}_n^i \neq e_n^i]$$
$$= \Pr[\hat{e}_n^i + e_n^i = 1] \tag{7-14}$$

The error rate can be written from inspection by using (7-14) and (7-10). When $e_n^i = 1$, there are four single-, four double-, and four triple-error patterns in the other four error symbols that make $\hat{e}_n^i = 0$. Likewise, when $e_n^i = 0$, there are four double-error patterns in the other four error symbols that make $\hat{e}_n^i = 1$. As a result, the bit error rate with definite decoding of this code is

$$P_{\text{dd}} = p[4p(1 - p)^3 + 4p^2(1 - p)^2 + 4p^3(1 - p)] + (1 - p)[4p^2(1 - p)^2]$$
$$= 8p^2(1 - p)^3 + 4p^3(1 - p)^2 + 4p^4(1 - p) \tag{7-15}$$

With feedback decoding, the calculation of bit error rate is much more complex. Consider the case of "perfect feedback." That is, assume that a magic genie always feeds back the correct channel error symbol. This eliminates error propagation. However, the actual decoder decision is not affected in any other way by the genie action. Then for the *genie-aided feedback decoder*, the estimated correction symbol is

$$\hat{e}_n^i = (e_n^i + e_n^p) \cdot (e_n^i + e_{n+1}^i + e_{n+1}^p) \tag{7-16}$$

since perfect removal of prior errors is assumed. Bit error rate can be determined in the same fashion as before from (7-16). When $e_n^i = 1$, there are three single-, two double-, and one triple-error patterns in the other three error symbols that result in $e_n^i = 0$. When $e_n^i = 0$, there are two double-error patterns in these error symbols that make $e_n^i = 1$. Thus, the error rate of the genie-aided feedback decoder is

$$P_{gd} = 5p^2(1 - p)^2 + 2p^3(1 - p) + p^4 \qquad (7\text{-}17)$$

Since the genie effectively provides perfect knowledge of the starting node for each decoder decision, one will always have

$$P_{gd} \leq P_{dd}$$

The relevance of the genie decoder concept is that the probability of *first decoding error* is exactly P_{gd}. If no prior decoding errors have occurred, then $\hat{e}_{n-1}^i = e_{n-1}^i$ and the estimate $\hat{e}_n^i$ assumes the form of (7-16) for the feedback decoder. This condition holds until the first decoding error occurs. Therefore, the probability of first decoding error, P_{fe}, is given by

$$
\begin{aligned}
P_{fe} &= P_{gd} \\
&= 5p^2(1 - p)^2 + 2p^3(1 - p) + p^4 \qquad (7\text{-}18)
\end{aligned}
$$

The overall bit error rate for feedback decoding is larger than P_{fe} because of the error propagation caused by erroneous feedback. This effect is rather difficult to analyze for long codes, but stochastic sequential machine methods have been used by Morrissey[65] to compute bit error rate in closed form for shorter codes. For this example he found that

$$P_{fd} = \frac{7p^2 - 12p^3 + 10p^4 - 4p^5}{1 + 3p^2 - 2p^3} \qquad (7\text{-}19)$$

For longer codes that are decoded either by table look-up or threshold decoding, the same error-counting technique can be used to determine P_{dd} and P_{fe}. This must be done individually for each code if table look-up is used. However, for a certain class of threshold-decodable codes a closed form expression for P_{dd} can be written that is dependent only on the code parameters and the channel error rate p. Determining P_{fd} is much more difficult. This quantity must usually be estimated by simulation.

Note that at high signal-to-noise ratio ($p \to 0$) with the $R = 1/2$, $v = 1$ example, $P_{dd} > P_{fd} > P_{fe}$. Of course, the error propagation phenomenon makes $P_{fd} > P_{fe}$ for any feedback decoder. The interesting point is that for this example and for most cases of interest $P_{fd} < P_{dd}$ at sufficiently small values of p. Since the use of feedback frequently offers a performance advantage with very little impact on complexity, it is widely used.

In general, P_{fd} is not always less than P_{dd}. The only class of codes for which it is known that $P_{fd} < P_{dd}$ as $p \to 0$ is the class of codes with $t_f > t_d$ [note that from (7-11) through (7-13) $t_f \geq t_d$].[65] This result becomes obvious by observing that as $p \to 0$

$$P_{\text{fd}} \approx K_f p^{t_f + 1}$$

and

$$P_{\text{dd}} \approx K_d p^{t_d + 1}$$

As long as the feedback decoder does not infinitely propagate errors and $t_f > t_d$, then

$$P_{\text{fd}} < P_{\text{dd}}$$

when $p \to 0$.

7.1.1.3. Error Propagation. The error propagation phenomenon of convolutional codes has been the subject of much attention in the research literature. There are several types of error propagation each with different causes. Error propagation can be either limited or unlimited. Limited error propagation cannot be entirely eliminated in a feedback decoder. It simply makes $P_{\text{fd}} > P_{\text{fe}}$, and it need not be particularly harmful. Some attention has been given in the literature to reducing its effect for particular codes by reducing the length of time that errors propagate. Unlimited error propagation is quite serious, and there are two types that can occur. *Type-I unlimited error propagation* is caused by poor code selection. These codes are the so-called catastrophic error-propagating codes discussed in the previous chapter. Massey and Sain[60] have determined the properties of the code generator polynomials that cause this problem. Note that such codes will perform poorly no matter what decoding technique is used. Systematic codes never have this problem (all infinite-weight information sequences produce infinite-weight code sequences). However, even when good codes are used, improper decoder design can cause *type-II unlimited error propagation* (a finite number of channel errors can cause the decoder to make an infinite number of information symbol errors). For table look-up feedback decoders, this phenomenon has been shown to appear when the decoding depth, L, is made too short.[66]

The $R = 1/2$, $v = 1$ code used a decoding depth of $L = 2$. It turns out that a reduction in decoding depth to $L = 1$ will have a disastrous effect on performance. Massey and Liu suggested analyzing the error propagation phenomenon by observing the autonomous behavior (i.e., with an all-zero input) of the syndrome register.[67] This register is a finite-state machine and an examination of the autonomous state diagram is most instructive.

Suppose that the decoding depth for the $R = 1/2$, $v = 1$ example is reduced to $L = 1$. The only alteration needed in the decoder of Fig. 7-2 is to change the decision device. Only the single syndrome symbol at the far right is available for producing the correction symbol $\hat{e}_n^i$. The only two sensible ways to use this symbol are to either set the correction symbol equal to it, i.e.,

$$\hat{e}_n^i = s_n^{(1)}$$

or avoid correction entirely, i.e.,

$$\hat{e}_n^i = 0$$

The autonomous state diagrams of the syndrome register for these two cases and the case with $L = 2$ are shown in Fig. 7-3. For the first case in Fig. 7-3a an information symbol error will result in a transition to state 3, which will produce the proper correction and a transition to state zero. Thus, no propagation occurs. However, the problem arises when a parity symbol error occurs. This causes a transition to state 2 and then to state 1 where false corrections are made. On each correction an incorrect feedback term is made, which causes the syndrome register to be "trapped" in state 1 resulting in type-II unlimited error propagation. To avoid unlimited error propagation there must always be a path from all nonzero states into state 0. The situation is better in Fig. 7-3b, and no error propagation occurs. However, this is only because no error correction is attempted. Thus, the output error rate is equal to the channel error rate, and a 3-dB penalty has been paid for adding the unused parity symbols. Finally, in Fig. 7-3c the properly designed decoder with $L = 2$ has a path from all nonzero states to state zero. This guarantees that after the last channel error occurs, the error propagation distance is *bounded*.

Another commonly used code is the $R = 1/2$, $v = 2$ nonsystematic code with $g_1(x) = 1 + x^2$ and $g_2(x) = 1 + x + x^2$. It has been shown that a table look-up feedback decoder for this code will have type-II unlimited error propagation with $L = 2$.[66] For $L = 4$ the error propagation distance after channel errors have ceased is unbounded, but it has a relatively small

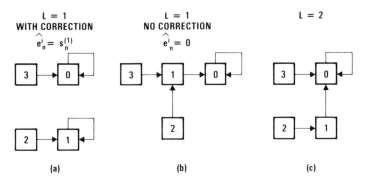

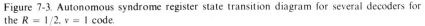

Figure 7-3. Autonomous syndrome register state transition diagram for several decoders for the $R = 1/2$, $v = 1$ code.

expected value. Finally, for $L = 6$ the propagation distance is bounded.

Some work has been done in bounding the value of L necessary to prevent type-II unlimited error propagation. For a noncatastrophic, $R = 1/2$, constraint-length-v code, it has been shown that a minimum-distance feedback decoder with decoding depth $L = v(2v + 1)$ will not suffer from type-II unlimited error propagation.[66] This result holds for both systematic and nonsystematic codes. For the example codes this bound indicates that the values $L = 3$ and 10 will be sufficient to guarantee no unlimited error propagation for $v = 1$ and 2, respectively. However, since the bound increases as v^2, it is probably quite loose at larger constraint lengths.

7.1.2. Table Look-up Decoding

Table look-up decoders have been employed where a very simple implementation is desired and only 1 to 2 dB of coding gain is required. Hard decision demodulator outputs are typically used, and moderate constraint lengths are necessary to keep the look-up table sizes manageable. The implementation of these decoders closely parallels that shown in Fig. 7-2 for the $R = 1/2$, $v = 1$ code.

7.1.2.1. Implementation Concepts. A generic block diagram of a $R = 1/2$ convolutional encoder and feedback decoder employing a table look-up device is shown in Fig. 7-4. This decoding technique has also been called

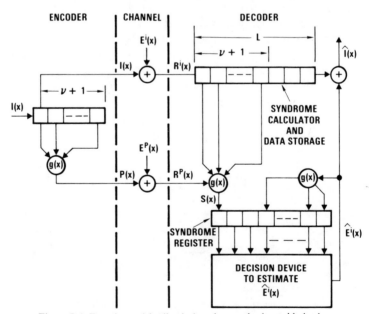

Figure 7-4. Encoder and feedback decoder employing table look-up.

simply feedback decoding in the literature.[68] The code constraint length can be moderately long (in the range of 10 to 12). As discussed previously, the decoder parameter L is made greater than v for proper operation. This parameter defines the span of syndrome symbols that are examined by the decision device to produce the correction symbols. Note that the syndrome calculator is a register with the first $v + 1$ stages connected as an encoder replica to calculate $S(x)$. The register is made L stages long so it can also serve as temporary storage for the received information symbols prior to correction. The syndrome register is another register of length L that holds the span of syndrome symbols that are to be utilized by the decision device. This register also has feedback determined by the code generator polynomial, $g(x)$, which allows removal of the effect on the syndrome span of all received information symbols that must be corrected.

The operation of the decoder is straightforward. Data symbols are encoded by a $R = 1/2$ systematic code into information and parity sequences, $I(x)$ and $P(x)$, which are transmitted over the digital channel. The received sequences are shifted into a syndrome calculator where $S(x)$ is computed. For each code branch one symbol from each sequence $R^i(x)$ and $R^p(x)$ is shifted in to compute one new symbol of the syndrome sequence $S(x)$ which is routed to the syndrome register. As each new syndrome symbol is shifted into this register, the last L syndrome symbols define an address for the decision device look-up table. The decision device, in effect, selects from all error patterns which could have produced this set of L syndrome symbols that pattern which will result in minimum probability of error. If this pattern contains an error in the symbol that is about to leave the data storage register, then the decision device outputs a 1 thereby correcting the error. The device outputs a zero otherwise. The correction symbol is fed back to the syndrome register to remove its effect on the span of syndrome symbols to be examined next by the decision device. Then as long as the decoder makes no errors, decoding of each subsequent symbol proceeds in exactly this fashion. However, a decoding error will cause incorrect feedback, which in turn causes a short burst of errors (i.e., the error propagation effect).

Code synchronization is not shown in this block diagram. As with other convolutional codes, the only level of synchronization needed is that of node or branch synchronization. This simply means that the sequences $I(x)$ and $P(x)$ must be separated if they have been multiplexed prior to transmission. This can be done in several ways. One of the simplest methods is to observe the rate at which the decoder corrects errors (the fraction of ones out of the decision device). If the fraction is much higher than the expected channel error rate, the decoder is probably out of sync and should switch sync positions. The hardware required to perform the counting and switching is trivial.

The decision device is usually implemented as a read-only memory (ROM). Thus, the assignment of syndrome patterns to most likely error patterns is done during the design process, usually through exhaustive search. The parity check matrix for a $R = 1/2$ systematic code is semi-infinite and may be written as

$$
\mathbf{H}^T = \begin{bmatrix}
g_0 & g_1 & g_2 & \cdots & g_v & 0 & 0 & 0\cdots \\
0 & g_0 & g_1 & \cdots & g_{v-1} & g_v & 0 & 0\cdots \\
0 & 0 & g_0 & \cdots & & g_{v-1} & g_v & 0\cdots \\
\multicolumn{8}{c}{\cdots\cdots\cdots\cdots\cdots\cdots\cdots\cdots\cdots\cdots\cdots\cdots} \\
\multicolumn{8}{c}{\mathbf{I}}
\end{bmatrix} \qquad (7\text{-}20)
$$

The syndrome and channel error sequence may be written as vectors

$$
\mathbf{S}^T = (s_0, s_1, s_2, \dots) \qquad (7\text{-}21)
$$

and

$$
\mathbf{E}^T = (e_0^i, e_1^i, e_2^i, \dots, e_0^p, e_1^p, e_2^p, \dots) \qquad (7\text{-}22)
$$

As shown previously, $\mathbf{S}$ is a function only of the error symbols and is given by

$$
\mathbf{S} = \mathbf{HE} \qquad (7\text{-}23)
$$

Since the decision device looks only at L syndrome symbols, one can consider the vectors $\mathbf{S}_L$ and $\mathbf{E}_L$ which are truncated to the first L branches and the truncated parity check matrix $\mathbf{H}_L$. Then

$$
\mathbf{S}_L^T = (s_0, s_1, \dots, s_{L-1}) \qquad (7\text{-}24)
$$

$$
\mathbf{E}_L^T = (e_0^i, e_1^i, \dots, e_{L-1}^i, e_0^p, e_1^p, \dots, e_{L-1}^p) \qquad (7\text{-}25)
$$

and

$$
\mathbf{H}_L^T = \begin{bmatrix}
g_0 g_1 \cdots g_v & & \\
& g_0 g_1 \cdots g_v & \\
\hline
& & g_0 g_1 \cdots g_v \\
& & \\
& \mathbf{I}_L &
\end{bmatrix} \qquad (7\text{-}26)
$$

The submatrix $\mathbf{I}_L$ is an $L \times L$ identity matrix which is multiplied by the parity subset of $\mathbf{E}_L$. Then the matrix equation

$$\mathbf{S}_L = \mathbf{H}_L\mathbf{E}_L \tag{7-27}$$

provides a set of L linear equations relating the L syndrome symbols to the $2L$ channel error symbols. Thus, for each syndrome vector $\mathbf{S}_L$, there are numerous channel error vectors $\mathbf{E}_L$ which satisfy (7-27). The probability of first decoding error will be minimized if one finds the minimum Hamming-weight $\mathbf{E}_L$ that satisfies (7-27) for each of the 2^L possible syndromes and uses these solutions as the look-up table for the decision device. For each minimum-weight solution found, the value of e_0^i in that solution would be used as the correction symbol that is produced by the decision device. The set of solutions for each $\mathbf{S}_L$ is typically found through exhaustive search. Note that they need be found only once, and that these solutions will specify the contents of the look-up table.

This suggested approach will result in a decoder that achieves minimum P_{fe} but not necessarily minimum P_{fd}. Determining the optimum look-up table for minimum P_{fd} is very difficult because of the effects of the limited error propagation that occurs with any feedback decoder. However, simulation results using a number of different look-up tables indicate that optimizing P_{fe} provides near-optimum performance on a P_{fd} basis.[68]

Note that the implementation complexity varies linearly with v and exponentially with L (look-up table size is 2^L). Thus, the decision device requires that moderate values of L be used. Assuming that the complexity of the decision device dominates, then codes should be selected which optimize performance at a given value of L. Define $d(L)$ as the minimum distance between the all-zero path and all paths in the L-branch truncated lower half code tree. Obviously, to optimize performance one should select code generators which maximize $d(L)$. However, for a given L, if there are several codes with the same $d(L)$, then the code that will achieve minimum P_{fe} as $p \to 0$ is that code that has the minimum number of paths with weight $d(L)$. Bussgang used just such a criterion in searching for optimum $R = 1/2$ codes, and several of the codes he found are listed in Appendix B, Table B-8.[69] These codes have odd minimum weight and are the codes that achieve this weight at the minimum value of L. Only one generator is given in this table because the codes are systematic. While this decoding technique can also be applied to nonsystematic codes, they do not offer any real advantage. For each of the codes in this table it is possible to find an equivalent nonsystematic code with a smaller value of v, but reducing v will have no appreciable impact on complexity.[69]

7.1.2.2. Performance. Performance analysis can be done in a manner similar to that demonstrated in Section 7.1.1.2. That is, one can find an expression for P_{fe} for a particular code by exhaustive search. While this approach may provide an underbound to P_{fd} which is in error by only a factor of 2 or 3, one is usually not satisfied until P_{fd} is known. Thus, computer simulation of the specific code and decision device must usually be used. In the region of interest the error rate is closely approximated by

$$P_{fd} \approx \begin{cases} Kp^{(d+1)/2}, & d \text{ odd} \\ Kp^{d/2}, & d \text{ even} \end{cases} \tag{7-28}$$

where K is a constant to be determined by simulation. Performance results for several codes of interest as a function of E_b/N_0 for coherent PSK modulation are shown in Fig. 7-5.[68,70] The actual coding gain at $P_b = 10^{-5}$ is rather modest. The $R = 3/4$, $d = 3$ code gains only 1 dB while the other codes gain about 2 dB. It is interesting that one does not gain a significant performance benefit by using a lower code rate. Also shown for each code is an error rate expression of the form (7-28) which can be used to evaluate performance on other hard decision channels. The table look-up approach is not well suited for utilizing demodulator soft decisions. Conceptually, one could utilize the syndrome symbols as well as an associated sequence

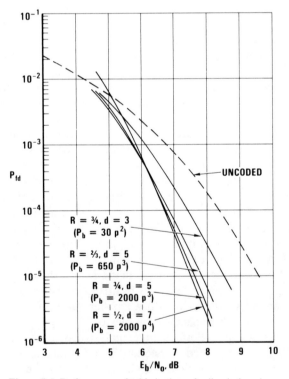

Figure 7-5. Performance of table look-up feedback decoders.

of reliability symbols as the address for a look-up table. However, this approach becomes exceedingly complex even for moderate volumes of L because the size of the look-up table increases exponentially with the number of bits in the address. As a result, only hard decision table look-up decoders have been used.

While table look-up decoding is not well suited to providing large coding gains on the Gaussian channel, it can be a very effective, simple technique for burst error channels. Some form of interleaving is needed when a random error-correcting code is used on a burst error channel. Fortunately, the implementation of any Meggitt-type decoder structure allows *internal interleaving* to be easily accomplished by simply substituting a D-stage shift register in place of each single stage in Fig. 7-4. This separates the transmitted data into D independent coded sequences which are then decoded as D independent sequences by the decoder. Any channel error burst less than D symbols apart will affect no more than one symbol in each of the D sequences. If the bursts have length less than D and occur independently, then the performance of this technique will be identical to that for the random error channel with the same value of p.

This discussion assumes that, as in Fig. 7-4, the $I(x)$ and $P(x)$ sequences are transmitted over separate channels that have independent error processes generating bursts of no more than D symbols in length. If the $I(x)$ and $P(x)$ sequences are to be multiplexed and transmitted over a single channel, then a slight modification is needed. One simply needs to delay either the $I(x)$ or $P(x)$ sequences by DL symbols prior to multiplexing, and then compensate for this delay after demultiplexing. Thus, when a burst occurs, the effects on received information and parity symbols are separated by a decoding constraint length. Otherwise, the operation of the decoder is as just described. With this procedure, bursts of length up to $2D$ can be accommodated without decoding errors. The application of interleaving to other types of decoder structures (e.g., BCH decoders, Viterbi decoders, and sequential decoders) often must be accomplished externally. This approach is discussed in Section 8.3.

7.1.3. Threshold Decoding

Threshold decoding is a practical technique discovered by Massey[13] which can achieve coding gains of 1 to 3 dB with relatively simple implementations. Considerable research interest has produced extensive lists of good codes.[13,71,72] Internal interleaving is easily accommodated in a fashion similar to that of the table look-up decoders. As a result, threshold decoding has seen application in a number of communication systems.

7.1.3.1. Threshold Decodable Codes. Threshold decoding of convolutional codes is performed in exactly the same manner as threshold decoding of block codes (see Chapter 3). The key point is that the code generators are chosen to enable one to solve the parity equations by taking a simple majority vote over a set of J orthogonal estimates, $\{A_k\}$, of an error symbol e_0^i. The estimates $\{A_k\}$ are formed by summing one or more syndrome symbols.

As shown in Chapter 3, the majority decoding rule is to set $e_0^i = 1$ when

$$\sum_{k=1}^{J} A_k > \lceil J/2 \rceil \qquad (7\text{-}29)$$

and to set $e_0^i = 0$ otherwise.

Convolutional codes which allow threshold decoding have a special structure which can be best understood from a simple example. Consider the $R = 1/2$ code with generator polynomial $g(x) = 1 + x^2 + x^5 + x^6$. Since these codes are typically decoded with $L = v + 1$, it is instructive to observe the truncated parity check matrix $\mathbf{H}_7$. This matrix[†] is

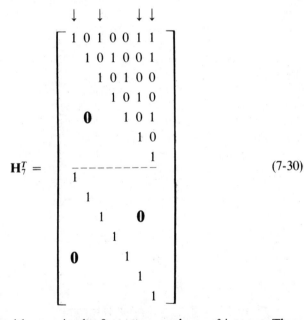

$$\mathbf{H}_7^T = \qquad (7\text{-}30)$$

The columns of $\mathbf{H}_7^T$ with ones in the first row are those of interest. These columns correspond to syndrome symbols that include e_0^i. In addition, note that no two of these columns both have ones in the same row for any other row of $\mathbf{H}_7^T$. Thus, this set of four columns is orthogonal on e_0^i. That is, one can use each one of the corresponding syndrome symbols as a

---

† The **0** indicates that all entries in the triangular portion of the matrix are zero.

separate estimate of e_0^i. This results in the set of estimates

$$s_0 = e_0^i + e_0^p$$
$$s_2 = e_0^i + e_2^i + e_2^p$$
$$s_5 = e_0^i + e_3^i + e_5^i + e_5^p \tag{7-31}$$

and

$$s_6 = e_0^i + e_1^i + e_4^i + e_6^i + e_6^p$$

Inspection of the set $\{s_0, s_2, s_5, s_6\}$ demonstrates that it is a set $\{A_k\}$ that is orthogonal on e_0^i and can be used in a threshold decoder. Note also that this set of estimates consists of single syndrome symbols rather than linear combinations of syndrome symbols. Thus, it is one of the so-called *canonical self-orthogonal codes* (CSOCs) with which one can use syndrome symbols directly as estimates. A more formal definition of these codes follows.

Definition. A *canonical self-orthogonal code* is a convolutional code for which no two parity equations include more than one symbol in common. That is, no two rows of $\mathbf{H}^T$ simultaneously have ones in more than a single column of $\mathbf{H}^T$.

Any code that satisfies this definition will have the property that a set of estimates orthogonal on e_0^i can be obtained simply from the single syndrome symbols corresponding to the positions in the first row of $\mathbf{H}^T$ in which there are ones. A procedure was suggested by Robinson and Bernstein[71] for constructing self-orthogonal codes using the concept of *difference triangles*.

Definition. A *difference triangle* of a row in $\mathbf{H}^T$ is a triangular array whose ijth element is the difference between the column indices of the $(i + j)$th 1 and the jth 1 in that row of $\mathbf{H}^T$.

For the example in (7-30) the difference triangle for the first row of $\mathbf{H}_7^T$ is

$$\begin{array}{l} \quad\; 2\ 3\ 1 \\ \mathbf{T} = 5\ 4 \\ \quad\; 6 \end{array} \tag{7-32}$$

Note that all the elements in the difference triangle are distinct. In fact, if all the elements are distinct, this is equivalent to requiring that any two columns of $\mathbf{H}^T$ which are simultaneously 1 in the first row cannot be simultaneously 1 in any other row [as in $\mathbf{H}_7^T$ of (7-30)]. Thus, given any difference triangle with distinct elements, a self-orthogonal code can be constructed in the following fashion. Use the elements of the first row of the difference triangle to define the first row of $\mathbf{H}^T$ by using the differences in the appropriate fashion. Then the other rows of $\mathbf{H}^T$ are determined by the appropriate shifts of the first row. The difference triangle is completely specified by its first row, which in turn can specify the first row of $\mathbf{H}^T$ and, of course, the code generator polynomial through (7-20). Thus, for a difference triangle which has $(d_{11}, d_{12}, \ldots, d_{1h})$ as its first row, the code generator polynomial is given by

$$g(x) = 1 + x^{d_{11}} + x^{d_{11} + d_{12}} + \cdots + x^v \tag{7-33}$$

with

$$v = \sum_{j=1}^{h} d_{1j} \tag{7-34}$$

If one desires to minimize constraint length for a specified number of elements of the difference triangle, then the sum of the elements in the first row should be minimized. This sum is equal to d_{h1}. Robinson and Bernstein[71] attempted to do this by trying to find difference triangles with the smallest maximum element. The difference triangles they found for several code rates are shown in Appendix B. The triangles for $R = 1/2$ codes are shown in Table B-9. Only even values of J are shown since they produce codes with $d = J + 1$ capable of correcting $J/2$ errors. Note that the example code used in (7-30) and (7-31) is the $J = 4$ code in this table.

The self-orthogonal codes have another interesting property in that they can be decoded by definite decoding. One can see this by examining the semi-infinite form of the parity check matrix in (7-30). For any row in this matrix one can examine the J columns in which ones occur. There will be no other rows (both above and below the row in question) in which more than a single 1 occurs in this set of J columns. This means that the effect of prior decoding decisions need not be removed in order to achieve a set of orthogonal estimates. Without feedback the set of estimates similar to (7-31) becomes

$$s_k = e_k^i + e_{k-2}^i + e_{k-5}^i + e_{k-6}^i + e_k^p$$
$$s_{k+2} = e_{k+2}^i + e_k^i + e_{k-3}^i + e_{k-4}^i + e_{k+2}^p$$
$$s_{k+5} = e_{k+5}^i + e_{k+3}^i + e_k^i + e_{k-1}^i + e_{k+5}^p \tag{7-35}$$

and

$$s_{k+6} = e^i_{k+6} + e^i_{k+4} + e^i_{k+1} + e^i_k + e^p_{k+6}$$

This provides a set of estimates orthogonal on e^i_k, and no feedback is required to guarantee orthogonality. When these codes are decoded via definite decoding, each estimate contains the same number of noise terms.

Code rates other than $1/2$ may be constructed by a similar method. In fact, there is an interesting duality between codes of $R = 1/n$ and those of $R = (n - 1)/n$. One can use $(n - 1)$ sets of difference triangles, and one such set will allow generation of both a $R = 1/n$ code and its dual $R = (n - 1)/n$ code. The key property required is that one needs $(n - 1)$ sets of difference triangles for which all elements from all the triangles are distinct. If this condition is satisfied, then this set of difference triangles can be used to form either a $R = 1/n$ code or a $R = (n - 1)/n$ code, both of which are self-orthogonal. The first row of each difference triangle can be used to compute a generator polynomial as in (7-33). This results in a set of $(n - 1)$ polynomials $g_1(x), g_2(x), \ldots, g_{n-1}(x)$. A $R = (n - 1)/n$ systematic self-orthogonal code is obtained as

$$T^j(x) = I^j(x), \qquad j = 1, 2, \ldots, n - 1$$

and

(7-36)

$$T^n(x) = \sum_{j=1}^{n-1} g_j(x)\, I^j(x)$$

That is, there are $n - 1$ transmitted information sequences with the single parity sequence $T^n(x)$. For the $R = 1/n$ code the same generator polynomials are used in a different fashion. The self-orthogonal code consists of a single information sequence and $(n - 1)$ parity sequences, i.e.,

$$T^1(x) = I^1(x)$$

and

(7-37)

$$T^j(x) = g_j(x)\, I^j(x), \qquad j = 1, 2, \ldots, n - 1$$

As an example, consider the pair of difference triangles that give the generator polynomials $g_1(x) = 1 + x$ and $g_2(x) = 1 + x^2$. The truncated parity check matrix for the $R = 2/3, J = 2$ code may be written

$$
\mathbf{H}^T =
\begin{array}{ccc}
s_0 & s_1 & s_2 \\
\downarrow & \downarrow & \downarrow \\
\end{array}
\left[
\begin{array}{ccc}
1 & 1 & 0 \\
0 & 1 & 1 \\
0 & 0 & 1 \\
\hline
1 & 0 & 1 \\
0 & 1 & 0 \\
0 & 0 & 1 \\
\hline
1 & 0 & 0 \\
0 & 1 & 0 \\
0 & 0 & 1 \\
\end{array}
\right]
\qquad (7\text{-}38)
$$

The first row has ones in the first and second columns, and no other rows have ones in both of these columns. Thus, $\{s_0, s_1\}$ form a pair of orthogonal estimates of e_0^1. Likewise, the fourth row has ones in the first and third columns, and no other rows have ones in both of these columns. Thus, $\{s_0, s_2\}$ form a pair of orthogonal estimates of e_0^2. Both e_0^1 and e_0^2 may be estimated correctly in the presence of single errors. The $R = 1/3$ code has a truncated parity check matrix

$$
\mathbf{H}^T =
\begin{array}{cccc}
s_0^1 \; s_1^1 & \quad & s_0^2 & \quad s_2^2 \\
\downarrow \;\; \downarrow & & \downarrow & \quad \downarrow \\
\end{array}
\left[
\begin{array}{ccc|ccc}
1 & 1 & 0 & 1 & 0 & 1 \\
0 & 1 & 1 & 0 & 1 & 0 \\
0 & 0 & 1 & 0 & 0 & 1 \\
\hline
1 & 0 & 0 & 1 & 0 & 0 \\
0 & 1 & 0 & 0 & 1 & 0 \\
0 & 0 & 1 & 0 & 0 & 1 \\
\end{array}
\right]
\qquad (7\text{-}39)
$$

Two syndrome sequences are computed for the $R = 1/3$ code

$$S^1(x) = g_1(x)\, R^1(x) + R^2(x)$$

and   (7-40)

$$S^2(x) = g_2(x)\, R^1(x) + R^3(x)$$

These two sequences correspond to the computations of the left and right halves of the $\mathbf{H}^T$ matrix, respectively. Note that the first row of $\mathbf{H}^T$ has four columns with ones and no other rows have more than a single one in these columns. Thus, $\{s_0^1, s_1^1, s_0^2, s_2^2\}$ form a set of orthogonal estimates on e_0^1. This code has $J = 4$ so double errors can be corrected.

Another useful class of codes is the class of so-called *trial-and-error codes* introduced by Massey.[13] These codes differ from the CSOCs in several respects. First, the J orthogonal estimates found for each code are not single syndrome symbols as with CSOCs, but some are formed as sums of two or more syndrome symbols. Also, the estimates must be used with feedback decoding to achieve the full error-correction capability. If definite decoding were used, a significant loss in performance would occur. An advantage of these codes is that one can achieve a given value of J with a shorter constraint length than with CSOCs. The $R = 1/2$ trial-and-error codes are shown in Table B-16.

A typical case is that of the $J = 4$ code. The truncated parity check matrix for this code is

$$
\mathbf{H}^T =
\begin{bmatrix}
1 & 0 & 0 & 1 & 1 & 1 \\
 & 1 & 0 & 0 & 1 & 1 \\
 & & 1 & 0 & 0 & 1 \\
\mathbf{0} & & & 1 & 0 & 0 \\
 & & & & 1 & 0 \\
 & & & & & 1 \\
\hline
 & 1 & & & \mathbf{0} & \\
 & & 1 & & & \\
\mathbf{0} & & & 1 & & \\
 & & & & 1 & \\
 & & & & & 1 \\
\end{bmatrix}
\qquad (7\text{-}41)
$$

Note that s_0, s_3, and s_4 are all orthogonal on e_0^i. However, when s_5 is added as an estimate, it is not orthogonal since both s_5 and s_4 have a 1 in the second row, i.e., they both include the e_1^i term. This term is canceled in s_5 by adding s_1. Thus, the estimate $s_1 + s_5$ provides the fourth orthogonal estimate.

7.1.3.2. Implementation Considerations. A block diagram of a typical hard decision threshold decoder for $R = (n - 1)/n$ codes is shown in Fig. 7-6. Conceptually, this design is very similar to that employed for table

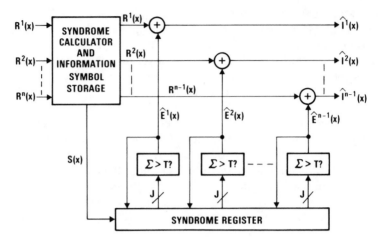

Figure 7-6. Hard decision threshold decoder for $R = (n-1)/n$ codes.

look-up decoders shown in Fig. 7-4. Syndrome calculation and data storage
are typically accomplished in the same register. The principal difference is
that the table look-up decision device has been replaced by decision devices
that simply sum J syndrome symbols and compare with the threshold,
$\lceil J/2 \rceil$. Thus, the decision device has complexity that grows only linearly
with J. Operation of the decoder is straightforward. As each symbol of the
received sequence $R^1(x), R^2(x), \ldots, R^n(x)$ is shifted into the syndrome
calculator, a single symbol of the syndrome sequence, $S(x)$, is calculated
and shifted into the syndrome register. The appropriate sets of J estimates
in the syndrome register are routed to the decision devices, and the esti-
mated error symbols $\hat{E}^1(x), \hat{E}^2(x), \ldots, \hat{E}^{n-1}(x)$ are produced and used to
correct both the received data and the contents of the syndrome register.
Assuming no error is made, then the next decoding cycle proceeds in exactly
the same manner. Somewhat more insight into this process is obtained
from the block diagram of Fig. 7-7 for the $R = 2/3$, $J = 2$ code. Note that
here the decision devices reduce to simple AND gates. Feedback to the
syndrome register is indicated, but it can be removed if definite decoding
is desired.

Soft decisions may also be utilized with threshold decoding, though at
a significant increase in complexity. The optimum technique for utilizing
soft decisions with threshold decoding is the APP decoding technique[13]
discussed in Chapter 4. Implementation considerations for convolutional
codes are identical to those discussed previously for block codes. A weight
w_k is assigned to each estimate, A_k. This weight is a function of the probability that A_k is correct. The APP decoding rule is to set $e_0^i = 1$ if and

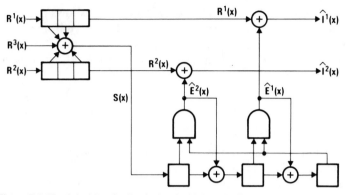

Figure 7-7. Hard decision feedback threshold decoder for $R = 2/3$. $J = 2$ code.

only if

$$\sum_{k=1}^{J} w_k A_k > \frac{1}{2} \sum_{k=0}^{J} w_k \qquad (7\text{-}42)$$

The weights, w_k, may be calculated either by the nonlinear mapping or the approximate APP technique[22,23] discussed in Chapter 4. A functional block diagram of an APP decoder is shown in Fig. 7-8. A comparison of

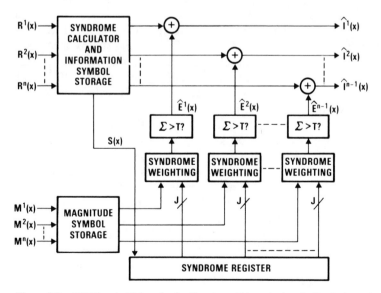

Figure 7-8. APP threshold decoder for $R = (n - 1)/n$ codes (no feedback shown).

this figure with Fig. 7-6 for a hard decision decoder shows that storage for magnitude symbols and a provision for weighting syndrome symbols by w_k have been added. This also makes the summations slightly more complicated. Note that (7-42) requires a variable threshold, but this equation may be modified to give

$$\sum_{i=1}^{J} w_k(1 - 2A_k) + w_0 < 0 \qquad (7\text{-}43)$$

A simpler implementation may be obtained by using this or a similar form with constants added to each side of (7-43).

Several observations can be made about decoder complexity at high data rates (several Mb/s). With hard decision decoders the principal contributors to complexity are the set of shift registers required for syndrome calculation and information symbol storage and the syndrome register. The code constraint length and hence complexity increases as J^2 while coding gain increases rather slowly with J. Thus, only small values of J are practical ($J \leq 6$). In a similar manner, with J fixed the complexity increases as n^2 for $R = (n - 1)/n$ codes. Note from Tables B-10 through B-15 that high-rate codes will require very long shift registers. When soft decisions are added, a set of registers for storing magnitudes is added, as well as more complex decision devices. (Typically, this results in about a factor of 2 increase in complexity.)

With careful code selection one can achieve significant coding gain with modest complexity. Several years ago the $R = 1/2$, $J = 6$ trial-and-error code was implemented to operate at 5 Mb/s utilizing four-level soft decisions with approximate APP decoding. Using standard SSI and MSI, less than 40 ICs were required to implement the encoder and decoder. This approach provided 2.5 dB of coding gain at $P_b = 10^{-5}$ with PSK signaling. These results compare very favorably with the table look-up decoders of the previous section.

Threshold decoding has the advantage that implementations at very high data rates (exceeding 50 Mb/s) are feasible. This is easiest if definite decoding is used with CSOCs. With no feedback to the syndrome register the difficult operations in the decision devices can be *pipelined* allowing an increase in the rate at which all the registers are shifted. Of course, there is a slight performance loss suffered in using definite decoding. If only a few shifts are required to propagate through the decision logic, then some of this loss can be recaptured by using *delayed feedback*. The CSOCs provide a large class of codes with which very high data rates are possible.

7.1.3.3. Performance Analysis. Massey[13] has discussed performance analysis techniques for threshold decoders using error-counting techniques similar to those presented in Section 7.1.1.2. The probability of first decoding error may be written

$$P_{fe} = (1 - p) \Pr\left[\sum_{k=1}^{J} w_k A_k > T \middle| e_0^i = 0 \right]$$

$$+ p \Pr\left[\sum_{k=1}^{J} w_k A_k \leq T \middle| e_0^i = 1 \right] \qquad (7\text{-}44)$$

This equation holds for all forms of threshold decoding. For hard decisions with simple majority decoding $w_k = 1$ for all k and $T = \lceil J/2 \rceil$.

Massey introduced the concept of the *size*, n_k, of an estimate A_k. This quantity refers to the number of error symbols included in A_k excluding e_0^i. This concept is important because at a fixed channel error rate p, the probability that an estimate A_k is incorrect increases monotonically with its size n_k. This is easily shown as follows. An extimate A_k of e_0^i is incorrect if and only if an odd number of error symbols in that estimate other than e_0^i are ones, and the probability that an odd number of errors occur is simply

$$P_k = \sum_{\substack{i \, odd}}^{n_k} \binom{n_k}{i} p^i (1 - p)^{n_k - i}$$

$$= \frac{1 - (1 - 2p)^{n_k}}{2} \qquad (7\text{-}45)$$

Note that the set of estimates, $\{A_k\}$, conditioned on $e_0^i = 1$ are the complements of the set, $\{A_k\}$, conditioned on $e_0^i = 0$. Therefore, (7-44) can be rewritten as

$$P_{fe} = (1 - p) \Pr\left[\sum_{k=1}^{J} w_k A_k > T \middle| e_0^i = 0 \right]$$

$$+ p \Pr\left[\sum_{k=1}^{J} w_k A_k \geq \sum_{k=1}^{J} w_k - T \middle| e_0^i = 0 \right] \qquad (7\text{-}46)$$

and only the probability distributions of the $\{A_k\}$ conditioned on $e_0^i = 0$ need be considered.

First consider the case of hard decisions with majority decoding. A generator for the random-variable $\sum_{k=1}^{J} A_k$ is simply

$$g(x) = \prod_{k=1}^{J} (1 - P_k + P_k x)$$

$$= \sum_{j=1}^{J} g_j x^j \qquad (7\text{-}47)$$

Each coefficient in this generating function is the probability that the random variable $\sum_{k=1}^{J} A_k$ is equal to the exponent of x in that term. With this technique the probability of first decoding error can be written

$$P_{fe} = (1 - p) \sum_{\substack{j > T}}^{J} g_j + p \sum_{\substack{j \geq \bar{T}}}^{J} g_j \tag{7-48}$$

where

$$\bar{T} = \sum_{k=1}^{J} w_k - T$$

The performance estimates assume a particularly simple form for the CSOCs with definite decoding and hard decisions.[73] In this case all estimates have the same size, σ. Also, since there is no feedback, every error symbol which is decoded depends only on a given set of neighboring channel error symbols and there is no dependency upon previous decoding decisions. Thus, the overall bit error probability is equal to P_{fe} and is given by (7-48). The generator for this case may be written

$$g(x) = [1 - P + Px]^J \tag{7-49}$$

where

$$P = \frac{1 - (1 - 2p)^{\sigma}}{2} \tag{7-50}$$

Since $w_k = 1$ for all k, (7-48) may be written

$$P_b = (1 - p) \sum_{i = T + 1}^{J} \binom{J}{i} P^i (1 - P)^{J - i} + p \sum_{i = J - T}^{J} \binom{J}{i} P^i (1 - P)^{J - i} \tag{7-51}$$

In this equation the thresholds are $T = J/2$ for J even and $T = (J + 1)/2$ for J odd. This equation may be used for both the $R = 1/n$ and $R = (n - 1)/n$ codes as determined from Tables B-9 through B-15. For the $R = (n - 1)/n$ codes the J given in the code tables is the J required in (7-51). In addition, each estimate has size $\sigma = J(n - 1)$ to be used in (7-50). For the $R = 1/n$ codes the J required in (7-51) is actually $J(n - 1)$ from the tables. The resulting parity check sizes are equal to the values of J in the tables.

This technique was used to compute the P_b for the $R = 1/2$ and $R = 2/3$ codes as shown in Figs. 7-9 and 7-10 when PSK modulation is used. Since hard decision quantization of the demodulator output is assumed, one could make performance estimates for other signal designs rather easily by determining the channel error rate p that results in the P_b of interest. The difference in E_b/N_0 required for the two signal designs is

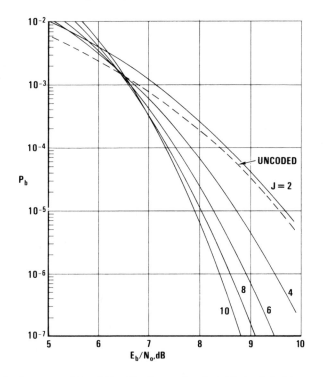

Figure 7-9. Performance of $R = 1/2$, self-orthogonal codes with hard decisions and definite decoding.

found by comparing them at the value of p corresponding to the desired operating point.

An interesting observation that can be made from these figures is that one quickly reaches a point of diminishing returns as J increases. Increasing J from 6 to 12 provides only about 0.5 dB more gain at $P_b = 10^{-5}$. Since complexity increases proportional to J^2, only the values of $J \leq 6$ are typically cost effective. Another interesting property is that at a fixed value of J, higher rate codes often outperform lower rate codes due to lower band-spreading loss. More will be said about this aspect shortly.

In order to determine the performance of APP decoding one must usually resort to simulation. Typical results for both block and convolutional codes indicate that an APP decoder provides a performance improvement of about 1.7 dB relative to a hard decision decoder when using eight-level demodulator soft decisions.

In the case of approximate APP decoding an analytical expression for P_{fe} can be found. Since the weights, w_k, take on a finite number of values, the expression (7-46) can be evaluated in much the same manner used with

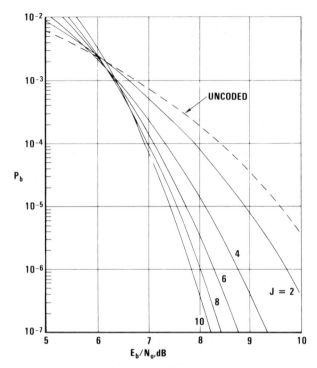

Figure 7-10. Performance of $R = 2/3$, self-orthogonal codes with hard decisions and definite decoding.

hard decisions (though it is much more tedious). This approach was used to calculate the performance curves of Fig. 7-11 for four-level quantization. Since the codes used are CSOCs, $P_b = P_{fe}$. The quantizer level was set at $0.4E_s^{1/2}$. Deviations of ± 3 dB in this setting cause only about 0.1 dB of performance degradation. The additional gain obtained by using four-level rather than two-level quantization is about 0.9 dB. A similar gain results for the higher-rate codes. It appears that the additional gain available with soft decisions using approximate APP decoding decreases as the size of each estimate increases [which is equal to $J(n - 1)$ for these codes].

Significantly more coding gain is obtained by using feedback decoding. For example, by using feedback with a $R = 1/2$, $J = 6$ code, the parity check sizes can be reduced from $\{6, 6, 6, 6, 6, 6\}$ to $\{1, 2, 3, 4, 5, 6\}$. With the reduced sizes, the probability of first decoding error, P_{fe}, can be calculated in the same fashion. This quantity is substantially less than P_b for definite decoding. Typically, P_b for feedback decoding is 2 to 3 times P_{fe}, but it must be determined through simulation. For most codes of interest, the use of feedback decoding has been shown through simulation to provide

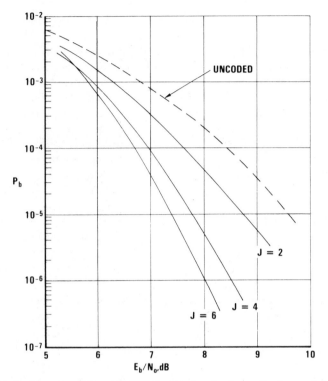

Figure 7-11. Performance of $R = 1/2$ CSOCs with four-level quantization and definite decoding.

about 0.6 dB more gain than definite decoding (at $P_b = 10^{-5}$) both for hard and soft decisions.

Performance calculations indicate that if J is kept fixed and code rate is varied, peak coding gain typically occurs for high code rates ($3/4 < R < 5/6$). The peak gain can be roughly 0.5 dB higher than that achieved with $R = 1/2$ codes. Unfortunately, the decoder complexity increases as n^2 so that increasing n significantly to achieve greater gain is not cost effective. The principal advantage of the higher-rate CSOC codes is that they can offer moderate coding gain in applications where only a very limited amount of redundancy is allowable.

7.2. Sequential Decoding Techniques

Sequential decoding was first introduced by Wozencraft,[64] but the most widely used algorithm to date is due to Fano.[74] Our discussions will focus principally on the Fano algorithm, but subsequently some comparisons with the "stack" sequential decoding algorithm[75,76] will be made.

While a Viterbi decoder extends and updates the metrics for all paths that can potentially be the best, a sequential decoder severely limits the number of paths that are actually updated. The fundamental idea of sequential decoding is that only the path that "appears" to be the most probable should be extended. We say "appears" since, because of the limited search, the decoder is never completely certain that this path is the best. This approach can be viewed as a trial-and-error technique for searching out the correct path in the code tree. It performs the search in a sequential manner always operating only on a single path. However, the decoder is allowed to back up and change previous decisions. Each time the decoder moves forward a "tentative" decision is made. The decision is made by extending the path being examined along its most probable branch. If an incorrect decision is made, subsequent extensions of the path will be wrong. The decoder is eventually able to recognize this situation by examining the path metric. However, it is quite possible to extend an incorrect path for many branches. When this occurs, a substantial amount of computation is required to recover the correct path. The decoder must search back and try alternate paths until it finally decodes successfully. It is very important to carefully select decoder parameters to allow quick recognition of an incorrect decision and quick recovery of the correct paths in order to mitigate the computational problem. The primary benefit of the sequential decoding approach is that each correct decision contributes to limiting the amount of computation that must be performed subsequently. At the same time, the path metric is providing an indication of the correctness of earlier decisions.

The branch metric at node i is defined by

$$\lambda_i = \sum_{j=1}^{n} \left\{ \log_2 \left[\frac{P(r_i^j \mid t_i^j)}{P(r_i^j)} \right] - B \right\} \tag{7-52}$$

where t_i^j is the transmitted symbol and r_i^j is the received symbol. (A binary $R = m/n$ code with n symbols per branch is assumed.) The path metric through node k is simply

$$L_k = \sum_{i=0}^{k} \lambda_i \tag{7-53}$$

The parameter B is a bias term to be explained later. Generally, this term is chosen such that the metric will be increasing in value over a correct path and decreasing in value over an incorrect path. The typical metric behavior is shown in Fig. 7-12. Although the metric for the correct path may temporarily show large decreases due to channel noise, over a longer

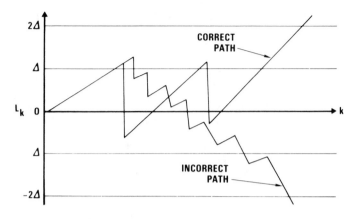

Figure 7-12. Typical metric behavior of correct and incorrect paths.

period of time it should be an increasing function. Also, if a burst of channel noise occurs, the metric on an incorrect path may increase temporarily making the path look like a good path, but it will usually start decreasing when the noise subsides. A sequential decoding algorithm must be able to detect a decreasing metric and quickly find a path with an increasing metric. To do this it uses a *running threshold T* which may be raised or lowered by increments of Δ, the *threshold spacing*. When the current metric falls below the running threshold T, it is an indication that the path being followed may be bad, and a search for a good path is initiated.

7.2.1. Sequential Decoding via the Fano Algorithm

The operation of a sequential decoder is just a trial-and-error search. The principal decoding rule for the Fano algorithm is that the decoder will never proceed either forward or backward on a path that falls below the current threshold. The 2^m branches emanating from the node under consideration are compared to the received branch and ordered according to their likelihood of occurrence. Normally, the decoder moves forward on the most likely branch at each node as long as the metric lies above the current threshold. Also, if $T + \Delta$ is crossed in the positive direction, the threshold is raised by Δ. When the decoder cannot move forward, it attempts to move back without violating the threshold and search all paths that have not been tried with the current threshold. When no paths can be found that satisfy the threshold, the threshold is lowered by Δ and the decoder begins searching forward again. A set of rules for implementing this procedure is shown in Table 7-1. This algorithm, due to Gallager,[3] is a restatement of Fano's sequential decoding algorithm.

Table 7-1. Sequential Decoding Rules for the Fano Algorithm

	Conditions		Action	
Rule	Previous move	Comparisons[a]	Final threshold	Move
1	F or L	$L_{k-1} < T + \Delta, L_k \geq T$	Raise (if possible)	F
2	F or L	$L_{k-1} \geq T + \Delta, L_k \geq T$	No change	F
3	F or L	any $L_{k-1}, L_k < T$	No change	B
4	B	$L_{k-1} < T,$ any L_k	Lower	F
5	B	$L_{k-1} \geq T,$ any L_k	No change	L or B

[a] By convention set $L_0 = 0$ and $L_{-1} = -\infty$.

The letters F, L, and B in Table 7-1 denote forward, lateral, and backward moves by the decoder. During a forward move the decoder moves forward to the most likely of the 2^m nodes emanating from the current node. A backward move is simply a move to the node preceding the current node. A lateral move is a move to a node differing from the current node only in the final branch and being the next most likely node in the order of their likelihood. When there is little noise, the decoder is normally using rule 1 and following the correct path. The decoder can proceed for many branches making only forward moves and extending only a single path.

The threshold is raised whenever possible to keep it less than Δ below L_k. If the noise becomes more severe and the metric falls below the threshold after a forward move, rule 3 is invoked and the decoder moves backward in order to find a path that lies above the current threshold. After a backward move either rule 4 or rule 5 is invoked depending upon whether L_{k-1} lies below or above the current threshold. If $L_{k-1} \geq T$, then a lateral move to the next most likely node is made if such a node exists, and a backward move is made if such a node does not exist. If $L_{k-1} < T$, no further backward moves are possible without violating the current threshold so the threshold is lowered by Δ and a forward move is made. The decoder now tries to find a path remaining above this reduced threshold. Rule 2 will now be used. As long as $L_{k-1} \geq T + \Delta$, the threshold cannot be raised, because if it were, the decoder would become trapped in an endless loop.

To better understand the Fano algorithm, consider the following simple example based on the $R = 1/2$, $v = 1$ code with the code tree shown in Fig. 7-1. Assume for the moment that hard decision decoding is used, and that the transmitted information sequence is 1 0 0 0 0.... The received sequence is $\mathbf{R} = 01\ 01\ 00\ 01\ 00\ 00...$. That is, there are single errors in the first and fourth branches. The branch metrics corresponding to this code with the received sequence $\mathbf{R}$ are shown in the tree of Fig. 7-13 using metric increments of $+0.5$ for each symbol that agrees and -4.5 for each

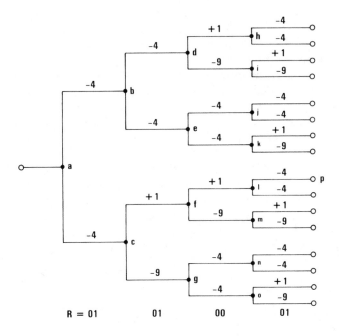

Figure 7-13. Branch metrics for the tree of Fig. 7-1.

Table 7-2. Tree Search of the Fano Algorithm for the Example of Fig. 7-13

Node	T	k	L_k	Final T	Move
b	0	1	-4	0	B
a	0	0	0	-5	F
b	-5	1	-4	-5	F
d	-5	2	-8	-5	B
b	-5	1	-4	-5	L
c	-5	1	-4	-5	F
f	-5	2	-3	-5	F
l	-5	3	-2	-5	F
p	-5	4	-6	-5	B
l	-5	3	-2	-5	B
f	-5	2	-3	-5	B
c	-5	1	-4	-5	B
a	-5	0	0	-10	F

symbol that disagrees. (The selection of these values for metric increments will be justified subsequently.) The steps taken by the Fano algorithm are listed in Table 7-2. The convention used when the decoder is presented with two branches with the same metric (as in the first step) is that the upper branch will always be tried first. Thus, the algorithm begins by trying to find a good path in the upper half tree. The first step forward results in a threshold failure which forces a step back and lowering of the threshold. However, in proceeding forward in the same direction to node d, another threshold failure occurs. This forces a movement back and into the lower half tree in an attempt to find a better path. Forward movement along the best path continues until node p, where another channel error causes a threshold failure. The decoder then must retrace its steps until it reaches node a and lowers the threshold to -10. Now there is a single path that always lies above this threshold. After searching the upper half tree, the decoder can move forward along the path a–c–f–l–p without violating the threshold. (These steps are left as an exercise for the reader.) The correct path has been found, and since there are no further channel errors, the metric will continually increase from this node along the correct path and only forward moves will be made.

7.2.2. Selection of Sequential Decoding Metric

The sequential decoding metric of (7-52) with $B = R$ (the code rate) was suggested by Fano based on heuristic arguments. Massey[77] has demonstrated that this metric allows the decoder to extend the most likely path based on the information available at each iteration of the decoding algorithm. These arguments will be briefly presented to provide further insight into sequential decoder behavior.

Consider a $R = m/n$ binary convolutional code and a sequential decoder which is attempting to decode the vector

$$\mathbf{r}_N = (r_0^1, \ldots, r_0^n, r_1^1, \ldots, r_1^n, \ldots, r_N^1, \ldots, r_N^n)$$

of the first $N + 1$ received branches. At any point the decoder will have examined M potential code tree sequences of varying lengths and compared them with $\mathbf{r}_N$. These sequences $\{\mathbf{t}_{k_1}, \mathbf{t}_{k_2}, \ldots, \mathbf{t}_{k_M}\}$ are each of the form

$$\mathbf{t}_k = (t_0^1, \ldots, t_0^n, t_1^1, \ldots, t_1^n, \ldots, t_k^1, \ldots, t_k^n)$$

A sequential decoder is restricted to extending one of these M sequences on its next move, and it is assumed to have no knowledge of the sequences in the unexplored part of the code tree. Thus, we shall select as the path $\mathbf{t}_k$ to be extended that path which maximizes $\Pr(\mathbf{t}_k | \mathbf{r}_N)$. Note that by Bayes' rule

$$\Pr(\mathbf{t}_k | \mathbf{r}_N) = \frac{\Pr(\mathbf{r}_N | \mathbf{t}_k) \Pr(\mathbf{t}_k)}{\Pr(\mathbf{r}_N)}$$

and this is equivalent to maximizing

$$\Pr(\mathbf{r}_N | \mathbf{t}_k) \Pr(\mathbf{t}_k)$$

or to maximizing

$$\Pr(\mathbf{r}_N | \mathbf{t}_k) \Pr(\mathbf{t}_k) \bigg/ \prod_{i=0}^{N} \prod_{j=1}^{n} \Pr(r_i^j)$$

Assuming information symbols that are independent and equally likely to be zeros or ones we have

$$\Pr(\mathbf{t}_k) = 2^{-(k+1)m}$$

In addition, since we have assumed that the decoder has no knowledge of the extensions of path $\mathbf{t}_k$ beyond the kth node, we model this situation as if the path $\mathbf{t}_k$ had a random tail added increasing its length to N nodes. Thus, we write

$$\Pr(\mathbf{r}_N | \mathbf{t}_k) = \prod_{i=0}^{k} \prod_{j=1}^{n} \Pr(r_i^j | t_i^j) \prod_{i=k+1}^{N} \prod_{j=1}^{n} \Pr(r_i^j)$$

Using these expressions we note that the decoder should maximize

$$2^{-(k+1)m} \prod_{i=0}^{k} \prod_{j=1}^{n} \frac{\Pr(r_i^j | t_i^j)}{\Pr(r_i^j)}$$

Finally, taking $\log_2(\cdot)$ of this expression gives (7-52) and (7-53) with $B = R$.

This result states that given the information that is available ($\mathbf{r}_N$ and the M previously examined code sequences of varying lengths), the decoder chooses to extend the most likely path (to within the quantization of the threshold spacing, Δ). The use of this metric results in a very "efficient" search for the correct path, and acts to mitigate somewhat the sequential decoder computational problem. Note that the M partial code words $\{\mathbf{t}_{k_1}, \mathbf{t}_{k_2}, \ldots, \mathbf{t}_{k_M}\}$ in general will have different lengths and the metric used will account for these differences. For each path $\mathbf{t}_{k_i}$ the metric (7-53) depends only on that path and the received symbols through the first $k_i + 1$ nodes. This is a result of the assumption that the decoder has no knowledge of the symbols in any extensions of that path. However, the use of this metric does not imply that the decoder will achieve maximum-likelihood decoding. The metric (7-53) is equivalent to a maximum-likelihood metric only when comparing paths with the same lengths. This metric is biased such that longer paths are favored over shorter paths to produce an efficient search. Because of this bias, a path that would have been selected by a maximum-likelihood decoder can fail to be extended by a sequential decoder. Fortunately, the impact of this nonoptimality is not severe.

7.2.3. Code Selection

Selection of codes for use with sequential decoding is not particularly critical as it is for the other techniques that have been considered. Any code can be decoded by any of the sequential decoding algorithms. In addition, decoder complexity is not strongly dependent on code constraint length. Thus, it is not necessary to optimize the code for a given value of v. Instead, one can increase v as much as necessary to achieve the desired level of performance. Since finding the optimum code (in the sense of the best weight structure) is computationally infeasible for large values of v, most researchers have been content to find codes with large values of d_f. The generators for a number of long codes suitable for sequential decoding are given in Appendix B in Table B-17. A long $R = 1/2$, systematic code found by Forney[78] and given in this table has been shown to have performance adequate for most applications. The resulting code generator is *nested*. That is, the generator may be shortened at any point to produce a good shorter code.

When it is desirable to minimize v, a nonsystematic code can be used. Several examples are also given in Table B-12. A very interesting nonsystematic code is the so-called "quick-look code" found by Massey and Costello.[79] The generators are related by

$$g_1(x) + g_2(x) = x \qquad (7\text{-}54)$$

This allows recovery of the transmitted information sequence from the received sequence by simply adding $R^1(x)$ and $R^2(x)$, i.e.,

$$R^1(x) + R^2(x) = I(x)g_1(x) + E^1(x) + I(x)g_2(x) + E^2(x)$$
$$= xI(x) + E^1(x) + E^2(x) \qquad (7\text{-}55)$$

These codes eliminate one of the disadvantages of nonsystematic codes [that $I(x)$ cannot be reliably estimated without decoding], and the resulting error rate of recovered estimate of $I(x)$ is only twice the channel error rate (an "error amplification factor" of 2). However, one needs a larger value of v to achieve the same performance as a nonsystematic code without this feature.

In summary, an adequate list of suitable codes is already available. However, if for some reason an application requires an even longer code it would not be difficult to generate such a code.

7.2.4. Sequential Decoder Computational Problem

A major problem with sequential decoding is that the number of computations required in advancing one node deeper into the code tree is a random variable. This characteristic strongly affects the complexity required to achieve a given level of performance. When there is little noise, the decoder is usually following the correct path requiring only one computation (i.e., one F, L, or B move) to advance one node deeper into the code tree. However, if the noise becomes severe, the metric along the correct path may decrease while the metric along an incorrect path may temporarily be larger. This causes the decoder to proceed along an incorrect path, and a large number of computations may be required before the decoder begins following the correct path. The number of computations required in getting through a period of severe noise is a rapidly increasing function of the number of times the threshold must be lowered. The most significant effect of the variability in the number of computations required in decoding a symbol is that a large memory is required to buffer the incoming data. Any finite buffer used with sequential decoding has a nonzero probability of overflowing, an event which must be considered in performance calculations.

This computational problem has been studied by researchers for a number of years. One of the more important results is the lower bound to the *distribution of computation* found by Jacobs and Berlekamp.[80] Their results are applicable to any sequential decoding algorithm. Such an algorithm is required to have only two properties:

1. Branches are examined sequentially so that, at any node, the decoder's choice among previously unexplored branches does not depend on received branches deeper in the tree.
2. the decoder performs at least one computation at each node of every path examined.

Rather than show the derivation of this bound, we shall discuss the simple example given by Jacobs and Berlekamp which illustrates the critical nature of the problem. Then the general result will be stated without proof.

Consider a binary erasure channel with erasure probability, δ (symbols are either erased or received correctly). Assume that an $R = m/n$ code is used, i.e., a code with 2^m branches per node and n symbols per branch. The probability of a burst of N erasures occurring in the first N symbols is δ^N. Assume that N is a multiple of n for simplicity. Note that after receiving erasures over these first N symbols, the decoder has absolutely no information with which to guide its selection of paths. The total number of paths (that are N symbols long) diverging from the origin is 2^{RN}. The decoder cannot use information beyond the first N symbols to order its initial search of these 2^{RN} paths. Thus, with probability $1/2$, at least $1/2$ of these paths must be examined before locating the correct path (with at least one computation per path). This means that with probability $1/2$, at least

$$L = 2^{RN-1}$$

total computations will be required before the correct path can be followed.

Now let C denote the number of computations required to decode the first M information symbols, and choose

$$N = \frac{1}{R} \log_2(2L)$$

Note that the only values of N allowed are integer multiples of n, and it is necessary that $NR < M$. Then if the first N symbols are erasures, L or more computations occur with probability at least $1/2$ giving

$$\Pr[C > L] > \tfrac{1}{2}\delta^N$$
$$= \tfrac{1}{2}2^{(NR \log_2\delta)/R}$$
$$= \tfrac{1}{2}(2L)^{(\log_2\delta)/R}$$
$$= aL^{-\alpha}$$

where the parameters $\alpha = -(\log_2\delta)/R$ and $a = 2^{-(1+\alpha)}$ have been defined. A distribution of the form $aL^{-\alpha}$ is called Pareto with exponent α. Unless α is very large, this distribution decreases slowly with L creating the necessity for a large buffer.

This example demonstrates several properties of sequential decoding algorithms. The parameters α and a in this bound depend only on the *channel* and the *code rate*. Obviously then, the computational problem is reduced by increasing the signal-to-noise ratio. Note that the Pareto distribution arises because, even with random noise, there is a sufficiently high probability that long bursts will occur with the result that they require a very large number of computations in order to find the correct path. The probability of a noise burst of length N decreases exponentially with N, while the number of computations required increases exponentially with N. The result of these two opposing effects is the $L^{-\alpha}$ form of the Pareto distribution. This example also demonstrates that sequential decoders are extremely sensitive to burst error channels, i.e., any channel for which the probability of a burst error of length N does not decrease exponentially with N.

Sequential decoder computational behavior can be accurately described in terms of the "Gallager function," $E_0(\rho)$, given by (1-64) and the exponential bound parameter R_0 which were both discussed in Section 1.4. Jacobs and Berlekamp[80] showed that, for a discrete memoryless channel, the distribution of computation with any sequential decoding algorithm is lower bounded by

$$\Pr[C \geq L] > L^{-\rho}[1 - o(L)] \tag{7-56}$$

where the Pareto exponent ρ for a code of rate R on this channel is given by the implicit solution of

$$R = \frac{E_0(\rho)}{\rho} \qquad (7\text{-}57)$$

In this result $o(L) \sim 1/(\log_2 L)^{1/2}$. The result is valid for $0 < \rho < \infty$ and $0 < R < C$, where C is the channel capacity. The interesting point that this bound shows is that no sequential decoding algorithm can have a distribution of computation that decreases more rapidly with L than a function of the form $L^{-\rho}$. The resulting exponent in this bound is a function only of the code rate R and the channel, and it is easily calculated through (7-57). The exponent determined in this manner provides a tighter bound than the exponent found in the preceding example. Experimental and theoretical results, particularly the upper bound to the distribution of computation found by Savage,[81] support the conclusion that the actual distribution is Pareto and that the actual exponent can be found from (7-57). Knowledge of the operating Pareto exponent ρ allows one to accurately estimate the buffer size required to achieve a given performance level (buffer overflow probability). Thus, (7-56) and (7-57) are important in the decoder design process.

Note that the exponential bound parameter R_0 as given by

$$R_0 = E_0(1) \qquad (7\text{-}58)$$

is the code rate at which $\rho = 1$ is the solution of (7-57). This rate is called the *computational cutoff rate* and is sometimes denoted by R_{comp}. This is a *practical limit* on the highest rate at which a sequential decoder can operate since a Pareto distribution with $\rho = 1$ is a distribution with an infinite mean. This indicates that any sequential decoder operating at $R > R_0$ will have a severe computational problem with frequent buffer overflows. The minimum feasible value of E_b/N_0 at which a sequential decoder can operate is usually predicted through calculating R_0. Since codes are usually chosen so that undetected error rates are extremely small under typical operating conditions, the computational problem has the dominant effect on system performance. Thus, R_0 is an important performance parameter.

7.2.4.1. Behavior with Coherent PSK. Consider the case of binary antipodal signaling over an additive Gaussian channel. The practical limit on sequential decoder performance for a code of rate R is the value of E_b/N_0 required to achieve $R_0 = R$. This parameter was evaluated in Section 1.4 for the case of PSK signaling with coherent detection. Required values of E_b/N_0 as a function of R are shown in Fig. 1-12. Demodulator quantizations of $Q = 2$, 8, and ∞ are shown. Operation at values of E_b/N_0 below those shown will result in an exceedingly severe computational load on the decoder. Note that with $R = 1/2$ codes, the required E_b/N_0 is 4.6 dB with hard decisions and 2.6 dB with eight-level quantization. The latter number is substantially better than can be achieved with Viterbi decoding of short codes.

In addition to the practical limit on E_b/N_0 one can also find the Pareto exponent at any E_b/N_0 above this limit. Consider the case of a binary symmetric channel with transition probability, p. Then the Gallager function can be written

$$E_0(\rho) = \rho - \log_2\{[(1-p)^{1/(1+\rho)} + p^{1/(1+\rho)}]^{(1+\rho)}\} \qquad (7\text{-}59)$$

The Pareto exponent, ρ, is given by the implicit solution of

$$R = 1 - \frac{1}{\rho}\log_2\{[(1-p)^{1/(1+\rho)} + p^{1/(1+\rho)}]^{(1+\rho)}\} \qquad (7\text{-}60)$$

For PSK signaling $p = Q[(2E_s/N_0)^{1/2}]$, and the resulting exponent for several code rates as a function of E_b/N_0 is shown in Fig. 7-14. In addition, similar results are shown for binary PSK signaling with eight-level demodulator decisions. Typical operating points in each case are the ranges of E_b/N_0 which produce $1 < \rho < 2$. Pareto exponents as predicted from Fig. 7-14 agree closely with those measured through simulation.

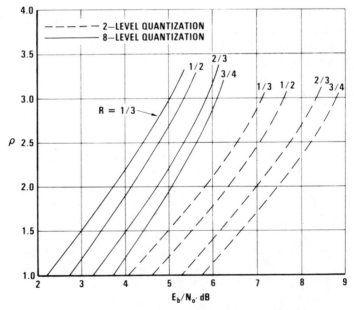

Figure 7-14. Pareto exponent as a function of E_b/N_0 and code rate.

7.2.4.2. Behavior with M-ary Orthogonal Signaling and Noncoherent Detection. We observed in Chapter 6 that a noncoherent system employing Viterbi decoding is most efficient when M-ary orthogonal signaling is used. Jordan has observed that this is also true in a system employing sequential decoding.[82] This is shown through the behavior of the R_0 parameter. We shall briefly summarize his results.

The demodulator for M-ary orthogonal signaling with noncoherent detection consists of a bank of M matched filters and envelope detectors. When these are sampled, the probability density function for the filter with signal present, $p_{s+n}(y)$, is given by (1-10).

For the case of signal absent, the probability density function, $p_n(y)$, is given by (1-11).

Jordan showed that with $Q = \infty$, the R_0 parameter can be expressed as

$$R_0 = \log_2 \left\{ M \left[1 + (M - 1) \left(\int_{-\infty}^{\infty} [p_{s+n}(y) \, p_n(y)]^{1/2} \, dy \right)^2 \right]^{-1} \right\} \quad (7\text{-}61)$$

Then R_0 is found by substituting (1-10) and (1-11) into (7-61) and performing the integration numerically. This has been done for several values of M, and the E_b/N_0 required to achieve $R_0 = R$ as a function of R is shown in Fig. 7-15. The code rate R is expressed in information bits per M-ary symbol.

Several aspects of this figure are deserving of comment. First, we note that, as expected, higher-level signaling alphabets are significantly more efficient in terms of E_b/N_0 than binary signaling. For example, the advantage of 16-ary signaling relative to binary is over 4 dB. Second, we note that for each value of M, there exists an optimum code rate which is typically slightly greater than $\log_2 M/2$. With binary signaling $R = 1/2$ codes are optimum. This is consistent with the results found in Chapter 6 with Viterbi decoding. In comparison with those results, a $R = 1/2$ sequential decoder could perform about 1 dB better than a Viterbi decoder. For larger values of M, the simplest way to utilize M-ary signaling is to use one M-ary symbol per code branch. Then the code rate selected will determine the number of branches emanating from each node in the tree (e.g., $R = 1$ gives a binary tree, $R = 2$ gives a quaternary tree, etc.). Using this approach, a binary tree ($R = 1$) is optimum for $M = 4$. At $M = 8$, the quaternary tree ($R = 2$) is most efficient, but one could use a binary tree at a cost of only 0.4 dB. Similarly, at $M = 16$, the quaternary tree is most efficient. This last system was implemented by Jordan,[82] and his measured performance results compare closely with the required E_b/N_0 as predicted by Fig. 7-15 after adjusting for nonideal filtering and quantization. Finally, this figure also implies that not only is there an optimum code rate to use with sequential decoding

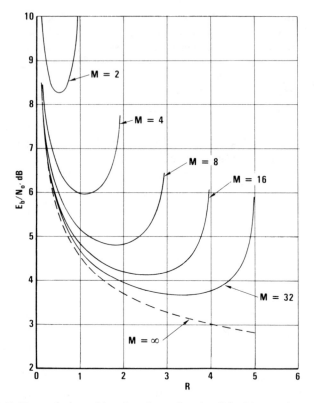

Figure 7-15. E_b/N_0 required to achieve $R = R_0$ as a function R for M-ary orthogonal signaling with noncoherent detection.

but there is also an optimum code rate to use in a system employing Viterbi decoding. Remember that the M-ary codes presented in the previous chapter utilized only $R = 1/2$ or $R = 1$ codes. The dual-3 code presented was a $R = 1/2$, 8-ary code. Figure 7-15 predicts that $R = 1/2$ codes are about 0.8 dB less efficient than $R = 1$ codes. The reason the $R = 1/2$ code was chosen is that the code constraint length is so short that the additional minimum distance is needed. Similarly, this figure also shows that $R = 1$ codes are near-optimum for 8-ary signaling. However, for larger alphabets ($M = 16$ or 32), one may wish to use $R = 2$ codes.

  7.2.4.3. Simulation Results. The decoder computational problem has a large influence on the decoder design since the decoder computational burden must be estimated. This is typically done by estimating the decoder cumulative distribution of computations and the average number of computations required per decoded branch. These quantities are influenced by code rate, E_b/N_0, and various decoder parameters such as metric assignments and threshold spacing.

First consider the influence of the metric assignment on the average number of computations. The form of the branch metric is given by (7-52). For the BSC the quantity added to the path metric when the received symbol agrees with the hypothesized channel symbol is

$$m_0 = \log_2 2(1 - p) - B \qquad (7\text{-}62)$$

and the quantity added when the received symbol disagrees with the hypothesized channel bit symbol is

$$m_1 = \log_2 2p - B \qquad (7\text{-}63)$$

The value of the bias should be set to $B = R$ to produce an efficient search. The general effect of using a bias term larger than R is to cause the decoder to enter a search more quickly as the result of a small amount of noise. This is desirable from an undetected error rate viewpoint since it increases the chance that the correct path will be examined (thereby providing an undetected error rate closer to that of a maximum-likelihood decoder). However, it is undesirable from a computational viewpoint because of the increased computational burden. Making B too small can also increase the computational burden because errors may not be recognized quickly enough.

To find the proper metric ratio, $\text{MR} = m_0/m_1$ for the BSC at $R = R_0$, (1-67) is solved for the corresponding value of p. Then by using this value of p and by setting $B = R$ in (7-62) and (7-63), the corresponding metric ratio is obtained. The resulting values of p and MR for $R = 1/2, 2/3$, and $3/4$ are shown in Table 7-3. (Note that the optimum metric values for a $R = 1/2$ code at $R = R_0$ are approximately $+0.5$ for each symbol agreement and -4.5 for each symbol disagreement, which were the values used in the example in Section 7.2.1.)

Simulation results showing the average number of computations per decoded branch, $\bar{C}$, as a function of E_b/N_0 for a $R = 1/2, v = 15$ systematic

Table 7-3. BSC Transition
Probability p and Optimum
MR at $R = R_0$

R	p	MR
1/2	0.045	$1/-9.15$
2/3	0.017	$1/-18.0$
3/4	0.009	$1/-27.7$

code with several metric ratios are shown in Fig. 7-16. At $R = R_0$ ($E_b/N_0 =$ 4.6 dB), the choice of MR $= 1/-9$ is far superior to the others. For higher values of E_b/N_0 this produces the minimum $\bar{C}$, but other choices such as MR $= 1/-11$ require only slightly more computations on the average. The latter choice may be preferable in this region due to an improved undetected error rate. [Note that MR $= 1/-11$ is optimum at $E_b/N_0 = 6.3$ dB as predicted by (7-62) and (7-63). The simulation results are consistent with this observation.]

The value of $\bar{C}$ is also a function of Δ, the threshold spacing. However, $\bar{C}$ exhibits a rather broad minimum with respect to Δ. For a metric ratio of $1/-n$ the value $\Delta = n + 1$ is nearly optimum and is also convenient to implement.

The distribution of computations is very important in estimating decoder buffer requirements. The tail of the cumulative distribution of computations per decoded branch has the form

$$\Pr[C \geq L] \approx AL^{-\rho}, \qquad L \gg 1 \qquad (7\text{-}64)$$

where A is a constant usually of the order of 1 or 2, and ρ is the Pareto exponent given by (7-57). This exponent can be found directly from Fig. 7-14 for a coherent PSK system. The constant A depends on the specific algorithm and decoder parameters. Optimization of decoder parameters such as the metric assignments and the threshold spacing can reduce the constant, A. Improper metric assignments or demodulator AGC setting has also been known to degrade the Pareto exponent with soft decision decoding. Simulation results for a Fano algorithm implementation with a

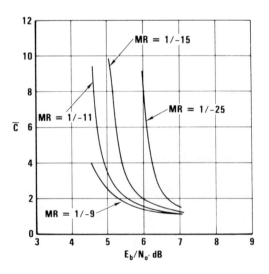

Figure 7-16. Average number of computations for a $R = 1/2, v = 15$ systematic code with hard decisions.

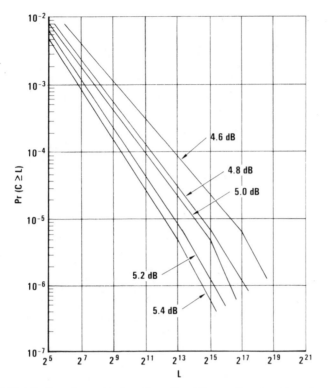

Figure 7-17. Distribution of computations for a $R = 1/2$, $v = 40$ systematic code with hard decisions.

$R = 1/2$, $v = 40$ code are shown in Fig. 7-17 (again hard decision quantization is assumed). This figure shows the measured $\Pr[C \geq L]$ when MR $= 1/-9$ and $\Delta = 10$ are used. The Pareto exponents given by these distributions correspond closely to those that would be predicted from (7-57). Thus, the tail of this distribution can be predicted by (7-64) with a constant A near unity.

7.2.5. Performance of Sequential Decoders

Exact analytical calculations of performance cannot be made for sequential decoders as is done with some other decoding techniques. However, performance can be estimated within a few tenths of a dB for most cases of interest by using certain approximations. Decoding errors can arise from two sources. The choice of code and metric ratio determines the undetected error rate, P_e. This is the bit error rate that would be observed if the decoder were allowed as many computations as it required to decode each symbol.

The second contributor to error rate arises from the finite size of the decoder buffer, and the resulting nonzero probability of overflow. This is a rather troublesome event, and the system can handle the problem in one of two ways. For example, one can provide an alternate method of decoding to estimate the span of symbols affected by the overflow. This can be done by simply delivering the received information symbols when a systematic code is used or by taking some linear combination of the received sequences (as was done with the "quick-look" codes) when a nonsystematic code is used. Naturally, this alternate decoding method can have a rather high error rate. The overall symbol error probability can then be estimated by

$$P_b = P_e + P_a \qquad (7\text{-}65)$$

where P_a represents the contribution to error rate due to alternate decoding of overflow events. A second technique for handling overflows is to simply erase the affected symbols. This is an excellent technique in many systems such as those with an ARQ capability because a sequential decoder provides a significant degree of error detection. In this case the coded sequence will consist of a sequence of moderately long "terminated" code blocks. That is, the convolutional code is turned into a block code in the following manner. Each block of L information symbols is encoded by a convolutional encoder (initialized at state 0). After the Lth symbol is shifted into the encoder, a sequence of v zeros is shifted in to return the encoder to state 0 (any other known state could also be used). This action "terminates" the code sequence in state 0. With this knowledge the decoder can determine when it has satisfactorily completed decoding a block. A fixed amount of time is allocated per block for decoding. If the buffer overflows, an erasure is declared, and a retransmission is necessary. The erasure probability is simply the buffer overflow probability per code block. The undetected error rate for this approach is P_e, which can be made as small as desired by increasing the code constraint length v.

7.2.5.1. Undetected Error Rate. Once a code is selected, the undetected error rate is significantly influenced by the metric ratio that is used. The metric ratio that produces the most efficient sequential decoder search does not always select the maximum-likelihood path.

Figure 7-18 shows some simulation results for a $R = 1/2$, $v = 15$ code for several metric ratios with hard decision quantization of the demodulator output. The optimum metric to minimize P_e is Hamming distance (which is closely approximated by MR $= 1/-25$). Note that metric ratios obtained by setting $B = R$ (those in the range $1/-9$ to $1/-11$) show a significant degradation in P_e. However, near-optimum decoding has a significant cost in terms of $\bar{C}$ as shown in Fig. 7-16. It is possible that the

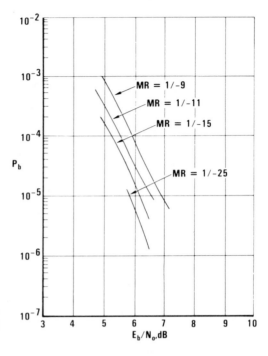

Figure 7-18. Performance curves for a $R = 1/2$, $v = 15$ code with hard decisions.

difference in performance between the metric with $B = R$ and the optimum metric is not so great with nonsystematic codes. We have observed this for shorter nonsystematic codes. This conclusion is supported by other theoretical results. It has been shown by ensemble coding techniques that setting $B = R$ in the metric results in an optimum error exponent for nonsystematic codes but a nonoptimum error exponent for systematic codes.[83,84]

Typically, the most cost-effective way to improve P_e is not by changing the metric ratio but by increasing v. Performance curves for several truncated versions of Forney's $R = 1/2$ code[78] are shown in Fig. 7-19. A metric ratio of $1/-9$ is used. Performance may be improved by about 0.5 dB by using $MR = 1/-11$. Constraint lengths in the neighborhood of 40 to 50 are necessary for operation in the $P_e = 10^{-5}$ or 10^{-6} range with R near $R_0(E_b/N_0 = 4.6$ dB). This code has been successfully employed in several decoder implementations.

The Massey and Costello $R = 1/2$ quick-look code[79] has also been implemented with excellent results. The $v = 31$ truncated version has $P_e < 10^{-5}$ at $E_b/N_0 = 2$ dB with soft decision decoding. This is more than adequate to operate near $R = R_0$ with negligible contribution to overall error rate due to P_e.

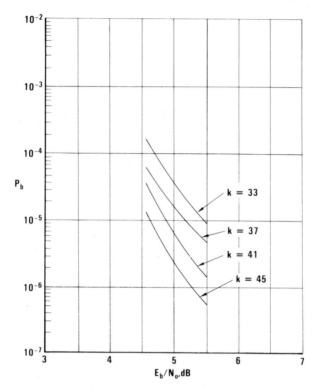

Figure 7-19. Performance curves for $R = 1/2$ systematic codes with hard decisions and $MR = 1/-9$.

In most cases the codes of Table B-17 will provide a value of P_e that is negligibly small compared to the contribution to error rate caused by buffer overflow. This is typically the most desirable mode of operation. If P_e is not negligible, it can be made negligible simply by increasing v. This could be done by adding an appropriate random "tail" to any of the generators of this table, and then checking to see if the desired performance improvement occurs.

Once P_e is made negligible, system performance may be accurately estimated by estimating the probability of buffer overflow.

7.2.5.2. Probability of Buffer Overflow. The probability of buffer overflow is usually the most critical performance parameter, and it is closely related to the distribution of computations per branch. There are only two decoder parameters that strongly affect the probability of buffer overflow. One is, of course, the buffer size, Γ, in code branches. The other is the decoder CPU speed factor, μ, i.e., the number of computations the CPU can perform during the time required to receive one branch.

Assume that a decoder with parameters μ and Γ has an initially empty buffer. Then this decoder must perform $\mu\Gamma$ computations without decoding a branch before a buffer overflow occurs. Thus, from (7-64) the initial overflow probability may be estimated as

$$P_0 \approx A(\mu\Gamma)^{-\rho} \qquad (7\text{-}66)$$

where the Pareto exponent, ρ, is given by (7-57). Simulations have shown that this is an accurate estimate of the buffer overflow probability when μ is somewhat larger than the average number of computations per branch, $\bar{C}$ (allowing the buffer to remain empty most of the time). The constant, A, in (7-66) depends upon the specific implementation and is usually on the order of 1 or 2. This estimate is less accurate when μ is close to $\bar{C}$ as has been observed in high-speed sequential decoder implementations. In this case, one must resort to simulation to reliably estimate P_0.

For systems in which the data are blocked (i.e., the code is terminated after L branches) one is more interested in the probability of failure to decode the L-branch block within the allowable time. This probability may be expressed as

$$P_{\text{ob}} \approx L P_0$$
$$\approx AL(\mu\Gamma)^{-\rho} \qquad (7\text{-}67)$$

The manner in which (7-66) and (7-67) are used in estimating system performance depends on whether the system employs block erasures or alternate decoding for processing buffer overflow events.

The most sensible approach in systems with the ARQ capability is to declare block erasures. The resulting system will have an undetected error rate, P_e, which is the undetected error rate of the sequential decoder. One may make P_e as small as desired by increasing ν. The block erasure rate is then given by the probability of failure to decode an L-branch block, i.e.,

$$P_{\text{eras}} = P_{\text{ob}}$$
$$\approx AL(\mu\Gamma)^{-\rho} \qquad (7\text{-}68)$$

Typical block sizes of about $L = 1000$ are common. This technique is well matched to applications such as packet-switched networks where ARQ techniques are used.

When alternate decoding is used, there are two methods of achieving decoder resynchronization after a buffer overflow. One method is to simply block the data with appropriate frame synchronization so that if a buffer overflow on one block occurs, the decoder can quickly move to the next block and begin decoding. In this way the decoder fails to decode only on a single block. The transmitted information symbols for this block are then estimated by "alternate means" and delivered to the user. For systematic codes this would involve simply delivering the received information symbols (with a fraction of errors equal to the channel error rate, p). For nonsystematic codes one would want to use a "quick-look" code to reconstruct an estimate of the transmitted information sequence. In this case the average error rate would be $2p$. Thus, the contribution to error rate due to alternate decoding can be expressed as

$$P_a = \beta p P_{ob} \qquad (7\text{-}69)$$

where $\beta = 1$ for systematic codes, $\beta = 2$ for quick-look codes, and P_{ob} is given by (7-67). Although the bit error rate is actually given by (7-65), P_b is usually dominated by P_a for long constraint lengths.

The other method is the so-called automatic resynchronization or "guess-and-restart" technique.[85] When the buffer becomes full, the decoder goes into the resynchronization mode and essentially gives up on the set of symbols it has been trying to decode. It is forced to move forward into the code tree (at least a constraint length) and is put into the state it would have been in had there been no errors in the previous constraint length of information symbols. If that happens to be the correct state, decoding will proceed normally. Often several attempts at resynchronization must be made before it is completed, but this technique is often preferable to blocking the data. While the decoder is in the resynchronization mode, uncorrected data are delivered to the user. These symbols are estimated in the same fashion as with blocked data. The average number of errors in the delivered data during the resynchronization process is dependent on E_b/N_0 and the decoder implementation. The probability of error due to alternate decoding may then be expressed as

$$P_a = N_a P_0 \qquad (7\text{-}70)$$

where P_0 is given by (7-66) and N_a is the average number of errors delivered to the user before resynchronization is achieved. For hard decision decoders N_a is a number on the order of 100. Unfortunately, this technique is much more difficult to use on soft decision decoders because of the higher channel error rates encountered.

7.2.6. Implementation Considerations

Sequential decoders have been successfully implemented a number of times at data rates ranging from low to very high for hard decision decoders and at moderate data rates for soft decision decoders. As a result, a considerable amount of practical experience with sequential decoder implementation has been obtained.

   7.2.6.1. Hard-Decision Decoders. Several high-speed (data rates of several Mb/s and higher) hard-decision decoders have been built in recent years (the first of these is described in detail in Forney and Bower[86]). These decoders all have similar architectures, as shown in Fig. 7-20 (though detailed implementations differ).

A $R = 1/2$, systematic code is assumed. Data flow through the decoder is as follows. When the information and parity sequences, $R^i(x)$ and $R^p(x)$, are received, a syndrome sequence, $S(x)$, is computed by an encoder replica. The received information sequence $R^i(x)$ then enters a delay which is equal to the delay of the decoder buffer plus the delay through the decoder CPU. The syndrome sequence is all that is required to reconstruct the error sequence $E^i(x)$, so it alone is operated on by the decoder. The sequence, $S(x)$, is written into the decoder buffer random access memory (RAM). Syndrome symbols are withdrawn, as required, by the decoder CPU in the lower portion of Fig. 7-20. The symbols that are returned from the CPU to the buffer are the correction symbols $\hat{E}^i(x)$. The withdrawal of syndrome symbols and replacement of correction symbols into the buffer occurs as required by the Fano algorithm decoder. The rate at which this occurs is highly variable because of the variation in the number of computations

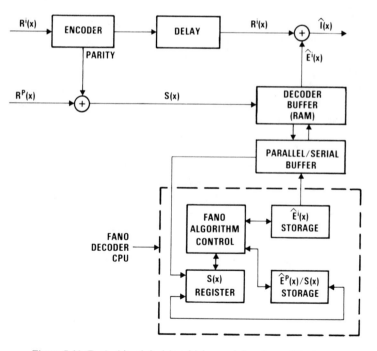

Figure 7-20. Typical hard decision, high-speed decoder architecture.

required to decode a symbol. Correction symbols, of course, are read from the buffer at a constant rate. The buffer is considered to be "full" and near overflow when it is full of undecoded syndrome symbols, and there are no correction symbols in it. The buffer is considered "empty" when it is full of correction symbols, and there are no syndrome symbols in it.

The decoder CPU is organized with enough temporary storage to avoid unnecessary accesses to the buffer RAM. That is, each syndrome symbol will be read from the RAM once, and later it will be replaced by the correction symbol determined by the Fano algorithm. Neither of these symbols will ever be used by the decoder CPU again. No backsearches required by the Fano algorithm will require additional data from the RAM. The only data required are held in the CPU storage. This type of organization in a high-speed decoder allows reasonable values for the word size and access time of the RAM.

The temporary storage required is enough to store all preceding "tentative" decisions for roughly the five constraint lengths preceding the decoder's deepest penetration. The depth of previous data stored is sometimes called the *backsearch depth* since the decoder would never be allowed to back up past this point. Storage for $S(x)$, $\hat{E}^i(x)$, and $\hat{E}^p(x)$ is provided. The $S(x)$ register contains the sequence of future syndrome symbols upon which the present estimate of the error symbols is to be based. This register is essentially an encoder replica that allows modification of v symbols of the $S(x)$ sequence whenever the hypothesized error symbols are changed (i.e., effects of preceding errors are removed).

When the decoder is moving forward in the code tree, the $S(x)$ register is shifted right. Estimated error symbols are also placed into the $\hat{E}^i(x)$ and $\hat{E}^p(x)$ storage. New error symbols are hypothesized using the contents of the $S(x)$ register under Fano algorithm control. When an error in the $\hat{E}^i(x)$ sequence is hypothesized, the code generator is added modulo-2 to the contents of the $S(x)$ register to remove the effect of that error. On backward moves the last symbols in the $\hat{E}^i(x)$ and $\hat{E}^p(x)$ sequences are used to restore to the $S(x)$ register the contents it had before that tentative decision was made. As backward moves [left shifts of the $S(x)$ register] are made, syndrome symbols are removed from the left-hand end of the $S(x)$ register and placed in $\hat{E}^p(x)/S(x)$ temporary storage. If the algorithm requires a search to the end of the backsearch depth, this storage will be filled with syndrome symbols. Then as the decoder moves forward again, these syndrome symbols are returned to the $S(x)$ register, and the $\hat{E}^p(x)/S(x)$ temporary storage again begins to fill with $\hat{E}^p(x)$ symbols.

The Fano algorithm control element simply implements the set of rules of decoder motion as given in Table 7-1, and it tabulates the relationship of the metric to the current threshold. The path metric hardware for $R = 1/2$ codes with hard decisions assumes a simple form. Define the metric increments for a branch as $+1$, $-\Delta + 1$, and $-2\Delta + 1$ for zero, one, and two errors on a branch, respectively. The threshold spacing Δ is assumed to be an integer. This choice produces a metric ratio of $1/(-2\Delta + 1)$. The metric can be accumulated in two counters. The first is a Δ-stage ring counter which counts modulo-Δ and shifts right or left depending upon whether the decoder is searching forward or backward (note that branch increments are always 1 mod-Δ regardless of whether the threshold is changed or not). Whenever the first stage of the ring counter is 1, the metric is 0 (modulo-Δ), and when the last stage is 1, the metric is Δ-1 (modulo-Δ). Another counter is used to accumulate multiples of Δ (the Δ counter). Then the actual metric is the sum of the contents of the mod-Δ counter and Δ times the contents of the Δ counter.

Nonsystematic codes can be used in a similar fashion. In fact, with "quick-look" codes the change is trivial. At the front-end of the decoder the syndrome is calculated in a slightly different fashion, i.e.,

$$S(x) = g_2(x) R^1(x) + g_1(x) R^2(x)$$
$$= g_2(x) E^1(x) + g_1(x) E^2(x) \tag{7-71}$$

As shown in (7-55) an estimate of the transmitted information sequence is easily obtained as $R^1(x) + R^2(x)$, and the necessary correction symbol is simply

$$\hat{E}(x) = \hat{E}^1(x) + \hat{E}^2(x) \tag{7-72}$$

Thus, there are only two modifications of the decoder CPU. The first involves allowing the $S(x)$ register contents to be modified by either $g_1(x)$ or $g_2(x)$ depending upon whether $E^2(x)$ or $E^1(x)$ was in error. The second is to simply form $\hat{E}(x)$ via (7-72) before returning it to the decoder buffer.

The branch synchronization function was not shown. This is usually implemented by measuring some parameter that indicates if the decoder is successfully decoding. One technique is to determine the fraction of corrections that are hypothesized to be errors through measuring a sufficiently long sequence. When the decoder is in sync, this fraction should be the channel error rate p while an out-of-sync condition produces a much higher fraction of hypothesized errors that is easily detected. Another method is to examine the behavior of the path metric. When the decoder is in sync, the rate of metric threshold tightenings will exceed the rate of threshold loosenings by a considerable margin. This is not the case when the decoder is out of sync, and this condition is easily detected.

A high-speed decoder architecture requires a considerable degree of complexity that is not necessary in lower-speed designs. One can design a lower-speed decoder with a substantial saving in hardware. For example, there are 2 to 4 compare operations per "computation" that could be done serially in a lower-speed design. The temporary storage for $\hat{E}^i(x)$ and $\hat{E}^p(x)$ does not need to be as large (an entire backsearch depth) if the computation cycle is slowed down. Finally, the decoder buffer RAM can usually be much smaller since the speed factor μ is usually large. This tradeoff will be discussed in more detail shortly. Decoders that operate in the 100-Kb/s range can be built for under 100 ICs including the decoder buffer.

7.2.6.2. Fano Algorithm Modifications. A considerable amount of effort has been expended in trying to improve the sequential decoder computation problem predicted by the Pareto distribution of computations. Several novel concepts have been investigated in an attempt to alleviate this problem. One such concept is the technique of quick threshold loosening. This technique has been implemented in a high-speed decoder to reduce the average number of computations, $\bar{C}$.[86] The rationale of the concept is the following. A decoder operating at rates below R_0 spends most of its time advancing along the correct path or backing up by a small amount to correct one or two errors. The quick threshold loosening technique reduces the amount of time spent on short searches thereby lowering $\bar{C}$ for $R = 1/2$ hard decision decoders. Metric values of $+1$, $-\Delta + 1$, and $-2\Delta + 1$ ($\Delta = 5$ or 6) for 0, 1, and 2 errors per branch are assumed. In addition, it is assumed that the two branches emanating from any node are complementary. In this situation there is a set of conditions under which a threshold can be loosened without requiring a backsearch. This occurs if, first, the last threshold adjustment was to raise it, and, second, if no errors are observed from that point to the present branch. In this case all alternate paths must fall below the current threshold (becuse of the choice of metric). If the present branch contains an error that causes a threshold violation, then the threshold may be lowered by Δ immediately because there are no other paths that lie above it. The decoder may then continue to advance. However, quick threshold loosening may not be employed again until the threshold is raised.

The effect of this strategy has been shown through simulation to have absolutely no effect on the tail of the distribution of computations, i.e., it does not help in long searches.[59] However, it achieves enough of a reduction in the number of computations required for short searches that a significantly lower value of $\bar{C}$ is obtained. For high-speed decoders that are operating with small decoder speed factors, μ, this technique gains a few tenths of a decibel. For low data rate applications where μ is large the benefit is very small. Fortunately, though, this technique is also very easy to implement.

Another interesting technique is that of *look-ahead decoding*. Note that in Fig. 7-20 the manner in which the Fano algorithm control utilized the contents of the $S(x)$ register to direct its search was not specified. However, in a standard implementation the Fano algorithm control bases its actions upon only a single syndrome symbol at a time. This means that any paths that eventually fall below the current threshold must be followed until the threshold violation actually occurs.

When using the look-ahead technique, the Fano algorithm bases each move upon a sequence of N syndrome symbols. This action is equivalent to looking forward N branches into the tree from the present node to determine if there is any path N branches ahead that satisfies the current threshold. If such a path exists, the decoder steps forward on the next branch if that can be done without violating a threshold. If no path exists N branches ahead that satisfies the threshold, then a backsearch is initiated. This technique can cause a reduction in the tail of the distribution of computations which amounts to about a factor of 2 for $N = 6$.[59] However, the slope of the distribution and hence the Pareto exponent is not changed; only the constant multiplying the $L^{-\rho}$ term is affected. As a result, the gain obtained is only about 0.1 dB. In addition, the implementation problems of this approach make it unattractive at high speeds. Implementation is no problem at lower speeds, but the expected benefit is rather small.

Thus, we observe that the basic Fano algorithm is very efficient and there are few modifications to it that will result in really significant improvements. The Pareto distribution of computations exists for *all* sequential decoder algorithms, and the best that any modification of the algorithm can do is to effect a small change in this distribution. The additional coding gain that can be obtained for a given level of complexity is only a few tenths of a decibel.

7.2.6.3. Decoder Parameters. The principal decoder parameters that affect performance are the speed factor, μ, and the buffer size, Γ. In fact, for most cases where μ is significantly greater than $\bar{C}$, the performance is only dependent on the $\mu\Gamma$ product (assuming that other decoder parameters are intelligently chosen). In this case one may compensate for smaller μ by increasing Γ or vice versa. However, there is not a significant degree of latitude in selecting μ in most cases. One would design the decoder CPU as efficiently as possible while maintaining a reasonably fast computation cycle. The desired degree of performance is then obtained by adding the necessary amount of buffer memory.

To gain more insight into the trade between $\mu\Gamma$ and performance, we shall develop an approximate expression for P_a from (7-68). For the cases of most interest the constant multiplying the $(\mu\Gamma)^{-\rho}$ term is roughly 200 (to within a factor of 2). Thus, assume that P_a can be estimated by

$$P_a \approx 200(\mu\Gamma)^{-\rho} \qquad (7\text{-}73)$$

Figure 7-21 shows the $\mu\Gamma$ product required to achieve values of P_a of 10^{-5}

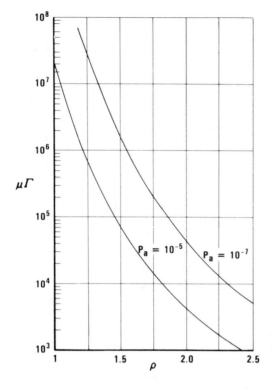

Figure 7-21. Value of $\mu\Gamma$ required to achieve a specified value of P_a $[P_a \approx 200(\mu\Gamma)^{-\rho}]$ as a function of ρ.

and 10^{-7} as a function of the Pareto exponent ρ. The $\mu\Gamma$ product required is very sensitive to ρ. By increasing ρ from 1.0 to 1.5, one can decrease the required $\mu\Gamma$ product by over two orders of magnitude. As shown in Fig. 7-14, this can be accomplished with an increase of less than 1 dB of E_b/N_0 with a coherent PSK system. Thus, to achieve that last 1 dB of coding gain and operate at $R = R_0$ is very costly in terms of the $\mu\Gamma$ product.

With present technology $\mu\Gamma$ products of 10^6 are easily obtained at data rates on the order of 100 Kb/s. At lower data rates the situation is even better since, for a fixed decoder CPU design, the speed factor μ increases as the data rate is decreased. Thus, as the data rate is decreased, the buffer size can be decreased proportionately while still maintaining the same amount of coding gain. If one does not desire to decrease Γ, then the $\mu\Gamma$ products increase proportionately allowing a decrease in the required Pareto exponent, ρ. This provides an amount of additional coding gain that can be determined from Figs. 7-14 and 7-21.

7.2.6.4. Soft Decision Decoders. Several soft decision sequential decoders have been implemented at moderate to low data rates. However, unlike the Viterbi algorithm, one does not obtain additional coding gain with little impact on complexity. The principal reason is that the decoder must now process m bits rather than 1 bit per channel symbol (for 2^m-level soft decisions), and storage is a large part of the decoder.

For example, consider the implementation of a decoder utilizing 2^m-level soft decisions with an architecture similar to that of Fig. 7-20. Rather than a single syndrome symbol per branch that is entered into the buffer, one must also store $2m - 2$ additional channel quality symbols per branch. Then the buffer RAM must actually be $(2m - 1)$ times the size of that for a hard decision decoder to achieve the same Γ [if the delay that stores the the $R^i(x)$ sequence is also part of this RAM, it is only m times as large]. In addition, the metric computation is considerably more complex with a larger range required. This results in a slower computation cycle thereby reducing the decoder speed factor μ at a given data rate. Thus, a soft decision decoder will typically have a $\mu\Gamma$ product that is less than that of a hard decision decoder by factors of 5 and 10 for four-level and eight-level quantization, respectively, for decoders of the same complexity. Conversely, a soft decision decoder that achieves the same $\mu\Gamma$ product as a hard decision decoder will require a much larger buffer size. These considerations act to reduce the gain achievable with soft decision decoding.

Example. The extent of this problem can be estimated from Figs. 7-14 and 7-21. Assume that a hard decision decoder achieves $\mu\Gamma = 10^6$. This $\mu\Gamma$ product can allow $P_a = 10^{-5}$ with $\rho = 1.22$ (from Fig. 7-21). Then with a $R = 1/2$ code the E_b/N_0 required to achieve $\rho = 1.22$ is 5 dB for hard decisions (from Fig. 7-14). If we wish to keep the same RAM size, then the $\mu\Gamma$ product for eight-level soft decisions will be decreased by a factor of 10 requiring $\rho = 1.46$ to achieve $P_a = 10^{-5}$. From Fig. 7-14 it is observed that $E_b/N_0 = 3.4$ dB is required to achieve this value of ρ. Thus, comparing decoders of roughly equivalent complexity, the actual additional coding gain provided by eight-level soft decisions is only 1.6 dB. A similar calculation may be made for four-level soft decisions assuming in this case that the $\mu\Gamma$ product is decreased by a factor of 5 relative to that with hard decisions. In this case the decoder must operate at $\rho = 1.37$, which is achieved at $E_b/N_0 = 3.65$ dB. Since this provides nearly as much gain as eight-level quantization, it may be preferred in many applications because of the reduction in implementation complexity in the decoder CPU.

There are other factors which also tend to reduce the gain available with soft decision decoding. One factor is that there is no "quick-threshold loosening" approach to reduce the computational problem. Another factor is that sequential decoders are sensitive to AGC inaccuracies which result in improper slicing threshold settings for the demodulator soft decisions. This effect reduces the achievable Pareto exponent, thereby accentuating the computation problem. These effects and complexity considerations tend to make soft decision decoders less attractive at high data rates (above several Mb/s). However, at low to moderate data rates, these problems are not severe. The potentially large coding gains (up to 7 dB at $P_a = 10^{-5}$) make soft decision sequential decoding quite desirable in these applications.

7.2.7. Stack Sequential Decoding

Another interesting class of sequential decoding algorithms are the so-called *stack algorithms* which were proposed and analyzed independently by Zigangirov[75] and Jelinek.[76] These algorithms have not found wide acceptance because the Fano algorithm is considered to be the most practical to implement. However, the stack algorithms are interesting because they are very simple conceptually, and they are more convenient to use in deriving certain analytical results concerning the properties of sequential decoders.

A flow diagram of a typical stack algorithm is shown in Fig. 7-22. The algorithm is extremely simple. The decoder creates a stack of previously searched paths of varying lengths, ordered according to their metric values (using the same metric as the Fano algorithm). At each step, the path at the top of the stack is extended to its 2^m successors (for $R = m/n$ codes) and replaced by these successors. Of course, a new metric is computed for each of these 2^m paths. Then the new stack is reordered according to metric

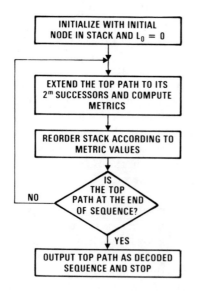

Figure 7-22. A stack decoding algorithm.

values, and the same process is repeated. This continues until the path at the top of the stack is at the end of the coded sequence.

As an example, consider the $R = 1/2$, $v = 1$ code with the code tree shown in Fig. 7-1. Furthermore, assume that the transmitted information sequence is 1 0 0 0 0..., and, with hard decisions, the received sequence is $\mathbf{R} = 01\ 01\ 00\ 01\ 00\ 00...$. Thus, there are single errors in the first and fourth branches. This is the same example considered previously for the Fano algorithm, and the corresponding branch metrics for this received sequence are shown in the tree of Fig. 7-13. At step 1 the initial node is extended to form paths 0 and 1 each with metric -4. The upper branch is arbitrarily placed at the top of the stack. At step 2 the top path is extended to give the paths 0 0 and 0 1 each with metric -8. After reordering the stack, the path 1 with metric -4 is at the top. At step 3 this path is extended to form paths 1 0 and 1 1 with metrics -3 and -13. The process continues in the same fashion, and after step 5 the path 1 0 0 0 is at the top of this stack. All upper extensions of this path will remain at the top of the stack so this path will eventually be chosen. This sequence of steps is shown in Table 7-4. The interesting point is that in comparing this table with the corresponding table for the Fano algorithm (Table 7-2) we see that the stack algorithm requires considerably fewer computations. However, this comparison is somewhat deceiving because a stack algorithm computation is much more complex than a Fano algorithm computation. Particularly troublesome is the operation of "reordering the stack."

There are several modifications one can make to this algorithm to make it more practical to implement. The most undesirable feature of the algorithm is the necessity for reordering the stack. The Jelinek algorithm partially alleviates this problem by quantizing the ordering. Paths are placed in "bins" each of which holds a stack of paths with metrics within Δ of each other. This considerably simplifies the complexity of the reordering process without significantly increasing the distribution of computation (it is still Pareto with variation as $L^{-\rho}$). (Note that this Δ has the same effect

Table 7-4. Stack Decoding for the Example of Fig. 7-20

Step	Stack contents after reordering (information sequence, metric)
1	0, -4; 1, -4
2	1, -4; 00, -8; 01, -8
3	10, -3; 00, -8; 01, -8; 11, -13
4	100, -2; 00, -8; 01, -8; 101, -12; 11, -13
5	1000, -6; 1001, -6; 00, -8; 01, -8; 101, -12; 11, -13

as the threshold spacing of the Fano algorithm.) A second serious disadvantage of stack algorithms is that the stack size (and hence memory) required to advance the correct path by one branch is proportional to the number of computations. Thus, the required stack size is also a Pareto distributed random variable. This problem is partially alleviated in the Zigangirov algorithm by discarding any path from the stack whenever its metric falls more than a fixed amount, Δ_0, below the metric of the top path. The probability of eliminating the correct path by doing this decreases exponentially with Δ_0, so that the effect on performance can be made negligible.

There are several advantages to the stack algorithms. One advantage is the ability of the stack algorithm to use idle periods effectively. When the buffer is empty, the Fano algorithm must wait for the next received branch before further calculations can be made. The stack algorithm, on the other hand, can use its idle periods to extend some of the paths further down in the stack. In this fashion the stack decoder need never be idle. Another apparent advantage is in the average number of computations required, an advantage that increases as $R \to R_0$. Jelinek's simulations showed that the two algorithms are equivalent computationally at $R = 0.8R_0$, and the stack decoder achieves an increasing advantage that reaches a factor of 7 better at $R = R_0$. Unfortunately, each of these stack algorithm computations is much more difficult than the Fano algorithm computations, and the stack decoder will require considerably more memory.

Problems

7-1. The $R = 1/2$, $v = 1$ code shown in Fig. 7-1 is used in a table look-up decoder with $L = 2$ (Fig. 7-2). The recieved sequence is 10 00 01 00 10. Compute the syndrome sequence and the decoded information sequence for both feedback and definite decoders.

7-2. The $R = 1/2$, $v = 5$ code with $g_1(x) = 1$ and $g_2(x) = 1 + x + x^3 + x^5$ is to be used in a table look-up decoder with $L = 6$. Find a decoding table that allows correction of all double and as many triple errors as possible with feedback.

7-3. The $R = 1/2$, $v = 1$ code with $g_1(x) = 1$ and $g_2(x) = 1 + x$ has an information sequence of $I(x) = \sum_{i=0}^{x} x^{2i}$. Find $T_1(x)$ and $T_2(x)$. Assume there are no channel errors, but that branch synchronization is lost [$T_1(x)$ and $T_2(x)$ are exchanged]. Find the resulting syndrome sequence, $S(x)$. Assuming a random information sequence, find the fraction of ones in the syndrome sequence and the fraction of correction bits that are one when definite decoding is used.

7-4. For the $R = 2/3$, $J = 4$ code in Table B-10 draw a block diagram of the encoder and a hard decision threshold decoder to the level of detail shown in Fig. 7-7.

7-5. For the $R = 1/2$, $J = 2$ code in Table B-9 draw a block diagram of the encoder and an approximate APP threshold decoder utilizing four-level demodulator soft decisions. Provide more detail than shown in Fig. 7-8.

7-6. Draw a block diagram of the hard decision feedback threshold decoder for the $R = 1/2$ code with $g(x) = 1 + x^6 + x^7 + x^9 + x^{10}$ which corrects all double errors and detects all triple errors.

7-7. Consider the use of $R = 1/2$ codes with sequential decoding on the BSC with transition probability, p. Compute the metric ratio, MR, with $B = R$ as a function of p and plot $1/MR$ vs. p for $p \leq 0.045$ ($R \leq R_0$).

7-8. Repeat Problem 7-7 for the binary-input–ternary-output channel with erasure probability, e, and error probability, p. Compute the standard metric ratio, $MR = m_0/m_1$, and the ratio of the metric for a correct bit to that of an erasure, $MR_e = m_0/m_e$. Plot $1/MR$ and $1/MR_e$ as a function of p for $e = 0.01, 0.1$, and 0.2.

7-9. Using the $R = 1/2$, $v = 48$ code found by Forney[78] (given in Table B-16) draw the code tree for the first four branches. Then, with a received code word 0 1 00 1 0 00 00..., construct the branch metric tree (as in Fig. 7-13). Using metric increments of $+0.5$ for each agreement and -4.5 for each disagreement and a threshold spacing of 5, apply the Fano algorithm (Table 7-1) until the all-zero code word is followed to depth 4 in the code tree. Show the results of the tree search in a table as in Table 7-2.

7-10. Repeat Problem 7-9 using the stack algorithm. Perform as many steps as necessary to extend the all-zero sequence through four branches and advance it to the top of the stack. List the steps in a table similar to Table 7-4.

7-11. Sequential decoding is to be used with a long-constraint-length $R = 1/2$ systematic code. Data are to be blocked into 1000-bit code words. Assume $A = 1$ and $\mu \gg 1$. Plot the probability of block erasure vs. E_b/N_0 for two-level hard decisions and eight-level soft decisions for $\mu\Gamma = 10^5, 10^6$, and 10^7. Repeat this process to find P_a. Repeat these calculations for a $R = 3/4$ code.

7-12. Assume that sequential decoding with continuous data transmission and alternate decoding of overflow events is used ($N_a = 100$). Plot P_a vs. E_b/N_0 under the same conditions as the previous problem for both $R = 1/2$ and $R = 3/4$ codes.

7-13. Consider the $R = 1/2$, $v = 2$ nonsystematic code generated by $g_1(x) = 1 + x^2$ and $g_2(x) = 1 + x + x^2$. (a) Using Euclid's algorithm (see Chapter 5), find two polynomials $a(x)$ and $b(x)$ such that $a(x) g_1(x) + b(x) g_2(x) = 1$. (b) Using $a(x)$ and $b(x)$ draw a circuit which, in the absence of errors, will recover the information sequence $I(x)$ from the received code sequences. (c) Using this result draw a block diagram similar to Fig. 7-4 which can be used to decode this code when errors are present. Show the details of the feedback connections. (d) Determine the appropriate entries for the look-up tables for decoding depths of $L = 2$, 4, and 6 if maximum-likelihood decoding is to be used. (e) For each value of $L = 2$, 4, and 6, draw a transition diagram similar to Fig. 7-3 and show that with $L = 2$ the error propagation distance is unbounded, that with $L = 4$ the distance is still unbounded but has a finite mean, and that with $L = 6$ the propagation distance is bounded.

8

System Applications

We believe that examination of a few typical application areas will provide the reader with much of the insight required to effectively utilize coding in communication systems. One such application occurs in systems that require a very powerful code. A practical technique for implementing such a code is that of concatenating two or more codes. With this technique one achieves a high level of performance by trading an increase in overall block length for a reduction in hardware complexity. A second important system application of coding occurs any time the predominant error-causing mechanism can be attributed to additive white Gaussian noise. Virtually all of the techniques discussed in this book are for correcting random errors and, consequently, are applicable to this problem. Another topic of considerable interest is that of coding for burst noise channels. In this case, the interesting questions are whether interleaving should be used and if so, what structure should be used. A related problem is that of coding for spread spectrum systems (because of the pulsed jammer problem). For these systems some definite conclusions can be made regarding the manner in which coding should be applied. Finally, a problem of a much different nature is that of communicating over limited bandwidth channels. In this case coding can provide substantial gains. This is an extremely interesting problem because the coding scheme should be selected in conjunction with the signal design to obtain the most efficient coding/modulation scheme.

8.1. Concatenated Codes

Concatenated coding was introduced by Forney[48] as a practical technique for implementing a code with a very long block length and a large error-correcting capability. This is accomplished by utilizing multiple levels of coding, and numerous configurations are possible. The most common approach uses two levels of coding. Reed–Solomon codes are commonly used as one of the codes, i.e., the so-called *outer code*; however, many different codes have been suggested for use as the *inner code*. Three specific configurations using orthogonal codes, short block codes, and convolutional codes as the inner code will be discussed.

8.1.1. Concatenated Coding Concepts

The basic concept of concatenated codes for two levels of coding is illustrated in Fig. 8-1. For simplicity, assume that the channel is a binary input channel. The outer code is a nonbinary code which utilizes K-bit symbols. These symbols enter the outer coder from the data source as shown. The outer code is assumed to be a block code which is n symbols long and where k of these symbols are information symbols. The K-bit symbols from the outer coder are further encoded by the inner coder. Here $N - K$ parity bits are appended to give an inner code block length of N bits. Although we show the N-bit symbol entering the channel, a parallel-to-serial conversion is typically used to allow binary transmission.

At the receive end, the channel may output either hard or soft decisions. In either case these decisions are presented in parallel to the inner decoder. The inner decoder performs its function by providing an estimate of each K-bit outer code symbol at a moderately low symbol error rate, p_s. The outer decoder then corrects as many symbol errors as possible and provides a very low output bit error probability, P_b.

The combination of inner coder, channel, and inner decoder is sometimes referred to as the *superchannel*. Likewise, the combination of outer and inner coders is referred to as the *supercoder*, and the combination of inner and outer decoders is called the *superdecoder*. Note that the resulting concatenated code has overall length $N^* = nN$ bits with $K^* = kK$ information bits per overall code word and with code rate $R^* = rR = kK/nN$, where $R = K/N$ and $r = k/n$. Although the overall length of the code is

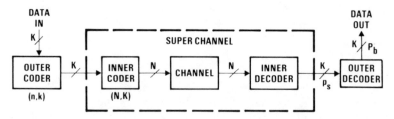

Figure 8-1. Concatenated coding approach.

nN, the structure imposed by the concatenation concept allows the decoding operation to be performed by two decoders for codes of length N and n, respectively. As we shall see, this allows a significant reduction in complexity over that which would be required to provide the same overall error rate with a single level of coding. The most natural choice for outer codes are the Reed–Solomon (RS) codes. These codes have been the most widely used because they are maximum distance separable codes ($d = n - k + 1$), and because they are relatively straightforward to implement.

The systems to be considered here all employ a Reed–Solomon outer code which is decoded by errors-only decoding. That is, the inner decoder produces only symbol decisions, and outputs neither symbol erasures nor a reliability indication for the symbol decisions. Assuming a symbol error rate p_s at the RS decoder input, the bit error rate at the RS decoder output may be estimated from the bound

$$P_b < \frac{2^{K-1}}{2^K - 1} \sum_{j=t+1}^{n} \frac{j+t}{n} \binom{n}{j} p_s^j (1 - p_s)^{n-j} \tag{8-1}$$

for a t-symbol-error-correcting RS code. This equation utilizes (5-71) with the factor $2^{K-1}/(2^K - 1)$ to account for the average number of information bit errors per symbol error. The performance of a specific concatenation scheme is determined by estimating p_s for the particular inner code of interest and then evaluating P_b through (8-1).

8.1.2. Reed–Solomon/Orthogonal Code Systems

The complexity of the RS outer decoder is determined by the symbol size (K bits), length n, and minimum distance d. If the outer code is fixed, one might then wish to find the most efficient inner code, i.e., that (N,K) code which, for a fixed K, allows operation at the minimum value of E_b/N_0. This is actually accomplished by the ($2^K - 1,K$) simplex code mentioned earlier. However, near-optimum performance is also provided by the closely related ($2^{K-1},K$) biorthogonal and ($2^K,K$) orthogonal codes. These codes may be implemented as binary block codes as discussed earlier or they may be implemented as M-ary modulation schemes (e.g., 2^K-ary FSK to provide orthogonal signals). In either case, the decoder complexity grows exponentially with K because a separate correlator or matched filter is required for each code word. Thus, only relatively small values of K are of interest ($K \leq 10$).

A very obvious characteristic of this class of inner codes is that they have very low rates. Thus, they would be useful only when very wide channel bandwidths are available (i.e., when the required data rate is much smaller than the available channel bandwidth). They have been used in deep space communication systems where the data rates are quite low and are of current interest in spread spectrum communication systems.

The word error rates for these inner codes using PSK modulation have been found by Viterbi [87] and are tabulated.[88] By using these results for the RS code symbol error rate p_s in (8-1), the bit error rate for the concatenated system is easily found. Some typical performance results are shown in Figs. 8-2 and 8-3 for RS codes with $r \approx 1/2$ and $r \approx 3/4$, respectively. These RS codes are concatenated with biorthogonal codes with $K = 5$, 6, and 7 (the biorthogonal codes perform roughly 0.3 dB better than orthogonal codes with equivalent complexity). Substantial performance increases are obtained as K is increased and operation at very low error rates for E_b/N_0 in the 2 to 3 dB range is feasible (Forney also indicates numerous code combinations that achieve this same result[48]). Another interesting point is that the $r \approx 3/4$ codes are slightly more efficient for these examples than the $r \approx 1/2$ codes. The decoders for these higher rate codes are also less complex since they have smaller values of d and correct

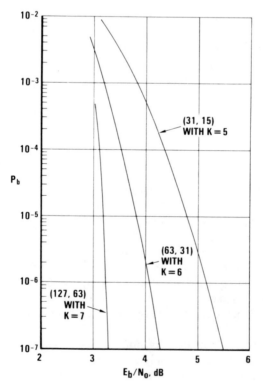

Figure 8-2. Performance of RS outer codes ($r \approx 1/2$) concatenated with biorthogonal inner codes.

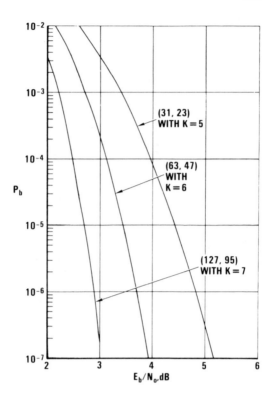

Figure 8-3. Performance of RS outer codes ($r \approx 3/4$) concatenated with biorthogonal inner codes.

fewer errors. Sometimes, however, there are other considerations that may dictate the use of a larger d (such as the presence of burst noise).

In principle, one can achieve even better performance than indicated in Figs. 8-2 and 8-3 by simply increasing K (and the RS code length) until the complexity becomes unreasonable. As this is done, however, the overall code rate decreases as $R^* \approx K2^{-K}$. This may prove to be one of the most severe limitations either because the bandwidth is not available or because of the difficulty of building good demodulators at the extremely low values of E_s/N_0 implied by these rates.

The case of noncoherent detection of the inner code is also quite interesting. We assume an M-ary orthogonal signaling scheme with noncoherent detection. The symbol error rate, p_s, at the inner decoder output is well known[89] and was given previously in (1-14). This expression can be used with (8-1) to predict performance. Typical results are shown in Figs. 8-4 and 8-5. The most interesting point is that these results degrade only about 1 dB relative to the coherent results presented in the previous figures and indicate an available coding gain of 9 to 10 dB (at $P_b = 10^{-5}$) relative to binary orthogonal signaling with noncoherent detection. This

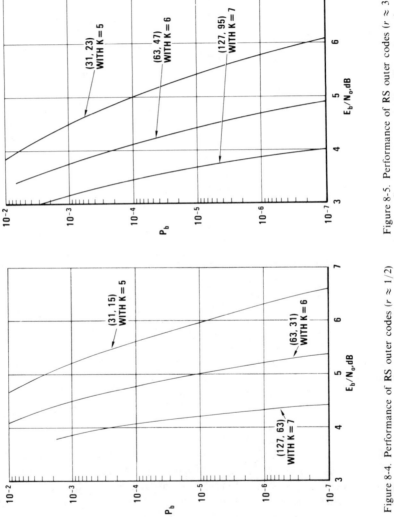

Figure 8-5. Performance of RS outer codes ($r \approx 3/4$) with 2^K-ary orthogonal signaling (noncoherent detection).

Figure 8-4. Performance of RS outer codes ($r \approx 1/2$) with 2^K-ary orthogonal signaling (noncoherent detection).

is consistent with our earlier observation (in Chapters 6 and 7) that the use of higher-level alphabets in a noncoherent system is quite beneficial. We also note that for this particular case the higher-rate codes ($r \approx 3/4$) are preferable.

8.1.3. Reed–Solomon/Short-Block Code Systems

While the use of orthogonal, biorthogonal, or simplex codes for inner codes offers excellent performance, there is the disadvantage that the overall code rate is very low. However, moderate code rates in the range $1/4 \leq R^* \leq 3/4$ are easily achieved by using short-block codes as inner codes. If these codes are decoded by an algorithm approximating maximum-likelihood decoding, then excellent performance can be obtained (though not quite as good as that of biorthogonal codes). A second benefit, which is not as widely recognized, is that such systems can be extremely effective in high data rate applications. RS codes are amenable to high data rate operation because no decoding operations need be performed at the data rate. Operations are performed either at the RS code symbol rate (a small fraction of the data rate) or a fixed number of times per code word (much less than the number of bits per code word). In addition, high data rate implementations of short block codes with minimum distances of 4 or less are straightforward. The structure of such a concatenated system facilitates "pipelining" the decoding operations, thereby easing the speed requirements on the logic. Coding gains of 5 dB with $R^* = 1/2$ are achievable with less complexity than using a stack of multiplexed lower rate Viterbi algorithm decoders. Such a system operating at an information rate of 40 Mb/s has been built and tested.[47, 90]

Performance curves for some typical $d = 4$ short block codes with 7 or 8 information bits per code word are shown in Fig. 8-6. These codes all have code rates slightly greater than $1/2$ to allow a concatenated code with $R^* \approx 1/2$ to be formed. These performance curves assume maximum-likelihood decoding and were determined via simulation. By using these results and (8-1) the coding gain for a number of concatenated systems can be obtained.[47] The coding gain at $P_b = 10^{-5}$ for a constraint on overall block length of 400 information bits per block is shown in Fig. 8-7. There are numerous combinations which can provide 5 dB of gain at $R^* \approx 1/2$. Similar curves can be shown for other block lengths with the slope of the curves increasing with block lengths. Thus, additional gain can be obtained with longer block lengths at the cost of correcting more errors per code word.

While these results assume maximum-likelihood decoding, implementation of a true maximum-likelihood decoder is unnecessary in many cases. The desired operating point (e.g., $P_b = 10^{-5}$) is such that the symbol

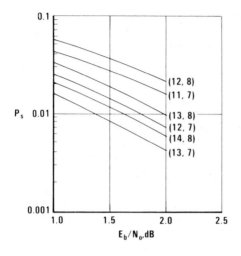

Figure 8-6. Short code performance (maximum-likelihood decoding).

error rate at the inner decoder output is roughly $p_s \approx 0.01$. It is of considerable interest to find a computationally efficient decoding algorithm for the inner code that approximates maximum-likelihood decoding (to within a few tenths of a dB) at $p_s = 0.01$. The Chase [26] decoding algorithm is one such approach (see Section 4.4). For these $d = 4$ inner codes, the Chase algorithm requires one to correlate with only four code words, and performance is typically degraded by about 0.2 dB.

Extension of this approach to higher code rates is straightforward. In fact, a significant reduction in complexity can be achieved, although

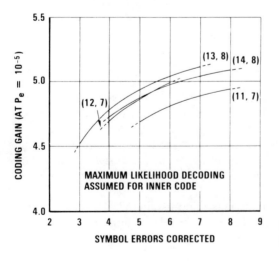

Figure 8-7. Concatenated code performance with 400-bit block.

the available coding gain is also reduced. One very attractive choice for the inner code is a simple $(K + 1, K)$, $d = 2$ single parity check code. With these codes the maximum-likelihood decoder is a device which selects either the received hard decision sequence, if parity is satisfied, or the hard decision sequence with the least reliable bit complemented. Using a (9, 8) inner code, coding gains of 4 dB or slightly less are possible even when correcting a small number of symbol errors with the RS code (on the order of 6 or less).

Implementation considerations discourage correction of a large number of symbol errors per RS code word. With 40 Mb/s implementations, correction of much more than about six symbol errors per code word is currently impractical [47]; however, at lower data rates serial implementations of the RS decoders allow the correction of many errors per code word. This and the potential for using longer inner codes can provide coding gains much higher than the 5 dB obtained at very high data rates. One can also deviate slightly from the original concatenation concept that each inner code word provides one RS code symbol. In fact, an inner code word could provide several RS code symbols, e.g., a (32, 16) inner code could be concatenated with an RS code using an eight-bit symbol. Each inner code word would represent two symbols. A slight problem would exist in that a word error in the inner code could cause two RS code symbol errors. This can be fixed by interleaving such that the two eight-bit information segments in each inner code word are used in separate RS code words. However, this approach allows the use of much longer inner codes, resulting in higher coding gains. A similar approach can be followed using convolutional codes as inner codes as discussed in Section 8.1.4.

8.1.4. Reed–Solomon/Convolutional Code Systems

One can use a convolutional code as the inner code and achieve excellent performance with relatively high overall code rates. Since the convolutional code would operate at a moderately high error rate, the use of short-constraint-length codes with Viterbi decoding (utilizing demodulator soft decisions) is the most promising approach. (Recall that the performance curves for sequential decoding typically have a much steeper slope than those for Viterbi decoding. As a result, sequential decoding is preferred at very low error rates, while Viterbi decoding is preferred at higher error rates.) Systems of this type were first investigated by Odenwalder.[91, 92]

The use of a convolutional inner code does not change the system block diagram significantly from that shown in Fig. 8-1. The K-bit symbols from the outer coder must be converted back to a serial format for convolutional encoding, but the block synchronization must be maintained, i.e., the start of outer code block and each RS code symbol must be identified. After Viterbi algorithm decoding the decoded data must be converted back to K-bit symbols for entry into the RS decoder. Unfortunately, there is one additional complication. The errors at the Viterbi decoder output are quite bursty so that the errors on successive RS code symbols are correlated. However, the degree of correlation beyond a decoding depth (5 or 6 constraint lengths) is insignificant. Since this correlation would cause performance degradation in the RS decoder, the problem is remedied through the application of interleaving. All that is necessary is to interleave in such a manner that no two symbols that occur within a decoding depth at the Viterbi decoder output belong to the same RS code word. (Several efficient interleaving techniques are presented in Section 8.3.)

Odenwalder[92] performed some Viterbi decoder simulations to determine the symbol error probability out of the decoder. The bound (8-1) was then used to determine the bit error probability at the RS decoder output. Ideal interleaving is assumed so that symbol errors at the RS decoder input are independent. Figure 8-8 shows some of these results for

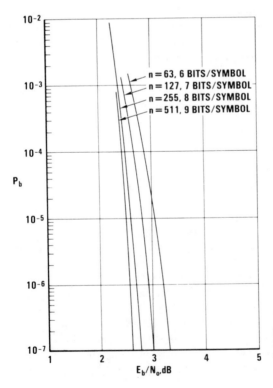

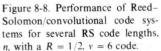

Figure 8-8. Performance of Reed–Solomon/convolutional code systems for several RS code lengths, n, with a $R = 1/2$, $v = 6$ code.

the widely used $R = 1/2$, $v = 6$ convolutional code and for RS codes with 6, 7, 8, and 9 bits per symbol ($n = 63$, 127, 255, and 511 per RS code word). The (n, k) parameters for each RS code are also shown. For each value of n the code selected is that with the best performance. Typically, there is a relatively broad minimum in P_b with respect to t (the number of errors corrected). For example, with $n = 255$, there is a broad minimum from $t = 16$ to $t = 24$, with the former value performing better at high P_b and the latter value performing better at low P_b. Similar results were obtained using other code rates with the same constraint length. The use of $R = 1/3$ provides about 0.6 dB more gain than $R = 1/2$ while $R = 3/4$ degrades about 1.2 dB.

In summary, the Reed–Solomon/convolutional code systems offer excellent performance. Operation with E_b/N_0 in the range of 2.0 to 2.5 dB to achieve $P_b = 10^{-5}$ is feasible with practical hardware for data rates of several hundred **Kb/s** and below. We also note that just as convolutional codes with Viterbi algorithm decoding perform better than biorthogonal codes, the concatenated system using convolutional codes as inner codes performs better than the system using biorthogonal inner codes. In addition, the performance curves are so steep that operation at several orders of magnitude below 10^{-5} can be achieved at the expense of only 0.5 dB more signal energy per bit. However, operation at extremely high data rates is not currently feasible both because of the difficulty of implementing very-high-speed Viterbi decoders and the larger number of errors which must be corrected by the RS decoder.

8.2. Coding for the White Gaussian Noise Channel

In recent years the utilization of forward error correction techniques has increased dramatically on channels which can be modeled as additive, white Gaussian noise channels. Typical applications include line-of-sight radio links such as satellite and space communication links. For this type of problem the objective is to reduce the required amount of E_b/N_0 relative to that of an uncoded system at a specified error rate. This coding gain can then be used by the system designer in the most economical fashion, e.g., to allow decreases in spacecraft power or in antenna size or to allow an increase in data rate. The types of coding techniques that are most appropriate for this problem, of course, are the random-error-correction techniques to which most of this book has been devoted.

Comparisons among these techniques will be made at $P_b = 10^{-5}$ (which is the most commonly specified operating point) and at $P_b = 10^{-8}$ (since there are differences in the steepness of the error rate curves as a function of E_b/N_0). Both coherent and noncoherent systems will be examined. For coherent systems it is assumed that BPSK or QPSK modulation is used with an uncoded bit error rate given by $Q[(2E_b/N_0)^{1/2}]$ and that errors occur independently from symbol to symbol. Similarly, for noncoherent systems, the performance will be compared with that of uncoded binary orthogonal signaling with noncoherent detection which has an error rate given by $\frac{1}{2}\exp(-E_b/2N_0)$.

Perhaps the most widely used technique for providing significant amounts of coding gain has been the use of short constraint length convolutional codes with Viterbi decoding. Specifically, the well-known $R = 1/2$, $v = 6$ code, which provides 5 dB of coding gain at $P_b = 10^{-5}$, has been applied in a large number of systems with varying data rates. This approach has typically outperformed block coding approaches for a given level of complexity. It has two natural advantages in that synchronization is simple and demodulator soft decisions are easily utilized. Though the decoding operations are not inherently much simpler than those of other techniques, it has been quite amenable to design optimization. Decoders from several vendors represent rather mature designs that utilize significant amounts of existing off-the-shelf LSI. The complexity comparisons discussed assume 1980 technology. Since LSI technology is rapidly changing, these comparisons could be substantially different in a few years. In the near future entire coding/decoding systems will be designed on one or a few custom LSI chips (e.g., Viterbi decoders, sequential decoders, etc.). Thus, an important emerging problem area is that of structuring decoding algorithms that are amenable to LSI design optimization.

A comparison of the coding gains available with the major random error correction techniques in a coherent system is shown in Table 8-1

Table 8-1. Comparison of Major Coding Techniques with BPSK or QPSK Modulation on a Gaussian Channel

Coding technique	Coding gain (dB) at 10^{-5}	Coding gain (dB) at 10^{-8}	Data rate capability
Concatenated (RS and Viterbi)	6.5–7.5	8.5–9.5	Moderate
Sequential decoding (soft decisions)	6.0–7.0	8.0–9.0	Moderate
Concatenated (RS and biorthogonal)	5.0–7.0	7.0–9.0	Moderate
Block codes (soft decisions)	5.0–6.0	6.5–7.5	Moderate
Concatenated (RS and short block)	4.5–5.5	6.5–7.5	Very high
Viterbi decoding	4.0–5.5	5.0–6.5	High
Sequential decoding (hard decisions)	4.0–5.0	6.0–7.0	High
Block codes (hard decisions)	3.0–4.0	4.5–5.5	High
Block codes—threshold decoding	2.0–4.0	3.5–5.5	High
Convolutional codes—threshold decoding	1.5–3.0	2.5–4.0	Very high
Convolutional codes—table look-up decoding	1.0–2.0	1.5–2.5	High

together with the data rate at which a reasonably practical decoder design can be operated. The quantization of data rates is rather crude: low (less than 10 Kb/s), moderate (10 Kb/s to 1 Mb/s), high (1 Mb/s to 20 Mb/s), and very high (greater than 20 Mb/s).

Compared with the other techniques in Table 8-1 (at $P_b = 10^{-5}$), the Viterbi decoding approach is seen to provide coding gains near the high end of the range. At moderate and high data rates, it also has significantly less complexity than other approaches with comparable coding gains. However, if more efficient algorithms for decoding long block codes with soft decisions are developed, they will undoubtedly be quite competitive. The only significant reductions in complexity occur when the very simple threshold decoding and table look-up decoding techniques are employed, but the achievable coding gains are several decibels less. Nevertheless, the very simple schemes may be useful in situations where extreme simplicity is needed (the internal interleaving capability may be of considerable interest).

At very high data rates, concatenated Reed–Solomon and short block code systems presently can provide roughly the same gain with less complexity than Viterbi decoding. Multiplexed or stacked Viterbi decoders have been used to achieve very high data rates, but this is a very costly alternative. However, the design of a Viterbi decoder on a chip could change this comparison.

At high data rates the Viterbi decoding algorithm has had the best cost/complexity tradeoff for several years. The techniques which can outperform Viterbi decoding have complexity which is proportional to data rate. It turns out that it is only for moderate rates that they become attractive. At these data rates the concatenated RS and Viterbi decoding system can perform about 2 dB better than the Viterbi decoder alone, but at the cost of doubling or tripling the complexity. A better case can be made for using sequential decoding with soft decisions. At data rates less than several hundred Kb/s it can outperform Viterbi decoding with roughly the same complexity. At low data rates the advantage shifts even more toward sequential decoding.

At very low error rates ($P_b = 10^{-8}$) the large minimum distance codes obtain an added advantage. For example, the concatenated and sequential decoding techniques achieve an additional 0.5 to 1.0 dB of gain relative to Viterbi decoding. Thus, stronger consideration should be given these techniques at very low error rates.

Similar results for noncoherent systems are shown in Table 8-2. A major difference in this table is the introduction of a higher-level modulation alphabet. It is difficult to obtain large coding gains when binary signaling is used. In fact, the coded system performance is substantially inferior (by as much as 4 to 5 dB) to that of a coherent coded system. How-

Table 8-2. Comparison of Major Coding Approaches for M-ary Orthogonal Signaling and Noncoherent Detection[a],[b]

Coding technique	M	Coding gain (dB) at 10^{-5}	Coding gain (dB) at 10^{-8}
Reed–Solomon	128	9.7	11.4
Reed–Solomon	64	9.0	10.4
Reed–Solomon	32	8.0	9.1
Sequential decoding (soft decisions)	8	7.5–8.5	9.5–10.5
Viterbi decoding	8	5.5–6.5	6.0–7.0
None	128	7.3	8.8
None	8	4.4	5.0
Sequential decoding (soft decisions)	2^b	4.5–5.0	6.5–7.0
Block codes (soft decisions)	2^b	4.5–5.5	6.0–7.0
Viterbi decoding	2^b	3.0–4.0	4.0–5.0
Sequential decoding (hard decisions)	2^b	3.0–3.5	5.0–5.5
Block codes (hard decisions)	2^b	2.5–3.5	4.0–5.0
Convolutional codes—threshold decoding	2^b	1.0–2.5	2.0–3.5
Convolutional codes—table look-up decoding	2^b	1.0–1.5	1.5–2.2

[a] Gain measured relative to performance of binary signaling.
[b] In some cases differentially coherent PSK can be employed to achieve an additional 3 dB gain.

ever, the use of M-ary orthogonal signaling for $M \geq 8$ provides a significantly greater "coherence time" in the system and allows the coded noncoherent system to perform much better relative to a coherent system. Note that the Reed–Solomon codes with M-ary modulation are equivalent to concatenated codes. Also, the Reed–Solomon code with 128-ary orthogonal signaling performs within 1 dB of a coherent concatenated coding system with the same RS code and a $K = 7$ biorthogonal inner code. Coding gains of 9 to 11 dB are observed relative to an uncoded binary system, but most of this gain is attributable to the M-ary modulation. This is indicated in the table for uncoded 8-ary and 128-ary orthogonal signaling. The benefit of additional coding is worth only 2.0 to 2.5 dB. Unfortunately, it is not always possible to employ M-ary orthogonal signaling because of the bandwidth requirements. In those systems where binary signaling must be used, performance is substantially inferior to a coherent system (unless differentially coherent PSK can be used).

It should be noted that a major part of these comparisons is influenced by today's digital integrated circuit technology. Advances in this technology could modify relative comparisons of complexity and achievable data rates, and the design of entire decoders on single LSI chips could significantly influence the manner in which these devices are applied.

8.3. Interleaver Structures for Coded Systems

All of the performance calculations discussed to this point have been for memoryless channels, i.e., the probability of error does not vary with time. In cases where burst errors are a problem, one potential solution involves utilizing one of the random-error-correcting coder/decoders discussed previously with a suitable interleaver/deinterleaver pair. Using this approach, the encoder output sequence is interleaved prior to transmission and deinterleaved prior to decoding so that errors are distributed more uniformly at the decoder input. A system block diagram is shown in Fig. 8-9. Note that if each code symbol is quantized to Q bits in the demodulator, then the deinterleaver requires a factor of Q more memory than the interleaver.

An *interleaver* is a device that rearranges (or permutes) the ordering of a sequence of symbols in a deterministic manner. Associated with an interleaver is a *deinterleaver* that applies the inverse permutation to restore the sequence to its original ordering. These devices can assume numerous configurations. Two principal classifications of interleavers are *periodic* and *pseudorandom*. The periodic interleaver is to be preferred in many applications since it is typically less complex than the pseudorandom interleaver. However, the pseudorandom interleaver offers a degree of robustness that is not present in periodic interleavers. Hence, it may be preferred in certain applications where there can be substantial variation in the burst characteristics of the channel.

The interleaver structures to be discussed here are applied *external* to the coder/decoder hardware. This is necessary for many decoding algorithms, e.g., Viterbi decoding, sequential decoding, and the standard BCH decoding algorithm. We mentioned previously that for decoders that have a Meggitt-type decoder structure, one can apply interleaving *internally* in a very simple fashion. The internal interleaving structure is preferable for those algorithms, and this structure needs no further discussion.

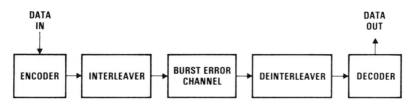

Figure 8-9. System block diagram for application of external interleaving/deinterleaving.

8.3.1. Periodic Interleavers

A periodic interleaver is an interleaver for which the interleaving permutation is a periodic function of time. Two types of interleavers are commonly used. *Block interleavers* accept symbols in blocks and perform identical permutations over each block of symbols. *Convolutional interleavers* have no fixed block structure, but they perform a periodic permutation over a semi-infinite sequence of coded symbols. The distinction between these two types of interleavers is much like the distinction between block and convolutional codes.

8.3.1.1. Block Interleavers. The typical case of a block interleaver involves taking the coded symbols and writing them by columns into a matrix with N rows and B columns. The permutation consists of reading these symbols out of the matrix by rows prior to transmission. Such an interleaver is referred to as a (B, N) block interleaver. The deinterleaver simply performs the inverse operation. Symbols are written into the deinterleaver by rows and read out by columns. Obviously, such an interleaver/deinterleaver could be easily implemented with current digital technology.

The most important characteristics of this interleaving approach are as follows:

1. Any burst of errors of length $b \leq B$ results in single errors at the deinterleaver output each separated by at least N symbols:
2. any burst of length $b = rB$ $(r > 1)$ results in bursts of no more than $\lceil r \rceil$ symbol errors separated by no less than $N - \lceil r \rceil$:
3. a periodic sequence of single errors spaced by B symbols results in a single burst of errors of length N at the deinterleaver output: and
4. end-to-end delay is $2NB$ symbols exclusive of the channel delay, and the memory requirement is NB symbols in both the interleaver and deinterleaver.

Typically, the interleaver parameters would be selected so that for all expected burst lengths, $b \leq B$. However, if there is substantial variation in the characteristics of the burst noise process, this type of interleaver lacks robustness, as demonstrated by the third characteristic.

Choice of the parameter N is dependent on the coding scheme used. In the usual application the effects of the channel memory are not observed over any span of N symbols at the deinterleaver output. Thus, N should be chosen to be larger than the decoding span. For block codes N should be larger than the code block length, while for convolutional codes N should be larger than the decoding constraint length. Thus, a burst of length $b \leq B$ can cause at most a single error in any block code word. Similarly, with convolutional codes there will be at most a single error in any decoding constraint length.

The use of a block interleaver introduces essentially the same synchronization problem that the use of a block code presents. Unless the start of each interleaver block is known, the deinterleaving process cannot be accomplished correctly. Of course, if this happens, the error-correction decoding process cannot proceed reliably. Synchronization of the interleaver/deinterleaver can be accomplished by using standard frame synchronization techniques. In this case, a sync word with good correlation properties is inserted periodically at the interleaver and then recovered by a frame synchronizer at the deinterleaver. Roughly 1%–2% overhead symbols are usually allocated to this process. An alternate approach (which requires no additional overhead symbols) involves replacing certain coded symbols by the sync word. These symbols are then erased at the decoder input. Care must be taken to ensure that these symbols are widely separated at the decoder input to avoid substantial degradation. The frame synchronizer design for a coded system is typically slightly more difficult than for an uncoded system because of the lower E_s/N_0 operating point.

8.3.1.2. Convolutional Interleavers. Convolutional periodic interleavers have been proposed by Ramsey[93] and Forney.[94] The structure we shall discuss here is shown in Fig. 8-10 and was proposed by Forney. The structures proposed by Ramsey have similar properties but will not be discussed. Defining the parameter

$$N = MB$$

this interleaver is referred to as a (B, N) interleaver and has properties quite similar to a (B, N) block interleaver.

The operation of this interleaver is straightforward. The code symbols are shifted sequentially into the bank of B registers with increasing lengths. With each new code symbol, the commutator switches to a new register, and the new code symbol is shifted in while the oldest code symbol in that register is shifted out to the channel. Naturally, the input and output commutators operate synchronously. The deinterleaver performs the

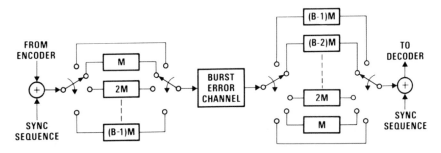

Figure 8-10. Shift register implementation of a convolutional interleaver/deinterleaver.

inverse operation. Note that for proper deinterleaving the deinterleaver commutator must be synchronized with the interleaver commutator. This synchronization problem will be discussed shortly. For many applications actual implementation of the interleaver can be done with a random access memory (rather than with shift registers) simply by implementing the appropriate control of memory access.

The most important characteristics of the convolutional interleaver are as follows:

1. The minimum separation at the interleaver output is B symbols for any two symbols that are separated by less than N symbols at the interleaver input;
2. the previous characteristic implies that any burst of $b < B$ errors inserted by the channel results in single errors at the deinterleaver output separated by at least N symbols;
3. a periodic pattern of single errors spaced by $N + 1$ symbols results in a burst of length B at the deinterleaver output; and
4. the total end-to-end delay is $N(B - 1)$ symbols and the memory requirement is $N(B - 1)/2$ in both the interleaver and deinterleaver. Note that this is half the required delay and memory in a block interleaver/deinterleaver.

The parameters B and N are selected in the same manner as for block interleavers. B is chosen to be larger than the length of the burst errors. N is chosen to be larger than the block length for block codes or the decoding constraint length for convolutional codes. By making these choices, the (B, N) block and convolutional interleavers will have very similar performance.

One advantage of the convolutional interleaver is that the synchronization problem is easier than that of a block interleaver. The reason is that the ambiguity is only of degree B for the convolutional interleaver while it is of degree NB in a block interleaver. Conventional frame synchronization techniques could be used, but there is another very interesting technique suggested by Viterbi et al.[95] This technique is indicated in Fig. 8-10, where a sync sequence is added modulo-2 to the coded symbols prior to interleaving and removed after deinterleaving. This sequence is chosen to be periodic with period B and to have a small autocorrelation function except at $\tau = 0$ (e.g., a maximal-length shift register sequence could be used). This sequence would be synchronized with the commutators in the interleaver and deinterleaver. Thus, when the interleaver and deinterleaver are not operating in synchronism, the sync sequence is not properly removed from the code symbols at the output of the deinterleaver. This results in an error rate of 0.50 at the decoder input. Such a condition is easily detected by any type of decoder. However, for a convolutional decoder which would normally be performing a branch synchronization function, virtually no additional hardware is required. The normal branch synchronization function will detect any out-of-sync condition in the interleaver, and all that is needed is a simple search strategy for searching all possible branch sync, symbol polarity, and interleaver sync ambiguities. The required search time increases linearly with B, which is normally small. We might also point out that a similar sync approach could be used with a block interleaver, but since the sync ambiguity is NB, the acquisition time will be a factor of N longer.

8.3.2. Pseudorandom Interleavers

The pseudorandom interleaver is a block interleaver which takes a block of L channel symbols after encoding and reorders, or permutes, them in a pseudorandom fashion. This can be implemented by writing the L symbols sequentially into a random access memory (RAM) and then reading them out pseudorandomly. One can store the desired permutation in a read-only memory (ROM) and then use this permutation to address the interleaver memory.

This technique provides a high degree of robustness to variability in the burst parameters, but at the cost of more complexity than a block or convolutional interleaver of the same size. The applications of most interest are antijam (AJ) systems utilizing some form of spread spectrum modulation. This type of application is emphasized in the following discussion. (Spread spectrum systems are discussed in Section 8.5.)

Note that if the same permutation is used on each interleaver block, there will exist certain interference patterns that can seriously degrade performance. In systems where such patterns are likely to occur (such as in AJ systems with intelligent jammers with knowledge of the permutation), then the permutation should be changed frequently to avoid this problem (e.g., for each interleaved block). One method for accomplishing this is to store a fixed number of permutations, say M, in an ROM and then randomly select a permutation from this group for each interleaved block. The parameter M needs to be large enough to eliminate any vulnerability to a jammer utilizing knowledge of the set of permutations to his advantage. Exactly how large M must be depends on the system performance criteria and estimates of jammer capability. This parameter would typically have a value in the range 10 to 1000.

A block diagram of a typical interleaver configuration (with interleaver operation slaved to a PN generator) is shown in Fig. 8-11. Channel symbols are written into the interleaver memory sequentially. After an entire block is written in, then these symbols are permuted by reading them out using the pseudorandom permutation stored in the address ROM. The actual

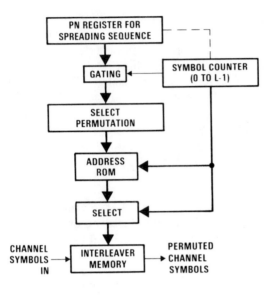

CHANNEL
SYMBOLS →
IN

PERMUTED
CHANNEL →
SYMBOLS

Figure 8-11. Pseudorandom interleaver configuration.

memory management could be accomplished using two RAMs in a ping-pong configuration; symbols can be written into one memory while symbols are read from the other memory. After completing this process, the roles of the two memories are interchanged, and a new pseudorandom permutation is selected.

The interleaver can be synchronized by establishing a fixed relationship between the state of the PN shift register used to generate the spreading sequence and the received symbol counter. Once lock is obtained for the PN sequence, the boundaries for the interleaver blocks are established. Also as shown in the figure, the state of the PN register at the block boundary is used to select the new permutation (one of M) for the succeeding inter-leaved block. For applications other than AJ systems, the deinterleaver can be synchronized by standard frame synchronization techniques (recognizing that channel errors will occur in bursts).

The only remaining question involves the selection of the permutations to be used. Numerous methods for generating sets of sequences are possible. However, such sets of sequences should be tested to verify if the required degree of "robustness" is achieved. Richer [96] has suggested the use of linear congruential sequences for the permutation. A pseudorandom sequence of integers from 0 to $L-1$ is produced by the relationship

$$A_{n+1} = (aA_n + c) \qquad \bmod L \qquad (8\text{-}2)$$

For a fixed L the sequence may be changed by changing a and c. To obtain a maximal length sequence (i.e., a sequence with period L), the parameters

a and c must satisfy the following conditions:

1. a and c must be less than L;
2. c must be relatively prime to L;
3. $(a - 1)$ must be a multiple of p every prime p dividing L;
4. $(a - 1)$ must be a multiple of 4 if L is a multiple of 4.

This approach then generates a sequence of numbers from zero through $L - 1$ that appears quite "random" if a is relatively prime to L. With proper choice of interleaver block length L, there will be many sequences which satisfy (8-2).

However, an interleaver based on a sequence generated by (8-2) is susceptible to certain periodic interference patterns with a period P which divides L.[96] To improve the randomness properties of the sequence we suggest an additional permutation. This involves taking the list of numbers produced according to (8-2) and permuting them using the sequence of states of a linear feedback shift register. This produces a sequence of addresses with excellent randomness properties. Implementation of this approach in a hardware pseudorandom interleaver has been accomplished.[97]

The sets of permutations thus produced are stored in the address ROM of Fig. 8-11. If the storage requirements are prohibitive, alternate implementations are possible. For example, one could store just one set of permutations from (8-2) and access these pseudorandomly using the states of a linear feedback shift register. The permutation can be changed by changing the connection polynomial of the feedback shift register.

As an example, suppose we wish to design an interleaver with $L = 2^m$. Then to produce a linear congruential sequence of length $L = 2^m$ from (8-2), we must select parameters a and c of the form

$$a = 4K + 1 < L$$

and

$$c = 2j + 1 < L$$

Obviously, there are many satisfactory choices for a and c. After this permutation is defined we suggest a further permutation using the states of a maximal-length feedback shift register. [Since the number of states is only $2^m - 1$, it is assumed that the zeroth number in the sequence (8-2) is always used first and all other numbers in the sequence are selected using the shift register states.] Typically, the number of potential connection polynomials for the feedback shift register is rather large. For example, for $L = 1024$ one needs degree-10 primitive polynomials. Peterson and Weldon[1] list 60 different primitive polynomials of degree 10. These polynomials could be stored in a 64×10 ROM. Accessing a given polynomial from the ROM produces an output which could then be used to control the feedback in the feedback shift register. This would allow 60 different permutations to be obtained. In addition, every cyclic shift (or a total of 1023) of each permutation is also available simply by changing the initial loading of the shift register. Both the choice of connection polynomial and the initial loading of the shift register can be slaved to the state of the PN shift register used to generate the spreading sequence.

To this point the deinterleaving operation has not been discussed. However, its structure is identical to that of Fig. 8-11 for the interleaver. The deinterleaver simply performs the inverse permutation. That is, if in a given block the permuted channel symbols are read out of memory pseudo-randomly, then at the deinterleaver these same symbols are written into memory with the same permutation. When these symbols are read out of memory the addressing will be sequential. Of course, using the same permutation at the deinterleaver presents no synchronization problem in AJ systems. Once the PN spreading sequence is synchronized, both the interleaver block boundary and the particular permutation used are known since they are slaved to the state of the PN sequence register.

8.4. Coding for Burst Noise Channels

There are two fundamental types of burst noise problems. The type that has received the most attention from researchers over the years is the problem which exists when soft decisions are not available from the demodulator. In this case, no information is available to identify single unreliable symbols. However, if the error mechanism which causes burst errors is understood, the resulting burst error statistics can be exploited by error-correction schemes. A second type of problem occurs when soft decisions are available from the demodulator to identify the unreliable symbols. When the noise is particularly severe during these symbols, they are sometimes simply "erased." Substantially better performance can be obtained in this case because random errors as well as bursts can be corrected efficiently.

The first type of problem arises naturally in certain communication media that are inherently digital. Examples are storage media such as film, magnetic tape, magnetic disks, and magnetic memories. Burst errors can occur through scratches, defects, etc. The nature of the error mechanism is such that reliability indications on each bit are difficult, if not impossible, to obtain. Typical approaches to solving this problem are techniques which correct long bursts of errors at the expense of missing some random error patterns. Cyclic codes such as Fire codes[98] can be used in this manner and decoded using a Meggitt-type decoder. Reed–Solomon codes can be used for correction of multiple short bursts, or they can be interleaved by symbols to correct multiple long bursts. One could also use random-error-correcting block or convolutional codes with appropriate interleaving. In addition, there are burst-trapping techniques found by Tong[99] and Gallager[3] which are effective in correcting long bursts provided that there is sufficient error-free guard space between bursts. Another approach is diffuse convolutional codes with threshold decoding,[100] which allows correction of long bursts with a guard space that contains a small number of errors. The greatest problem in applying these techniques is acquiring an accurate characterization of the burst statistics of the channel. Such statistics are often quite difficult to obtain. With the plethora of available coding approaches for this type of problem, we feel that it is not necessary to discuss it further.

The second type of burst noise problem is more interesting, and the use of coding can be more effective. Often the channel is basically a Gaussian noise channel that is occasionally corrupted by large bursts of noise or interference such as in an RFI or intentional jamming environment. PSK modulation is assumed. A near-optimum strategy in providing likelihood information when the interference bursts are present is to simply blank the affected symbols so that symbol erasures are produced.[101] This procedure will be assumed, and it will also be assumed that the burst lengths are longer than a symbol time and only complete symbols are blanked. In addition, if interleaving is used, there is no need for accurate knowledge of the channel burst statistics (other than the maximum burst length). Several burst erasure processes will be considered. The principal results will apply to the use of convolutional codes with Viterbi decoding on Gaussian noise and burst erasure channels. However, some comparisons with block codes will be made. Finally, one can draw some interesting conclusions regarding the use of interleaving for this type of channel. The performance in an actual system will depart somewhat from these results because of difficulties in implementing perfect blankers or because the blanker may be approximated by a clipper or limiter.

8.4.1. Burst Noise Processes

Several types of burst noise processes will be considered. In each case, the channel is modeled simply as a Gaussian noise channel with a specified value of E_b/N_0 when the burst noise is not present. Then when the burst noise is present, it is considered to have such a large value that it can easily be detected and blanked by the demodulator. This results in a burst of symbol erasures. It will be assumed that either the burst length is fixed or its maximum value is known.

The spacing between bursts can assume several forms. One such process is the *periodic burst erasure process*. In this case a burst of B symbol erasures occurs every P symbols (the parameter P is the interference *period*, B is the *burst length*, and $\delta = B/P$ is the *duty factor*). A typical example of such an interference process is RFI from an interfering pulse radar. We will find that in many cases this type of burst erasure process does not significantly degrade the performance of random-error-correcting codes even when no interleaving is used. The burst length must be quite long before a sizable degradation occurs. A much more troublesome process is the *random burst erasure process*. With this process the burst length is assumed to be fixed at B symbols with an average duty factor of δ and with exponentially distributed interarrival times. Because of the unpredictable nature of spacing between bursts, only very short bursts can be tolerated without resorting to the use of interleaving. The most damaging type of interference process is *intentional jamming*. With this process the jammer utilizes knowledge of the signal structure to place erasures in the most damaging locations. This type of interference requires the most sophisticated interleaving scheme.

The *random erasure process* will also be treated. This process is of interest because it is the process that determines coded performance when several types of interleaving are used. These results can be used as a base line with which to compare performance of a system without interleaving. This allows one to evaluate the tradeoff between the performance gain achieved by adding the interleaver and the additional complexity added to the system.

8.4.2. Convolutional Code Performance with Random Erasures

Convolutional code performance with a random erasure process of duty factor δ is quite straightforward to compute.[102] In this case, each symbol has probability δ of being erased, and all erasures are statistically independent. Thus, a given code word of weight j can have its weight reduced to $j - i$, $0 \leq i \leq j$, by erasures, and the probability of weight reduction to $j - i$ is given by $\binom{j}{i} \delta^i (1 - \delta)^{j-i}$. Accordingly, the event error probability for a weight-j code word, when averaged over the random erasure process, becomes

$$\overline{P_e(j)} = \sum_{i=0}^{j} \binom{j}{i} \delta^i (1 - \delta)^{j-i} Q\left[\left(\frac{2(j - i)RE_b}{N_0} \right)^{1/2} \right] \qquad (8\text{-}3)$$

and the average union bound on P_b as given by (6-11) for an $R = m/n$

code is therefore

$$\bar{P}_b \le \frac{1}{m} \sum_{j=0}^{\infty} w_j \sum_{i=0}^{j} \binom{j}{i} \delta^i (1-\delta)^{j-i} Q\left[\left(\frac{2(j-i)RE_b}{N_0}\right)^{1/2}\right]$$

which can be rearranged as

$$\bar{P}_b \le \frac{1}{m} \sum_{j=0}^{\infty} \hat{w}_j Q\left[\left(\frac{2jRE_b}{N_0}\right)^{1/2}\right] \tag{8-4}$$

where

$$\hat{w}_j = \left(\frac{1-\delta}{\delta}\right)^j \sum_{k=j}^{\infty} \binom{k}{j} w_k \delta^k \tag{8-5}$$

is an average weight structure which reflects the deterioration of the code weight structure due to the random erasure process.

Of special significance in (8-4) is the appearance of a nonvanishing $j = 0$ term, which indicates that as $E_b/N_0 \to \infty$, the bound on $\bar{P}_b$ approaches a "floor" at $\hat{w}_0/2m$, or

$$P_{\text{floor}} = \frac{1}{2m} \sum_{k=0}^{\infty} w_k \delta^k \tag{8-6}$$

This error rate floor occurs because of the nonzero probability of erasures occurring in all the positions in which two code words differ. In this case an error rate of $1/2$ occurs.

Some interesting performance curves for specific codes can be generated using these results. Assuming random erasures with probability of occurrence of δ, the probability of bit error is given by (8-4) and (8-5). Thus, the performance of the familiar $R = 1/2$, $v = 6$ code is shown in Fig. 8-12 (assuming a coherent PSK system).

Performance degradation (i.e., E_b/N_0 increase required to maintain $P_b = 10^{-5}$) as a function of δ for a random erasure process is shown in Fig. 8-13 for several interesting codes. The $R = 1/4$ code is derived by two repetitions of each channel symbol of the $R = 1/2$ code. The advantage obtained in using the $R = 1/4$ code is a doubling of the code free distance from 10 to 20 and the resulting reduction of P_{floor} as given by (8-6). The $R = 2/3$ and $R = 3/4$ codes are punctured codes. Comparing the $v = 6$ curves it is seen that there is a significant performance advantage associated

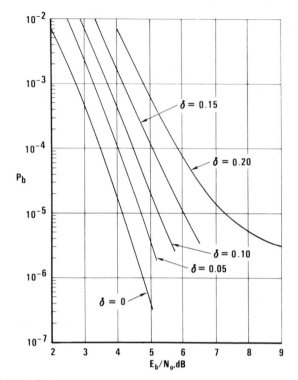

Figure 8-12. Performance of $R = 1/2$, $v = 6$ code with random erasures.

with reducing the code rate. This is particularly evident when operating near the error rate floor. For example, one cannot operate at $P_b = 10^{-5}$ with $\delta = 0.20$ and $R \geq 2/3$. Even with a $R = 1/2$ code one is operating near the BER floor and over 3 dB of degradation is encountered. However, by reducing the code rate to $1/4$, the degradation is reduced to 1.5 dB, and this is accomplished with a decoder of equivalent complexity.

One can accomplish a similar result maintaining a constant code rate by increasing the code constraint length. Similar results are shown for $v = 8$. Performance improvements of 1 to 2 dB for $R = 2/3$ and $3/4$ codes are observed for duty factors in the range $0.05 \leq \delta \leq 0.10$. Also shown is the lower bound on degradation. A fraction δ of erased symbols implies that the received signal energy is only a fraction $(1 - \delta)$ of the energy that is received with no erasures. Thus, the received signal energy is reduced by $10 \log_{10}(1 - \delta)^{-1}$. Note that for $R \leq 1/4$ the degradation is very close to this bound at interesting values of δ.

We should point out that the values shown in Fig. 8-13 and the subsequent figures show the degradation for that code measured relative to that code's performance in the absence of erasures. Thus, to compare different codes, the difference in performance in the absence of erasures must be counted as well as the difference in degradation.

The results of Fig. 8-13 may be applied to any burst erasure process when used with pseudorandom interleaving (for a sufficiently long interleaver). In addition, if one considers a random burst erasure process, the results apply when either random or periodic interleaving is used.

Similar results may be obtained with other coding/decoding techniques. Long block codes have large minimum distances and suffer little degradation over this range of δ if an approximation to maximum-likelihood decoding is used. For example, the (48, 24) and (128, 64) codes have $d = 12$ and 22, respectively. Union bounds on performance can be obtained in exactly the same manner, but the computation is complicated by the difficulty of obtaining the weight structure of most block codes. With complex concatenated systems, degradation relative to the lower bound shown in Fig. 8-13 should be almost nonexistent (because of the extremely large minimum distances). However, in many cases such sophistication should not be necessary.

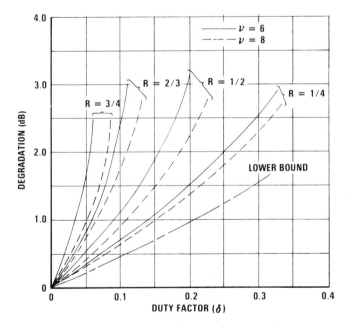

Figure 8-13. Performance degradation (at $P_b = 10^{-5}$) for a random erasure process.

8.4.3. Convolutional Code Performance with Periodic Burst Erasures

The effect on the performance of a convolutional code with Viterbi decoding caused by a periodic erasure process can be determined by calculating the effect of periodic erasures on the code weight structure. Suppose that the input to the decoder is perturbed by a periodic erasure process in which a burst of B consecutive symbols is erased every P symbols. There are P different phasings of this erasure process with respect to the decoder input sequence. For a given phasing of the erasure bursts, a modified weight structure of the code can be determined by carrying out the same exhaustive search process used to obtain $\{w_j\}$, but omitting the code word positions in which an erasure appears. (Observe that the decoder knows the location of the erased symbols but cannot use any code distance accumulated in the erased positions to discriminate among code words.) By substituting this modified weight structure into the union bound, a union bound on P_b is obtained for this given erasure phasing. The net bound on error probability is then obtained by averaging over all phases of the erasure process. Equivalently, one can derive an average weight structure $\tilde{w}_j$ by determining the modified weight structure for each phase of the erasure process and then averaging over phase. Insertion of $\tilde{w}_j$ into (8-4) just as with random erasures yields a union bound on $\bar{P}_b$.

The average weight structure is given by

$$\tilde{w}_j = \frac{1}{P} \sum_{p=1}^{P} w_j(p) \tag{8-7}$$

where $w_j(p)$ is the weight structure determined by the usual exhaustive search, but with the erasure process beginning at symbol p (that is, we omit distance accumulated in symbol $p, p + 1, \ldots, p + B - 1$: $p + P$, $p + P + 1, \ldots$: $p + P + B - 1$:..., etc.). As an example, Fig. 8-14 illustrates the first several terms of the original weight structure of the $R = 1/2$, $v = 6$ code, and the average weight structure with periodic erasure processes with $B = 1$, $P = 10$ and $B = 10$, $P = 100$. Note that while erasure bursts can significantly reduce the minimum distance of the code (e.g., to 7 for $B = 1$, and to 3 for $B = 10$), small multiplying factors appear in the low-weight terms of the union bound. Because of this, performance does not degrade as much as would be expected simply from considerations of the distance reduction. For comparison, the average weight structure for a random erasure process with the same average duty factor as given by (8-5) is also shown.

Performance degradation for a $R = 1/2$, $v = 6$ code and a periodic burst erasure process, as determined by (8-4) and (8-7), is shown in Fig.

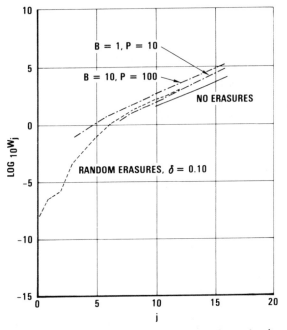

Figure 8-14. Weight structures for the $R = 1/2$, $v = 6$ code.

8-15.[103] These curves show degradation as a function of burst length for 5%, 10%, and 20% duty-factor periodic erasure processes. This is representative of the performance degradation that would be encountered when *no interleaving* is used. As expected, the best performance for a given δ is achieved at $B = 1$. For a *known* periodic burst erasure process one could configure a periodic interleaver/deinterleaver that transforms the known process into a process with maximally separated single erasures (with performance equivalent to the $B = 1$ case of Fig. 8-15). Note that for bursts of length up to five symbols, the degradation increases surprisingly slowly, and in fact is comparable to or less than the degradation due to random erasures of the same duty factor. Thus, for bursts up to five symbols, pseudorandom interleaving does not help, but rather hurts performance, and no form of interleaving can improve performance significantly. In addition, for the 5% and 10% duty factors, at least, the degradation is modest even for rather long erasure bursts (less than 2 dB at $B = 8$).

An interesting aspect of the curves is the anomalous behavior at 10% duty factor in which performance is slightly better with $B = 4$ than with $B = 3$. The explanation for this is that the longest minimum weight code

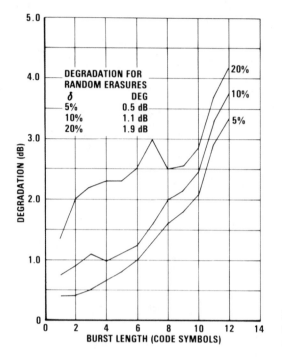

Figure 8-15. Performance degradation (at $P_b = 10^{-5}$) for the $R = 1/2, v = 6$ code.

word is 32 symbols long, which is just long enough to suffer two erasure burst hits of length 3 and period 30 symbols. The distribution of weight in this code word is such that the weight is reduced by 5 with this worst-case phasing. However, with $B = 4$ and $P = 40$, the weight-10 code word is too short to suffer multiple hits, and thus can have weight reduced by at most 4.

Observe also that as B increases, the curves for all three duty factors tend to become parallel, while for smaller values of B there are substantial differences in the shapes of the curves. The explanation for this effect is that for larger values of B with fixed δ, the periods become too long to permit multiple hits on the low-weight code words, which are the significant contributors to the union bound. Thus, for example, at $B = 10$, regardless of whether $\delta = 5\%$, 10%, or 20%, the degrading effect is attributable only to a single erasure burst, and thus the duty factor enters the result only by way of the averaging over a period. On the other hand, for small B, periods are short enough to allow *multiple hits* for larger duty factors, and the curves can differ significantly.

Finally, note that the curves of Fig. 8-15 are valid only at $P_b = 10^{-5}$. Asymptotically, i.e., as $E_b/N_0 \to \infty$, degradation is determined only by the worst-case distance reduction. It is significant that the levels of degradation at $P_b = 10^{-5}$ are substantially less (up to 3 dB over this range) than the asymptotic degradation. The reason is, as we noted earlier, that although the weight reduction with worst-case phasing can be sizable, the impact is mitigated substantially by averaging over the erasure period.

Similar curves can be shown for the $R = 2/3$ and $3/4$ codes. To demonstrate another interesting effect these curves will be shown with the burst length normalized to an information symbol time. That is, degradation is shown as a function of the BR product (the burst length in information symbols). The results for $\delta = 0.05$ are shown in Fig. 8-16 for $R = 1/2, 2/3$, and $3/4$. (If lower rate codes were constructed by repetitions of the channel symbols of the $R = 1/2$ code, the curves would be almost identical to that of the $R = 1/2$ code in Fig. 8-16.) Note that if one holds the information rate constant and uses the higher code rate to achieve a bandwidth reduction, then we are comparing the performance in the proper manner. The interesting point is that the performance degradations for the three cases are so close. Obviously, this behavior is significantly different from what one observes with random erasures. From these and other observations we have made (see Fig. 6-25), it appears that the higher-rate codes have longer paths with a significantly lower density of ones than the $R = 1/2$

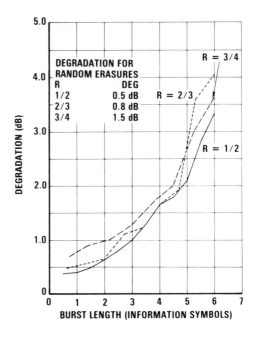

Figure 8-16. Performance degradation (at $P_b = 10^{-5}$) for $v = 6$ codes and $\delta = 0.05$.

code, and a burst of erasures does not produce the weight reduction and hence the performance degradation that one might expect.

As an indication of dependence upon the constraint length of the code, Fig. 8-17 shows degradation at $P_b = 10^{-5}$ with $\delta = 10\%$ for $R = 1/2$ codes of several constraint lengths. We observe that the effect of increasing the code constraint length is much more beneficial than reducing the code rate (at constant information rate) when the burst erasure process is periodic and no interleaving is used. This is in contrast to the results of Fig. 8-13 for random erasure processes.

One can think of this periodic erasure process as a process that causes an erasure burst of length B followed by an erasure-free guard space of length $G = P - B$. Thus, these results are also quite meaningful in cases where the process is specified to have bursts of length B or less with a guard

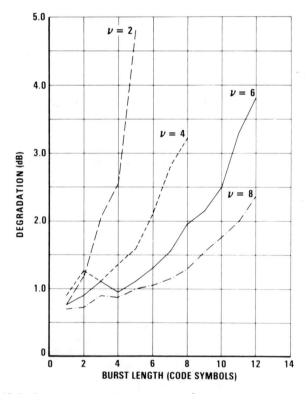

Figure 8-17. Performance degradation (at $P_b = 10^{-5}$) for $R = 1/2$ codes with $\delta = 0.10$.

space of $G \geq P - B$. The predicted performance degradations should provide an upper bound to the degradation expected in the latter case.

Similar results can be obtained with block codes. For example, Berlekamp has shown that the (60, 30) block code with $d = 12$ can allow one burst of 6 or 12 erasures per code word with performance degradations of only 0.8 and 1.9 dB, respectively (at $P_b = 10^{-5}$ assuming maximum-likelihood decoding).[104] This result is slightly better than one obtains with the $R = 1/2$, $v = 8$ code which also has $d_f = 12$. The use of long block codes has somewhat more potential than convolutional codes when used without interleaving because very long large-distance codes can be used. Long convolutional codes can be decoded via sequential decoding, but a significant computational problem will result in the presence of a burst erasure process.

8.4.4. Performance Degradation Due to Random Erasure Bursts

The results of the preceding sections have established that a short burst of erasures is not particularly harmful, provided that it is not followed closely by another burst. However, closely spaced bursts can occur with significant probability for a random burst erasure process. Thus, it is of much interest to estimate the resulting performance degradation. Unfortunately, it is much too tedious to estimate the performance degradation directly for specific codes as we did for the random and periodic burst erasure processes. However, indirect techniques for estimating performance degradation using the R_0 parameter are often quite meaningful. Recently, Viterbi[95] has shown that R_0 for the random burst erasure channel is given by

$$R_0 = \frac{1}{B} \log_2 \left[(1 - \delta) \left(\frac{1 + e^{-E_s/N_0}}{2} \right)^B + \delta \right] \qquad (8\text{-}8)$$

with B and δ as defined previously. Performance degradation is computed for specific values of B and δ by finding the increase in E_s/N_0 (compared with the $\delta = 0$ case) required to maintain the same R_0 parameter. Using this measure, one sees that the rate at which degradation increases with burst length is much more severe than was observed with a periodic erasure process. As a result, an interleaver is needed unless the bursts are very short ($B \leq 2$) or the duty factor is very low ($\delta < 0.05$). The severe degradation observed is due to a significant probability that multiple bursts will occur within a decoding constraint span.

8.4.5. Interleaving Implications

The results of the previous sections can be used to predict performance with no interleaving, periodic interleaving, or pseudorandom interleaving for each of the erasure processes considered.

When no interleaving is used, a convolutionally coded system performs well when the erasures are either random erasures or periodic short burst erasures. Performance degradation versus δ is given in Fig. 8-13 for the random erasure process. Degradation for periodic burst erasure processes for a number of cases is given in Figs. 8-15, 8-16, and 8-17. These results demonstrate that with *no interleaving* significant erasure bursts can be accommodated with very little performance degradation when compared to maximally spaced erasures with the same duty factor. With the $R = 1/2$, $v = 6$ code, for example, use of an interleaver is probably not justifiable if the erasure burst lengths are no greater than $B = 5$ (see Fig. 8-15). Conversely, the performance degradation due to random burst erasures indicates that, except for very short bursts or very small δ, interleaving is usually needed.

The two principal classes of interleavers are periodic (either block or convolutional) and pseudorandom. The periodic strategy attempts to produce maximally spaced erasures while the pseudorandom strategy attempts to produce random erasures at the decoder input. It has been widely assumed that the periodic approach will result in superior performance when it is exactly matched to the erasure parameters (B and P), but that it is much less robust to changes in these parameters than the pseudorandom interleaver. With the results of the previous sections, we can quantify the degree of robustness of the periodic interleaver.

Suppose that erasure bursts of length B and duty factor $\delta = B/P$ are expected. Then if a block interleaver is designed for this erasure process (with the number of interleaver rows large relative to P/B to allow an increase in P), the decoder will see *single* erasures spaced by P/B symbols. The performance for this case is found from our results with $B = 1$ and $\delta = B/P$, and it appears to be superior to that of the pseudorandom interleaver (with the same δ) in all cases. Now suppose that the burst length increases to rB, but with the same δ. In this case the interleaving strategy is no longer matched to the erasure process. As a result, the erasures will occur in bursts of r erasures separated by rP/B symbols at the decoder input. This is identical to the case we have analyzed. The results show that as long as r is small (say 5 or less for the $R = 1/2$, $v = 6$ code), there will be little degradation from either the perfectly matched periodic interleaver or the pseudorandom interleaver. Thus, a fairly significant degree of robustness to a change in B is demonstrated.

However, the periodic interleaver is not suitable for intentional jamming because knowledge of the deinterleaver structure can be used advantageously. As noted in Section 8.3, a jammer could produce single erasures spaced by small integer multiples of B symbols. Any desired duty factor could be achieved by turning this process on and off. At the deinterleaver all the erasures will lie in the same column and will thus enter the decoder as a long burst resulting in severe degradation.

If the process is a random burst erasure process, then interleaving is usually required and a periodic interleaving strategy will give the best performance providing the maximum burst length can be bounded. The interleaver parameter B should be chosen to be the *maximum* burst length. Unlike the previous example, one cannot afford for a burst to cover several interleaver rows. It does not hurt to be conservative in selecting B (burst lengths that are much less than B are no problem with *random* bursts). The number of rows in the interleaver should be at least as large as a decoding span. In this case the erasures will appear as random single erasures and the performance degradation may be predicted from Fig. 8-13.

The preceding discussion assumed a block interleaver for ease of presentation. All of the principles of the preceding discussion apply also to a convolutional interleaver although some of the details may differ.

Pseudorandom interleaving is of interest when the interference environment is not well characterized or is unknown (e.g., intentional jamming). With true random interleaving the performance is a function only of E_b/N_0 and the average duty factor δ as given in Fig. 8-13. Since an actual interleaver is not truly random but must have a finite length, the performance will depart somewhat from Fig. 8-13. The required interleaver length is always greater than that of a periodic interleaver for a well-defined periodic burst erasure process, but it may be either more or less than that required for a random burst erasure process depending on the decoding span and the duty factor δ.

Our results are summarized in Table 8-3. For random erasure bursts with bounded maximum burst length the periodic interleaver is usually chosen. However, the periodic erasure burst process is much more benevolent, and our results show that for erasure bursts of moderate length (up to about half the constraint length in transmitted symbols) and low duty factor, degradation is modest even without interleaving. For longer bursts where interleaving is required, a periodic interleaver matched to the nominal burst parameter is robust against sizable variation in the burst parameters (with constant δ). However, there are some situations where periodic interleaving strategies do not offer sufficient robustness (e.g., intentional jamming). The random interleaving approach offers the greatest degree of "robustness" and is recommended when the interference environment is uncertain.

Table 8-3. Interleaver Application Matrix for Several Burst Erasure Processes[a]

	No interleaver	Periodic interleaver	Random interleaver
Random erasures	Yes (Fig. 8-13)	No	No
Known periodic burst erasures	Yes, for moderate B and δ (Figs. 8-15, 8-16, 8-17)	Use for $BR \geq 3$ (Figs. 8-15, 8-16, 8-17)	No
Random burst erasures	Poor performance except for small B and δ	Best performance (Fig. 8-13)	Occasionally requires less storage than periodic (Fig. 8-13)
Intentional jamming	No	Nonrobust to certain periodic burst erasure processes	Maximum robustness (Fig. 8-13)

[a] Figure denoted in parentheses provides performance results.

8.5. Coding for Spread Spectrum Systems

Spread spectrum techniques can be used for interference suppression, ranging, or transmitted energy density reduction. We shall be concerned with the first of these applications — protection against intentional interference or jamming. Antijamming (AJ) protection is the most extensive use of spread spectrum communications. The degree with which these techniques are being used has grown rapidly in the past several years because of recent advances in integrated circuit technology which has allowed higher achievable spreading factors and practical implementations of complex systems.

The spread spectrum techniques involve signaling techniques which greatly expand the transmit spectrum relative to the data rate. They include direct sequence PN spreading, frequency hopping, time hopping, and hybrid techniques. These techniques attain AJ protection basically by forcing the jammer to deploy his power over a much wider bandwidth than would be necessary for an unspread system. Thus, for a fixed jammer power, the jamming spectral density is reduced in proportion to the ratio of spread to unspread bandwidth.

We shall be concerned mainly with direct sequence PN spread and frequency hop systems. For each type of spread spectrum signaling waveform there is typically an optimum jamming strategy that takes advantage of the *waveform structure* and maximizes error rate. For example, with an uncoded direct sequence system all the signal energy in any information symbol occurs over one symbol time. Thus, a pulse jamming strategy can capitalize on this structure by transmitting high peak power pulses that are at least one symbol time in length thereby significantly increasing the probability of bit error. Coding offers a solution to this problem by making the waveform unstructured. The gains that can be obtained from coding in these situations can be much larger than can be obtained on a random error channel such as the BSC or Gaussian channel.

8.5.1. Direct Sequence PN Spread Systems

A block diagram of a coded direct sequence (DS) PN spread AJ system is shown in Fig. 8-18. The data modulator could use any one of a number of standard techniques (PSK, FSK, etc.). However, one of the advantages of DS spread systems is that coherent demodulation can be used ($\Delta\omega = 0$ and $\phi \approx 0$ in the figure). Thus, PSK is utilized most often. The resulting waveform is spread by multiplying it by a carrier that has been modulated by a pseudorandom NRZ sequence with a chip rate, R_c, that is much larger than the information rate, R_b. The spreading ratio, R_s, is defined as the ratio R_c/R_b. The data waveform is recovered by multiplying the received signal by a synchronized replica of the pseudorandom sequence. This operation also has the benefit of spreading the jammer energy over a bandwidth much wider than the data bandwidth.

The pseudorandom sequence is usually generated by utilizing a maximal-length feedback shift register (MLFSR). Such a register with n stages generates a sequence of length $2^n - 1$ symbols before repeating. The security of the AJ system is enhanced by choosing n large to prevent the jammer from ever observing a complete cycle of the sequence. However, one can observe $2n$ symbols of such a sequence and compute the feedback connections of the MLFSR (using the Berlekamp algorithm). To further

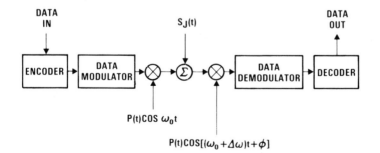

Figure 8-18. Coded direct spread AJ system.

enhance security the output sequence from the MLFSR is not used directly. Instead, the outputs from several stages in the MLFSR are combined in a nonlinear logic network to produce the output symbol. This process has the advantage of making it computationally infeasible for a listener to observe a portion of the sequence and to then determine the network that generates the sequence. These codes are typically changed frequently to further enhance security.

8.5.1.1. Performance of Uncoded DS Spread Systems. The data modulation technique most widely used in direct sequence systems is PSK. One can perform an approximate analysis of this system by calculating an "effective" E_b/N_0 (where N_0 denotes an "effective" noise spectral density due to the jammer). Assume a received signal power S and a tone jammer with average power J. Then the effect of the despreading process is to collapse the spread signal back into its original bandwidth and to multiply the interfering signal by the PN code. As a result, the interference process now has a $\text{sinc}^2(x)$ power spectral density shape identical to that of the PN spread carrier reference. For large spreading ratios the noise spectrum is flat in the vicinity of the data spectrum and has $N_0 = J/R_c$. The effective E_b/N_0 for such a system is simply the signal-to-noise ratio in a bandwidth equal to the information rate (after despreading). Thus, the effective E_b/N_0 is

$$E_b/N_0 = R_c S/R_b J = R_s S/J$$

If we make the assumption that the resulting noise is Gaussian (valid for large R_s), the error rate is given by

$$P_b = Q\left[(2R_s S/J)^{1/2}\right] \qquad (8\text{-}9)$$

The spreading ratio, R_s, can also be called the *processing gain* of the system since it makes the effective E_b/N_0 larger than the actual S/J by a factor R_s. (The processing gain is sometimes defined as the ratio of the spread bandwidth to the data bandwidth.) Knowing the desired E_b/N_0 (say 9.6 dB for $P_b = 10^{-5}$), one can define the system *jamming margin* (J/S) as

$$\frac{J}{S} = \frac{R_s}{(E_b/N_0)_{\text{req'd}}} \qquad (8\text{-}10)$$

This quantity gives the maximum jammer-to-signal power ratio at the demodulator input that can be tolerated before P_b falls below the desired operating point. Thus, the J/S increases linearly with R_s.

The above analysis is approximate and deviations can occur due to several causes. First, the noise at the demodulator output is not Gaussian but a filtered spread cw jammer. However, one can show (based on the central limit theorem) that if a very broadband signal with virtually any statistics is passed through a very narrowband filter, the output is approximately Gaussian. Since the spread jammer has bandwidth that is a factor R_s larger than the bandwidth of the integrate-and-dump filter in the PSK demodulator, this condition is satisfied for large spread ratios, thus justifying the Gaussian assumption.

The second difficulty is that the P_b computation in (8-9) assumes that the Gaussian noise contains components both in phase and in quadrature with the desired signal. The quadrature component (which is half of the noise) of course does not affect P_b. However, if the jammer is coherent in the phase and frequency with the carrier, then all the jammer noise contribution is in phase with the desired signal. Since (8-9) counts only half the noise, performance will be 3 dB worse than indicated. This is a real problem if the jammer frequency is much closer to the carrier frequency than R_b. In this case a significant fraction of symbols will have all the noise in phase, and this worst case will govern performance. Only if the jammer frequency is sufficiently far away from the carrier frequency will the noise phase average and provide the performance of (8-9). (A potentially more severe problem with a coherent jammer occurs if the received waveform is limited before despreading.)

This 3-dB difficulty with the worst-case cw jammer (and the potential problem with limiting) can be alleviated by using a QPSK spreading sequence. By this we mean that the same data are transmitted on the I & Q channels, but that a different PN code is used on each channel. In this case the noise phase "averages" on each bit and the original result of (8-9) remains valid. Thus, a QPSK spreading sequence has roughly a 3-dB advantage over BPSK for the worst-case cw jammer.

The performance with noise jamming can be determined in the same manner. Since the noise spectrum will be wider than that of cw, it will be still wider after the despreading process. How much wider it will be will depend on the noise spectrum itself. However, the performance will be a little better but closely approximated by that predicted by (8-9).

The direct spread approach can produce a signaling waveform that is highly unstructured relative to any parameter but time. That is, the transmitted symbols all have equal length and occur with no guard space. A jammer can make much more efficient use of this power by adopting a strategy that takes advantage of this structure. Pulsed jamming is such a strategy. Assume first that the jammer power is concentrated into pulses with peak power J_p. Again assume that when the jammer is on, the demodulator output statistics are Gaussian. If the jammer pulses are many symbols long and the jammer duty cycle is δ, then only a fraction δ of the symbols are jammed, and for these symbols the E_b/N_0 is $R_s S/J_p$. Thus, the bit error probability is

$$P_b = \delta Q \left[(2R_s S/J_p)^{1/2} \right]$$

The average jammer power, J, is δJ_p. [Note that, at $\delta = 1$, this expression is identical to (8-9).] For a fixed average jammer power one can select δ such that P_b is maximized. This represents the worst-case pulse jammer. Thus,

$$P_{max} = \max_{0 \leq \delta \leq 1} \left\{ \delta Q \left[(2\delta R_s S/J)^{1/2} \right] \right\}$$

This maximum is easily found to be

$$P_{max} = \frac{0.083 J}{R_s S}$$

resulting in P_{max} decreasing only inversely with $R_s S/J$. P_{max}, as well as P_b, is shown in Fig. 8-19 for several values of δ.

8.5.1.2. *Performance of Coded DS Spread Systems.* We observed in the previous section that although the DS technique can provide excellent AJ protection against the continuous noise or tone jammers, there is a *worst-case* jammer (pulsed) that seriously degrades performance. For the worst-case jammer, the error rate behaves as $P_e \approx K (E_b/N_0)^{-1}$. This sensitivity to a worst-case jammer can be eliminated through intelligent application of diversity and/or error correction coding. However, one must be very careful. The use of coding places structure (in the form of time) in the signaling waveform. For example, if a random-error-correcting block code of length n is used, the jammer gains an advantage by using a pulsing strategy with pulses on the order of n symbols long. The diversity provided by the code is no longer useful in this case. If one used a burst-error-correcting code, then the jammer could still gain an advantage by using a pulse longer than the burst-correction capability or by inserting a series of isolated single errors. Thus, while coding provides the promise of eliminating susceptibility to worst-case jammers, one must be very careful to try to make the coding approach *unstructured with respect to time.*

An excellent technique for making the waveform unstructured with respect to time is pseudorandom interleaving. As discussed in Section 8.3, the technique involves simply taking a long block of coded symbols that are about to be transmitted and scrambling in a pseudorandom fashion

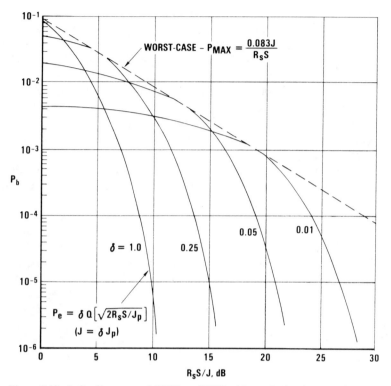

Figure 8-19. P_b for direct spread BPSK or QPSK with a pulsed noise or cw jammer.

the order in which they are to be transmitted. An inverse operation will be performed at the receiver which has the effect of making the channel errors entering the decoder appear random even if generated by a pulse jammer.

When using pseudorandom interleaving, a good random-error-correcting code and a decoding technique that utilizes soft decisions will provide excellent performance. It is very important that the demodulator be able to identify the very noisy symbols and that the decoder be able to utilize this soft decision information in order to provide the best performance with a pulsed jammer.

Implementation of an optimum PSK demodulator for use with coding or diversity in both continuous and pulse-jamming environments requires a more elaborate receiver structure (e.g., a different AGC) than would be needed for an uncoded system. A coded system achieves coding gain because of the minimum distance or number of symbol positions in which all code words disagree. The optimum receiver structure for applying coding can be found by finding the optimum receiver structure for a message consisting of M repetitions of the same symbol. This problem is the general Gaussian problem discussed in detail in Van Trees.[105] A block diagram of the optimum receiver structure is given in Fig. 8-20. Because of the nonstationary character of the noise with pulse jammers, it becomes very important to estimate the noise variance on each symbol to allow an optimum decision to be made in the decoder. Note that scaling the integrate-and-dump output voltage by $2E^{1/2}/\sigma_k^2$ (shown in the block diagram) results in a *signal suppression* effect on the value of m_k for high noise conditions as shown in the figure. This is exactly the desired effect which allows the decoder to treat those symbols with high noise as very unreliable symbols. In addition, it is very important that the AGC strategy be robust to varying jammer strategies. In general, this is a rather difficult problem because one must be certain that there do not exist jammer strategies utilizing knowledge of the AGC time constants as this can cause serious errors in the attempt to properly scale the integrate-and-dump output.

We can gain some insight into how good the random-error-correcting code must be by examining the use of diversity. Diversity of order M is a very simple $R = 1/M$ coding scheme accomplished with M repetitions of each symbol. Consider the "on–off" pulse jammer with $J = \delta J_p$. When used with random interleaving and the optimum receiver, an error is made only if the jammer hits all M repetitions of a given symbol. The error rate becomes

$$P_e(M) = \delta^M Q\left[(2\delta R_s S/J)^{1/2}\right]$$

It is easily shown that the "worst-case" performance is given by

$$P_{\max} \le \frac{K_M}{(R_s S/J)^M}$$

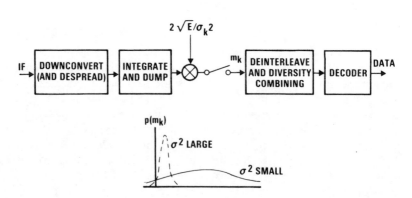

Figure 8-20. Optimum receiver structure for PSK with coding in nonstationary Gaussian noise.

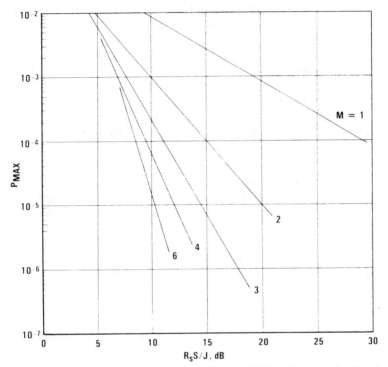

Figure 8-21. Worst-case P_b for direct spread BPSK and QPSK with Mth-order diversity in the presence of a pulsed noise jammer.

where K_M is a constant that depends on M. Figure 8-21 demonstrates the behavior of the worst-case P_b curve for Mth-order diversity for several values of M. Obviously, most of the severe effects of pulse jamming can be eliminated through the intelligent use of diversity and pseudorandom interleaving. The worst-case P_b curve of any random-error-correcting code with distance d behaves very similarly to a diversity scheme with $M = d$. However, the performance is several decibels better because the code rate is much higher than $R = 1/M$ (resulting in less bandspreading loss). Figure 8-21 indicates that code distances ≥ 6 are desirable to achieve good pulse jamming performance.

Now consider the use of convolutional coding with Viterbi algorithm decoding (again assuming a random interleaver). For a single "on–off" pulse jammer with duty cycle δ the union bound on P_b for a $R = m/n$ code may be written as

$$P_b \leq \frac{1}{m} \sum_{k=d}^{\infty} w_k \delta^k Q \left\{ [2Rk\delta(R_s S/J)]^{1/2} \right\} \tag{8-11}$$

where d is the code free distance and J is the average received jammer power. This union bound was used to compute the performance curves shown in Fig. 8-22 as a function of $R_s S/J$ (for $R = 1/2$, $v = 6$ with $d = 10$). For any desired P_b there exists a jammer duty cycle which makes the most effective use of average jammer power, J. Note that at $P_b = 10^{-5}$ the optimum jammer is roughly $\delta = 0.6$. However, even at this value of δ the performance degrades by only 0.2 dB relative to the performance with a continuous jammer. The situation is even better at higher error rates. Note also that at any P_b there is a critical value of jammer duty cycle below which the error rate caused by the jammer will always be below the desired P_b. Similar results are obtained at other constraint lengths. For example, the $R = 1/2$, $v = 2$ code has $d = 5$ and behaves roughly as a fifth-order diversity scheme with less bandspreading. The worst-case performance is roughly 1.4 dB worse than that obtained with continuous jamming.

The use of convolutional coding provides a very large increase in AJ margin against the worst-case jammer. To achieve $P_b = 10^{-5}$ requires $R_s S/J = 38$ dB for the uncoded system while the coded system requires only $R_s S/J = 4.4$ dB. Thus, a coding gain of 33.6 dB is achieved. We should

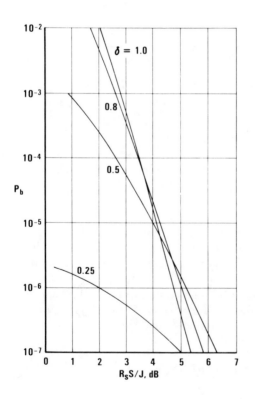

Figure 8-22. Performance of a $R = 1/2$, $v = 6$ code with direct spread PSK and pulsed jamming.

also point out that other random-error-correcting codes could have done almost as well. For example the $R = 1/2$, $v = 2$ code provides a gain of 31.9 dB, and even $M = 2$ diversity provides a gain of 18 dB.

Had we chosen to use a hard decision decoder, the performance would degrade substantially (much more than the 2-dB degradation seen on white Gaussian noise channels). The reason for this behavior is that the decoder can correct errors much more effectively if the reliable and unreliable symbols are identified as such. As an example, the $R = 1/2$, $v = 6$ code requires a channel error rate of 0.02 to achieve a decoded bit error probability of 10^{-5} with hard decision decoding. With worst-case jamming a value of $R_s S/J = 6.2$ dB is required to achieve $p = 0.02$ in an uncoded system (from Fig. 8-19). Thus, the $R = 1/2$ coded system will require $R_s S/J = 9.2$ dB to achieve $P_b = 10^{-5}$. This is 4.8 dB worse than the performance with the same code with soft decisions. Even so, this approach provides a gain of 33.2 dB relative to the uncoded system with worst-case jamming. In some cases this approach is desirable since it eases the problem of designing a robust AGC in the demodulator.

There is a rather important practical problem that involves sizing the interleaver block length. There are several important considerations. One involves the pulse length that a jammer can produce and still produce very high peak powers. This question involves the technology of power amplifier tubes in use, and today the limit on pulse lengths is less than a millisecond. However, it could change in the future. The other consideration involves how long any system degradations last that may occur due to a pulsed jammer. As an example, consider the situation when adaptive antennas are used to place nulls on jammers. Then the time required to achieve a null on the pulse jammer is of interest since a very low signal-to-noise ratio will be encountered while the antenna is adapting. The interleaver length should be chosen to be very long relative to the length of these times when the signal-to-noise ratio is very low. If these events are confined to 5% to 10% of the total interleaver length, then only a small degradation will occur (assuming the use of soft decisions).

Although these results have been presented using convolutional codes, other random-error-correcting techniques could also have been used. Long block codes or concatenated codes potentially provide even better performance.

8.5.2. Frequency Hop Systems

Another important spread spectrum modulation technique is frequency hopping. In this technique the data-modulated carrier is sequentially hopped under the control of a PN code into a series of slots into which the available bandwidth is split. As with the direct spread technique the data modulation format may be any of a number of standard techniques. A block diagram of an uncoded frequency-hop system is shown in Fig. 8-23. As indicated, the frequency hopping operation is accomplished by controlling a frequency synthesizer by a PN code generator. In order to recover the data an identical PN code generator must be synchronized at the receiver. The hopping operation may occur several times per symbol (fast hop) or at the symbol rate or slower (slow hop).

   8.5.2.1. Performance of Uncoded Frequency Hop Systems. Several papers have appeared that provide analysis of the performance against "worst-case" jammers of both coded and uncoded frequency hop systems with several modulation formats. [106, 107] First, consider the case of binary FSK (with orthogonal signals) with noncoherent detection and slow frequency hopping. Noncoherent detection is often used because it is difficult for the frequency synthesizer to maintain phase coherence between hops. Assume that the jammer is a noise jammer with power J and with all its energy within a fraction δ of the available frequency slots as shown in Fig. 8-24. This would be called a partial band-jamming strategy. It is well known that the error rate for orthogonal signaling with noncoherent detection is

$$P_b = \tfrac{1}{2}\, e^{-E_b/2N_0}$$

where E_b/N_0 will be measured within the particular signaling slot of interest. For the partial band jammer, the error rate will be zero for a fraction $(1 - \delta)$ of the time and will be P_b for a fraction δ of the time (where $N_0 = J/\delta W$). Thus, the average error rate with a partial band jammer is

$$P_b = \frac{\delta}{2} e^{-\delta W S/2 J R_b} \tag{8-12}$$

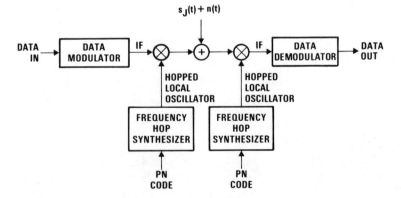

Figure 8-23. Uncoded frequency hop AJ system.

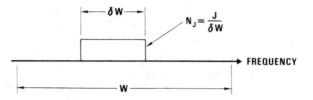

Figure 8-24. Partial band noise jammer.

where $WS/R_b J$ is the average or "effective" E_b/N_0. A series of curves for different values of δ is shown in Fig. 8-25. It can also be readily shown that the worst-case bit error probability occurs at $\delta = \min(2R_b J/WS, 1)$ and is given by

$$P_{\max} = \left(\frac{eWS}{R_b J}\right)^{-1} \tag{8-13}$$

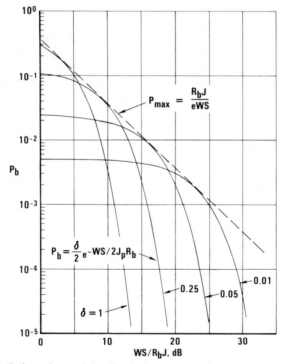

Figure 8-25. P_b for orthogonal signaling in the presence of a partial band noise jammer.

for $WS/R_bJ \geq 2$ and by (8-12) with $\delta = 1$ otherwise. Note that this "worst-case" partial band-jammer result is very similar to that obtained with pulse jamming (partial time jamming) of direct spread QPSK, but the jammer does not need to pay the cost of higher peak power. In fact, the result we have just derived is also applicable to pulse jamming of frequency hop FSK. For a pulse jammer with duty cycle δ and average power $J = \delta J_p$, performance is given by (8-12) with the worst-case bit error probability given by (8-13). Thus, the curves of Fig. 8-25 still apply. The partial band jammer is a very real threat since it appears at the receiver as a pulse jammer without paying the cost of a higher peak power signal.

Note that in these results the "effective" E_b/N_0 can be written as

$$\frac{E_b}{N_0} = \frac{W}{R_b} \frac{S}{J}$$

Thus, the ratio W/R_b plays the same role as R_s for direct sequence systems and can be called the processing gain.

The poor "worst-case" performance of frequency hop can be dramatically improved through intelligent application of diversity or coding. For example, fast frequency hopping (several times per symbol) is a form of diversity and will save 20 to 40 dB at error rates in the range of 10^{-3} to 10^{-5} against partial band-jamming strategies. However, one must be very careful because such a strategy now becomes vulnerable to a partial time (or pulse-jamming) strategy unless special precautions are taken. Coding represents a more efficient form of diversity and its use will be discussed in the following section.

8.5.2.2. Performance of Coded Frequency Hop Systems. As noted previously, coding considerations are quite different in noncoherent systems compared with coherent systems. The principal difference is the advantage gained by using M-ary orthogonal signaling. Since bandwidth is no problem in spread spectrum systems, M-ary orthogonal signaling is an ideal choice in many noncoherent systems. Worst-case coded system performance for such a system can be calculated in a direct fashion.

As a specific example, consider the use of M-ary orthogonal signaling and noncoherent demodulation with convolutional coding and Viterbi decoding. An AGC strategy similar to that in Fig. 8-22 is also optimum, i.e., each matched filter output should be scaled by a factor $E^{1/2}/\sigma_k^2$. Then the analysis of the partial band jammer is nearly identical to that of the pulse jammer discussed previously. In a comparison by the decoder of two paths that differ in k positions, the jammer must hit all k of these positions in order to cause an error. Then as in (8-11), the union bound on P_b for a $R = m/n$ code operating against a partial band jammer with duty factor δ may be written as

$$P_b \leq \frac{1}{m} \sum_{k=d}^{\infty} w_k \delta^k P_k \qquad (8\text{-}14)$$

where P_k is the probability of error for noncoherent combining of k transmissions as given by (1-42) with

$$\rho = k\delta RWS/R_bJ$$

Performance in the presence of a worst-case partial band jammer is obtained by maximizing (8-14) over δ.

Performance for several codes of interest is shown in Fig. 8-26. (The performance of these codes on a Gaussian channel was previously shown

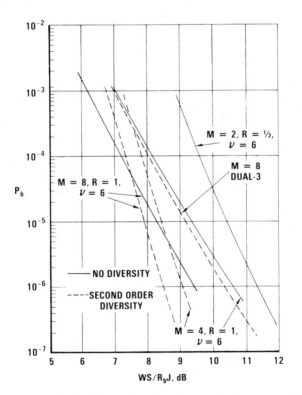

Figure 8-26. Worst-case performance of selected convolutional codes with frequency-hopped orthogonal signaling and partial band or pulsed noise jamming.

in Fig. 6-18.) Results are shown for alphabet sizes of $M = 2, 4$, and 8. Obviously, the use of higher-order alphabets is quite beneficial (by more than 2 dB for $M = 8$ relative to $M = 2$ at $P_b = 10^{-5}$). In addition, when DPSK can be used the $M = 2, R = 1/2, v = 6$ performance curve improves by 3 dB. Also note that the use of second-order diversity can be of substantial benefit. The curve for the dual-3 code with no diversity was not shown because of its poor performance. It is 2 dB worse than the $M = 2, R = 1/2, v = 2$ code which performs about 3 dB worse than the $v = 6$ code. Of this degradation about 2.2 dB is the degradation caused by the worst-case partial band jammer relative to a continuous jammer. However, second-order diversity can allow a performance improvement of 1.2 dB for the worst-case jammer. As with coherent systems, a substantial performance degradation occurs when hard decision decoding is used. For the case of the $R = 1/2, v = 6$ code, this degradation is about 5.2 dB.

In comparing these results with those of Fig. 8-22, we note that in a worst-case jamming environment the noncoherent system degrades by 3.5 to 6 dB relative to a coherent system depending on the alphabet size. An excellent comparison of coded direct sequence systems with frequency hop systems using the R_0 parameter was recently presented by Viterbi.[108] This comparison leads to very similar conclusions. Viterbi showed that binary signaling in a noncoherent frequency hopped system is about 6 dB worse than in a coherent direct sequence system. The problem occurs in trying to utilize large distance codes. Noncoherent combining loss prevents maximum utilization of the code distance.

The E_b/N_0 required as predicted from R_0 is about 2.5 dB less than that required to achieve $P_b = 10^{-5}$ at each value of M as predicted in Fig. 8-26. These results indicate that more sophisticated coding schemes may allow somewhat better performance than that shown in Fig. 8-26.

We note that coding can offer very significant gains with partial band jamming (on the order of 35–40 dB for worst-case jamming at $P_b = 10^{-5}$). If the hopping is done at least once per symbol, then this gain is accomplished with *no interleaving*. However, if coding is applied to such a system with no interleaving, then the coded system now becomes vulnerable to a worst-case full-band, partial-time jammer. Thus, depending on the hopping rate and the estimated jammer capability for long, high-peak-power pulses, interleaving may be necessary. Fortunately, in many cases the jammer cannot produce a high-peak-power pulse that is longer than a hop time.

Finally, we note that although direct spread systems are 3 dB more efficient than frequency hop systems, they are not necessarily always preferable. With present technology the achievable bandspreading is more than an order of magnitude better with frequency hopping compared with direct sequence spread. This gives frequency hopping an advantage in many applications.

8.6. Coding for Bandwidth-Constrained Channels

A problem of increasing interest is that of obtaining a moderate amount of coding gain on channels with very limited bandwidth. Implicit in this problem is that a higher-level signaling alphabet (or M-ary modulation scheme) is used to achieve increased spectral efficiency. However, the use of a higher-level alphabet causes a significant complication in the problem of performance calculation and in the search for optimum codes. Design of good codes is receiving increasing attention from researchers, and some interesting results have been reported.[109,110,111]

For this type of problem the system requirements such as information rate, bandwidth, etc. will determine the desired signal designs (or potential designs, i.e., PSK, PAM, etc.) and the required symbol rate and number of information bits per symbol, m. Thus, an uncoded system would require a 2^m-ary modulation scheme. The best approach in applying coding with higher-level alphabets is to use a high code rate. The usual method is to expand the signal design to include 2^{m+1} points and then apply a $R = m/(m+1)$ code. Since there is no change in the symbol rate, the coded system will have the *same bandwidth* as the uncoded system and carry the same number of information bits per symbol (m). However, assuming a well-chosen code, the coded system will allow a reduction in the required E_b/N_0 by an amount equal to the coding gain. The following discussion will not consider actual bandwidth or effects of filtering but will only consider the problem of designing efficient codes for several well-known M-ary modulation schemes. The coding techniques discussed will employ convolutional codes with Viterbi decoding, but other techniques could also be used with similar results.

There are some rather substantial differences between this problem and the binary signaling problem treated in Chapter 6 in terms of performance calculations and in terms of the manner in which path likelihoods are calculated. Consider a received vector ρ which is to be compared with two equally likely code words $\mathbf{x}_0$ and $\mathbf{x}_1$, each with n orthogonal components. As shown in Section 1.3.3, the optimum decision rule is to select code word $\mathbf{x}_0$ whenever

$$p(\rho|\mathbf{x}_0) > p(\rho|\mathbf{x}_1) \tag{8-15}$$

and $\mathbf{x}_1$ otherwise. Assuming Gaussian noise that is independent and has variance σ^2 on each code word component, (8-15) may be written

$$\prod_{i=1}^{n} \exp\left[-(\rho_i - x_{0i})^2/2\sigma^2\right] > \prod_{i=1}^{n} \exp\left[-(\rho_i - x_{1i})^2/2\sigma^2\right]$$

This optimum decision rule is equivalent to selecting $\mathbf{x}_0$ whenever

$$\sum_{i=1}^{n} (\rho_i - x_{0i})^2 < \sum_{i=1}^{n} (\rho_i - x_{1i})^2 \tag{8-16}$$

Therefore, the metric required by the decoder for the ith orthogonal component of $\mathbf{x}_0$ is $(r_i - x_{0i})^2$. Note that this is a different metric than was used previously with binary PSK signaling since in that case $|x_{0i}| = |x_{1i}|$ and (8-16) could be reduced to

$$\sum_{i=1}^{n} \rho_i x_{0i} > \sum_{i=1}^{n} \rho_i x_{1i} \tag{8-17}$$

The type of metric indicated by (8-17) for binary PSK is normally implemented in Viterbi decoders, but there is no significant problem in implementing the metric indicated by (8-16) for higher-level signaling alphabets. This would require relatively minor changes in the decoder front end.

Note that (8-16) can also be written as

$$\| \boldsymbol{\rho} - \mathbf{x}_0 \|^2 < \| \boldsymbol{\rho} - \mathbf{x}_1 \|^2 \tag{8-18}$$

Consequently, the error rate in the comparison between x_0 and x_1 is determined by the Euclidean distance between $\mathbf{x}_0$ and $\mathbf{x}_1$. The squared distance is

$$d^2(\mathbf{x}_0, \mathbf{x}_1) = \| \mathbf{x}_0 - \mathbf{x}_1 \|^2 \tag{8-19}$$

and the error rate in deciding between $\mathbf{x}_0$ and $\mathbf{x}_1$ is

$$P_e(\mathbf{x}_0, \mathbf{x}_1) = Q\left[\frac{d(\mathbf{x}_0, \mathbf{x}_1)}{2\sigma} \right] \tag{8-20}$$

Assuming that code word $\mathbf{x}_0$ is transmitted, a union bound on word error rate may be obtained by summing $P_e(\mathbf{x}_0, \mathbf{x}_j)$ over all other code words. However, because of the multilevel or multiphase nature of the signal design, the error rate can change with a different transmitted code word. Thus, to obtain the union bound on sequence error rate one must average over all M possible code words, i.e.,

$$P_s \leq \frac{1}{M} \sum_{i=0}^{M-1} \sum_{\substack{j=0 \\ (j \neq i)}}^{M-1} P_e(\mathbf{x}_i, \mathbf{x}_j) \tag{8-21}$$

This computation indicates that the error rate calculations will be substantially more tedious than was the case with binary PSK signaling. In the binary case, the squared Euclidean distance, $d^2(\mathbf{x}_0, \mathbf{x}_1)$, is proportional to the Hamming distance between binary code words, and (8-20) is simply given by $Q\left[(2dE_s/N_0)^{1/2} \right]$, where d is the corresponding Hamming distance. This simplification with linear codes allows one to assume the all-zero transmitted code word so the sum over i in (8-21) is not needed and the sum over j is obtained directly through knowledge of the significant terms of the code weight structure. The complexity of the calculations indicated by (8-20) and (8-21) has hindered the search for truly optimum codes for higher-level modulation schemes.

The nature of the calculation in (8-20) suggests that the code properties alone (e.g., the code weight structure) do not necessarily determine error rate. What is important is the set of Euclidean distances achieved by the combination of the coding and modulation scheme. That is, it is very important to achieve the proper match between the code and the signal design. Thus, codes are designed specifically for certain higher-level modulation schemes.

In the search for good codes, researchers have typically used *asymptotic coding gain* rather than exact bit error probability in comparing codes. The important parameter is the free Euclidean distance achieved by the combined coding/modulation scheme. The code free distance is defined as the minimum of the distances between different code words as given by (8-19), i.e.,

$$d_{\text{free}} = \min_{\text{all } i \neq j} [d(\mathbf{x}_i, \mathbf{x}_j)] \tag{8-22}$$

Then a lower bound on error-event probability is given by

$$P_s \geq Q(d_{\text{free}}/2\sigma)$$

and this bound is approached asymptotically at high signal-to-noise ratios. The asymptotic coding gain is thus defined as

$$G_a = 20 \log_{10}\left(\frac{d_{\text{free}}}{d_{\text{ref}}}\right) \tag{8-23}$$

where d_{ref} is the minimum Euclidean distance between different sequences of the uncoded reference system with the same average or peak power. The objective, of course, in code selection is to maximize asymptotic coding gain.

As a specific example, consider coding for M-ary PSK modulation using convolutional codes and Viterbi decoding to carry two information bits per transmitted symbol. This can be accomplished by using a $R = 2/3$ convolutional code with 8-PSK modulation. This signal design is shown in Fig. 8-27 and is assumed to have unit energy per transmitted symbol. Consider the $R = 2/3$, $v = 2$ code with trellis structure as shown in Fig. 6-7. This code is selected only because it has been discussed previously. (The advantages of punctured codes are of no utility in this application.) The important step in the code design process is to carefully match the code branches to the signal points. A Gray code mapping of code branches onto signal points is shown in Fig. 8-27. This mapping is intuitively satisfying since all branches that differ by Hamming distances of 1, 2, and 3 will differ in Euclidean distance by at least d_0, d_1, and d_2, respectively. The reader can verify that the minimum free Euclidean distance is

$$d_{\text{free}} = (d_0^2 + d_1^2)^{1/2} \qquad (8\text{-}24)$$

which is the distance between the all-zero path and the path 001 011 000.... However, the reference uncoded system that achieves two bits per symbol is 4-PSK (or QPSK). This signal design contains just the four points on the axes. Assuming the same average (and peak) power, the reference system has Euclidean distance equal to d_1, so the resulting asymptotic coding gain is

$$G_a = 20 \log_{10} \left[(d_0^2 + d_1^2)^{1/2} / d_1 \right]$$

$$= 1.1 \text{ dB} \qquad (8\text{-}25)$$

This rather meager gain can be improved substantially by using coding in a more intelligent fashion. This can be done by taking advantage of the wide disparity in distance between points in the signal set (note that d_3/d_0 is 8.4 dB). Consider the mapping shown in Fig. 8-28. The first two bits associated with each point (underlined) represent the output of the familiar $R = 1/2$, $v = 2$ code with trellis structure shown in Fig. 6-3. The third bit is an uncoded data bit. Thus, this encoding scheme still carries two information bits per branch, and it has the advantage that the $R = 1/2$ decoder is particularly simple to implement. Note that the mapping is such that the maximum separation between signal points is used for pairs where the coded branches are equal but the uncoded data bits are different (i.e., the

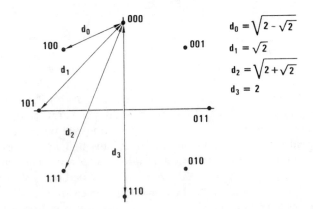

Figure 8-27. Gray code mapping of code branches onto 8-PSK signal set for $R = 2/3$, $v = 2$ code.

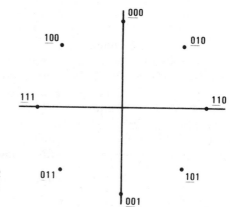

Figure 8-28. Mapping of $R = 1/2$, $v = 2$ code branches (underlined) onto 8-PSK signal set.

separation between $\underline{000}$ and $\underline{001}$ is d_3). A trellis diagram that represents this mapping is shown in Fig. 8-29. This is the standard $R = 1/2$, $v = 2$ trellis, but there are parallel transitions between states. The parallel transitions correspond to $R = 1/2$ code branches that are identical but with different uncoded third bits. The mapping of Fig. 8-28 assures that parallel transitions always have a Euclidean distance of d_3 between them.

For each received signal point, the decision as to which of the parallel transitions to retain may be made by a comparison which is "outboard" of the Viterbi decoder. This decision may be made with d_3/d_1 (or 3 dB) less

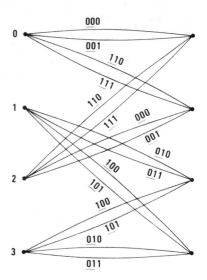

Figure 8-29. Trellis structure of the code used for 8-PSK (from mapping of Fig. 8-28).

signal energy than the uncoded reference system. The metrics corresponding to the squared distances between the received point and the nearest 0 0, 0 1, 1 0, and 1 1 points would then be used in a standard Viterbi decoder. The uncoded (or third) bit corresponding to each of these branches would also be retained. After a path has finally been selected by the Viterbi algorithm, then both data bits on each symbol are available. The Euclidean distances achieved on $R = 1/2$ branches that differ in one or two positions are d_0 or d_1, respectively. The $R = 1/2$ code minimum distance path is 1 1 0 1 1 1, which leads to a minimum Euclidean distance of $(d_0^2 + 2d_1^2)^{1/2}$. As a result, the free Euclidean distance of this coding scheme is

$$d_{\text{free}} = \min\left[d_3, (d_0^2 + 2d_1^2)^{1/2}\right] \tag{8-26}$$

which results in a coding gain of 3 dB.

    This approach provides substantially more coding gain than is obtained with the mapping of Fig. 8-27 using a decoder of roughly the same complexity. While this technique may seem highly ad hoc, it offers a significant level of coding gain for many signal designs. The reason is that in many situations the most efficient approach will involve coding only a portion of the information bits in each symbol. The wide disparity in distance between signal points can be used to carry some information without the necessity for coding. However, this technique has its limits. Note that with 8-PSK, $d_3 = 2$ so that the parallel transitions limit the free Euclidean distance to be no larger than 2. Thus, no more than 3 dB of gain can be achieved. Achievement of more gain with 8-PSK will require elimination of parallel transitions and subsequently the use of an actual $R = 2/3$ code. A constructive approach for achieving the best mapping of code branches onto signal points will be discussed shortly.

    Another example using a $R = 1/2$ code with the same mapping approach, but with an 8-PAM signal set is shown in Fig. 8-30. This mapping will carry two information bits per symbol so the reference system with the same peak power is 4-PAM with $d_{\text{ref}}^2 = 4/9$. The trellis for this code may

Figure 8-30. Mapping of $R = 1/2$ code branches (underlined) onto 8-PAM signal set.

also be drawn similar to Fig. 8-29 with parallel transitions, and the squared distance between parallel transitions is $d_{par}^2 = 64/49$. Note that with this assignment the maximum asymptotic coding gain is bounded by

$$\text{gain} \leq 20 \log_{10}(d_{par}/d_{ref})$$

$$= 4.7 \, \text{dB}$$

This happens to be larger than the corresponding figure for 8-PSK (3 dB), but the distance accumulated by the nonparallel code transitions is smaller. From Fig. 8-30 we see that all antipodal comparisons (e.g., 1 1 vs. 0 0) have squared distance of $d_{anti}^2 = 16/49$ while all orthogonal comparisons have squared distance of $d_{orth}^2 = 4/49$.

Recognizing that the antipodal and orthogonal branch comparisons generate these Euclidean distances, knowledge of the composition of the paths in terms of number of orthogonal and antipodal branches can be used to determine asymptotic coding gain. For example, with the $R = 1/2$, $v = 2$ code the minimum free Euclidean distance is achieved by two antipodal comparisons (one when the two paths diverge and the other when they merge) and one orthogonal comparison. This gives

$$d_{free}^2 = 2d_{anti}^2 + d_{orth}^2$$

$$= 36/49 \tag{8-27}$$

which results in a total asympotic coding gain of 2.2 dB. At longer constraint lengths the gain increases significantly. Assume that the $R = 1/2$ codes of Table B-2 are used. Again each code will have at least two antipodal comparisons contributing to its free Euclidean distance, and the worst-case Euclidean distance results when the rest of the code free Hamming distance, d_f, occurs in orthogonal comparisons. Thus, the overall code free distance is lower bounded by

$$d_{free}^2 \geq \min \left[d_{par}^2, 2d_{anti}^2 + (d_f - 4)d_{orth}^2 \right] \tag{8-28}$$

Using this relationship the minimum asymptotic coding gain of the $R = 1/2$ codes from Table B-2 have been computed and are shown in Table 8-4. The column d_{free}^2/d_{ref}^2 and the first gain column refer to a comparison of coded 8-PAM with uncoded 4-PAM with the *same peak power*. If the comparison were made on the basis of systems with the *same average power*, slightly more coding gain is available as shown in the last column. The difference is obtained by noting that the peak-to-average power ratio for M-PAM is

$$\frac{P_{peak}}{P_{av}} = \frac{3(M-1)}{(M+1)} \tag{8-29}$$

We observe that for systems using the same peak power, the 4-PAM system

**Table 8-4. Asymptotic Coding Gain Obtained with the $R = 1/2$ Codes
of Table B-2 and the 8-PAM Signal Design of Fig. 8-30[a]**

v	d_f	d_{free}^2/d_{ref}^2	Gain (peak power) (dB)	Gain (average power) (dB)
2	5	1.65	2.18	3.31
3	6	1.84	2.64	3.77
4	7	2.02	3.05	4.18
5	8	2.20	3.43	4.56
6	10	2.57	4.10	5.23
7	10	2.57	4.10	5.23
8	12	2.94	4.68	5.83

[a] Gain is computed relative to an uncoded 4-PAM signal design.

uses 1.13 dB *more average power* than 8-PAM. This is the amount of additional coding gain if an average power comparison is made.

While the ad hoc coding schemes just presented offer excellent performance in many cases, one might wonder how codes should be designed
to obtain improved performance. Specifically, one would like an efficient
algorithm for finding near-optimal codes. Ungerboeck has described such
an approach.[109,110] His approach involves the concept of partitioning
the signal constellation into subsets with distance between the elements of
the subset that increases as the size of the subset decreases. This approach
is called *mapping by set partitioning*. This procedure, using an 8-PSK signal
design, is shown in Fig. 8-31.

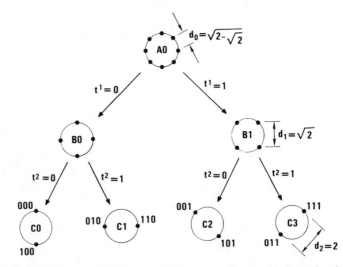

Figure 8-31. Partitioning of 8-PSK signal design into subsets with increasing free subset
distances ($d_0 < d_1 < d_2$).

The set partitioning is used in the following fashion in designing trellis codes. When parallel transitions are allowed in the trellis (as in Fig. 8-29), the two signals that would be used are taken from one of the subsets with the largest free subset distance (C0, C1, C2, or C3). This is exactly what was done in the previous examples. For transitions that originate at the same state and lead to different states, channel signals are assigned with the next-largest distance, d_1 (i.e., the signals are chosen from B0 or B1). This rule is also used for transitions that merge at a given state but originate at different states. Within this set of constraints, signal points can be assigned to transitions in the trellis to maximize the code free Euclidean distance. This procedure has been carried out by Ungerboeck to find a code at each value of v which maximizes free distance for a number of signal designs.[109, 110] One such code trellis is shown in Fig. 8-32 for $v = 3$. Note that parallel transitions are not used. (We found previously that parallel transitions limit the achievable gain to 3 dB, and this value of gain can be obtained with $v = 2$.) As a result, paths that diverge from and merge into a given state are assigned transitions from sets B0 and B1. (The numbers on the left by each state refer to the signal design points used for each of the four paths diverging from that state.) The minimum free Euclidean distance occurs with the two paths that are shown in the trellis, and the free distance is

$$d_{free}^2 = 2d_1^2 + d_0^2$$

This results in a coding gain of 3.6 dB relative to uncoded 4-PSK. By using this approach with larger constraint lengths, Ungerboeck demonstrated achievable gains comparable to those in Table 8-4. In addition, similar

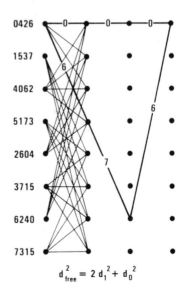

Figure 8-32. Trellis code with $v = 3$ for 8-PSK modulation with two information bits per symbol.

$$d_{free}^2 = 2 d_1^2 + d_0^2$$

techniques were used to find good codes for various M-PAM, M-QAM, and M-AMPM signal designs. He noted that for higher-order signal designs the extensive use of parallel trellis transitions is desirable from a performance as well as an implementation viewpoint. These results demonstrated that for v in the range of 6 to 8, coding gains in the range of 5 to 6 dB were achievable. He also performed a number of simulations which indicated that the asymptotic coding gains predicted were about 0.5 dB optimistic for interesting values of bit error probability.

A similar type of problem has been considered by Anderson and Taylor[111] for continuous-phase FSK (CPFSK) modulation. They designed special trellis codes that were matched to the signal design to maximize free Euclidean distance. Their results indicate that codes can be designed with performance that is 2 to 4 dB better than uncoded BPSK and with a narrower spectrum (the bandwidth between the first nulls is 3/4 that of BPSK).† The codes presented are very interesting but beyond the scope of this book. We should point out that there are also other approaches for achieving equivalent performance with the same spectral width to the first null. One such approach would be to use a QPSK signal design with symbol rate that is 3/4 that of the uncoded BPSK and with $R = 2/3$ coding. As indicated in Fig. 6-14, coding gains of 2 to 4 dB can be obtained with constraint lengths in the range $2 \leq v \leq 5$. However, the comparison between this and CPFSK may change if a significant amount of filtering is introduced.

Problems

8-1. An (n, k) RS code with symbols defined over $GF(2^K)$ is used as an outer code and concatenated with a Hamming code with $N = 2^m - 1$ and $K = N - m$. Find the overall code block length N^*, code rate R^*, and minimum distance d^* for the concatenated code. Assuming hard decision decoding of the Hamming code, what is the minimum number of bit errors in an overall block that could cause a decoding error? Compare this with d^* and comment.

8-2. For the concatenated system in the preceding problem, let $n = 2^K - 1$ and $k = n - 2t$, where t is the number of errors corrected by the RS code. Assuming soft decision maximum-likelihood decoding of the inner code, give an expression for *asymptotic coding gain* of the concatenated system as a function of m and t. Plot gain versus t for $m = 3, 4, 5,$ and 6. Repeat for an inner code that is a double-error-correcting BCH code.

---

† The reader is cautioned that reduced spectral occupancy does not necessarily mean that the coded sequence can be passed through a narrow filter with less performance degradation. In some cases the opposite may be true. An interesting example is provided in the paper by Pelchat and Geist.[112]

8-3. Consider a $R = 1/2$ systematic convolutional encoder with $g(x) = 1 + x^D + x^{2D} + x^{3D+1}$. Give the block diagram of a double-error-correcting threshold decoder (diffuse threshold decoding[100]). Assuming a periodic burst error process, what is the maximum burst length and minimum error-free guard space that can be allowed for error-free operation? How can the coder/decoder be modified to allow single error correction in the guard space? Compare with table look-up feedback decoding of interleaved $R = 1/2$, $d = 3, 5$, and 7 codes.

8-4. Consider a binary burst erasure channel (erasure bursts of length up to $2D$ are allowed, but no errors can occur). A $R = 1/2$ systematic convolutional code with $g(x) = 1 + x^D$ is used. Construct a decoder for this code that allows correction of burst erasures of lengths up to $2D$. What error-free guard space is needed (see Kohlenberg and Forney[100])?

8-5. For a convolutional code used over a random erasure channel, the average weight structure $\{\hat{w}_j\}$ is given by (8-9). Express this quantity in terms of partial derivatives of the code-generating function $T(D, L, N)$ and the erasure probability, δ.

8-6. A (48,24) block code is to be decoded with infinitely quantized soft decision maximum-likelihood decoding. The code weight structure is given by the following nonzero terms: $w_0 = 1$, $w_{12} = 17296$, $w_{16} = 535095$, $w_{20} = 3995376$, $w_{24} = 7681680$, and $w_{48-i} = w_i$ for $i \leq 24$. Find an expression for the union bound on word error rate for PSK signaling on a Gaussian noise channel with random erasures occurring with probability δ. Plot P_s vs. E_b/N_0 for $\delta = 0, 0.05, 0.1$, and 0.20. Find an expression for the floor on P_s as a function of δ and plot the result.

8-7. An (8,4) extended Hamming code is to be decoded via maximum-likelihood decoding using infinitely quantized demodulator soft decisions. Repeat the previous problem and plot P_s vs. E_b/N_0 for $\delta = 0, 0.01$, and 0.10. The terms in the weight structure of this code are $w_4 = 14$, $w_5 = w_6 = w_7 = 0$, and $w_8 = 1$.

8-8. Assume a $R = 1/2$, $v = 6$ code is used with maximum-likelihood decoding using infinitely quantized soft decisions. PSK signaling on a Gaussian channel is used. A *periodic* burst erasure interference process with $\delta = 0.10$ is also present. Assuming the burst length is $B = 4$, what E_b/N_0 is required to achieve $P_b = 10^{-5}$ for the three cases of no interleaving, a matched period interleaver, and a random interleaver? Repeat for $B = 8$ and $B = 12$.

8-9. The error rate for DS spread PSK with pulsed jamming is given by $\delta Q\left[(2R_s S/J_p)^{1/2}\right]$. Find P_{max} using the bound $Q(\alpha) \leq \frac{1}{2}e^{-\alpha^2/2}$. What is the worst-case δ? Compare the result with that of Section 8.5.1.2.

8-10. The (7,4) block code is used with DS spread PSK and hard decision decoding. Find the word error probability, P_s, assuming a worst-case pulsed jammer. Compare with the $R = 1/2$, $v = 6$ code (Fig. 8-22). Repeat for the (24,12) code.

8-11. The error rate for DS spread PSK with diversity of order M is given by $\delta^M Q\left[(2\delta R_s S/J)^{1/2}\right]$. (We assume a random interleaver.) Using the bound

$Q(\alpha) \leq \frac{1}{2}e^{-\alpha^2/2}$, find the worst-case δ, and P_{max}. Plot for $M = 1, 2, 3, 4, 5$, and 6, and compare with Fig. 8-21.

8-12. Assume that DS spread PSK is used and that the BER of the $1/2$, $v = 6$ convolutional code is adequately approximated by the first term of its weight structure, $w_{10} = 36$. Using the bound $Q(\alpha) \leq \frac{1}{2}e^{-\alpha^2/2}$ give an expression for P_b for a pulse jammer with duty factor δ (assuming random interleaving). Find the worst-case δ and P_{max} from this expression. Plot P_b for $\delta = 1$ and P_{max} in the range from 10^{-5} to 10^{-10} and compare.

8-13. Using a 4-PAM signal design and a $R = 1/2$, $v = 2$ code, design a coding scheme that transmits one information bit per symbol and performs 2.5 dB better (asymptotically) than uncoded BPSK on an average power basis. How does the scheme compare on a peak power basis? How much more gain could be obtained with the $R = 1/2$, $v = 6$ code?

8-14. Using a 16-PAM signal design and the $R = 1/2$, $v = 2$ code, design a coding scheme that transmits three information bits per symbol. Compute the asymptotic coding gain relative to uncoded 8-PAM on both peak and average power bases.

8-15. Using a 4-QAM signal design (four levels on both I and Q channels for a total of 16 signal points) and the $R = 1/2$, $v = 2$ code, design a coding scheme that transmits three information bits per symbol. Compute the asymptotic coding gain relative to both 8-PAM and 8-PSK on both peak and average power bases.

Appendix A. Code Generators for BCH Codes

As described in Chapter 2, code generator polynomials can be constructed in a straightforward manner. Here we list some of the more useful generators for primitive and nonprimitive codes.

Generators for the primitive codes are given in octal notation in Table A-1 (i.e., 23 denotes $x^4 + x + 1$). The basis for constructing this table is the table of irreducible polynomials given in Peterson and Weldon.[1] Sets of code generators are given for lengths 7, 15, 31, 63, 127, and 255. Each block length is of the form $2^m - 1$. The first code at each block length is a Hamming code, so it has a primitive polynomial, $p(x)$, of degree m as its generator. This polynomial is also used for defining arithmetic operations over $GF(2^m)$. The next code generator of the same length is found by multiplying $p(x)$ by $m_3(x)$, the minimal polynomial of α^3. Continuing in the same fashion, at each step, the previous generator is multiplied by a new minimal polynomial. Then the resulting product is checked to find the *maximum* value of t for which $\alpha, \alpha^2, \ldots, \alpha^{2t}$ are roots of the generator polynomial. This value t is the error-correcting capability of the code as predicted by the BCH bound. The parameter k, the number of information symbols per code word, is given by subtracting the degree of the generator polynomial from n. Finally, on the next iteration, this generator polynomial is multiplied by $m_h(x)$ [with $h = 2(t + 1) + 1$] to produce a new generator polynomial. This process produces a sequence of BCH code generators of length $n = 2^m - 1$ with increasingly larger degree and error-correction capability. (In some cases the actual minimum distance will exceed this bound.)

The minimal polynomials necessary for constructing this table may be found in Peterson and Weldon. In addition, these tables may be used for constructing codes of length greater than 255. The tables in Peterson and

Weldon are complete through degree 16, which would allow construction of all primitive BCH codes through length 65535 using the same technique used here.

Table A-2 provides a list of selected code generators of nonprimitive BCH codes. These particular codes were selected because one cannot shorten a primitive BCH code to the same n and k and achieve the same or larger minimum distance. The arithmetic operations for these codes are defined over $GF(2^m)$, where m is the value indicated in the table. In this case the actual error-correcting capability is indicated since it is larger than the BCH bound in a number of cases.

Table A-1. Generator Polynomials of Primitive BCH Codes

n	k	t	Generator polynomial, $g_t(x)$ (octal)
7	4	1	$g_1(x) = p(x) = 13$
	1	3	$g_1(x) \cdot (15) = 177$
15	11	1	$g_1(x) = p(x) = 23$
	7	2	$g_1(x) \cdot (37) = 721$
	5	3	$g_2(x) \cdot (7) = 2467$
	1	7	$g_3(x) \cdot (31) = 77777$
31	26	1	$g_1(x) = p(x) = 45$
	21	2	$g_1(x) \cdot (75) = 3551$
	16	3	$g_2(x) \cdot (67) = 10765\ 7$
	11	5	$g_3(x) \cdot (57) = 54233\ 25$
	6	7	$g_5(x) \cdot (73) = 31336\ 5047$
	1	15	$g_7(x) \cdot (51) = 17777\ 77777\ 7$
63	57	1	$g_1(x) = p(x) = 103$
	51	2	$g_1(x) \cdot (127) = 12471$
	45	3	$g_2(x) \cdot (147) = 17013\ 17$
	39	4	$g_3(x) \cdot (111) = 16662\ 3567$
	36	5	$g_4(x) \cdot (15) = 10335\ 00423$
	30	6	$g_5(x) \cdot (155) = 15746\ 41655\ 47$
	24	7	$g_6(x) \cdot (133) = 17323\ 26040\ 4441$
	18	10	$g_7(x) \cdot (165) = 13630\ 26512\ 35172\ 5$
	16	11	$g_{10}(x) \cdot (7) = 63311\ 41367\ 23545\ 3$
	10	13	$g_{11}(x) \cdot (163) = 47262\ 23055\ 27250\ 155$
	7	15	$g_{13}(x) \cdot (13) = 52310\ 45543\ 50327\ 1737$
	1	31	$g_{15}(x) \cdot (141) = 77777\ 77777\ 77777\ 77777\ 7$
127	120	1	$g_1(x) = p(x) = 211$
	113	2	$g_1(x) \cdot (217) = 41567$
	106	3	$g_2(x) \cdot (135) = 11554\ 743$

Table A-1 (continued)

n	k	t	Generator polynomial, $g_t(x)$ (octal)
127	99	4	$g_3(x) \cdot (367) = 34470\ 23271$
	92	5	$g_4(x) \cdot (277) = 62473\ 00223\ 27$
	85	6	$g_5(x) \cdot (325) = 13070\ 44763\ 22273$
	78	7	$g_6(x) \cdot (203) = 26230\ 00216\ 61301\ 15$
	71	9	$g_7(x) \cdot (357) = 62550\ 10713\ 25312\ 7753$
	64	10	$g_9(x) \cdot (313) = 12065\ 34025\ 57077\ 31000\ 45$
	57	11	$g_{10}(x) \cdot (345) = 33526\ 52525\ 05705\ 05351\ 7721$
	50	13	$g_{11}(x) \cdot (301) = 54446\ 51252\ 33140\ 12421\ 50142\ 1$
	43	14	$g_{13}(x) \cdot (323) = 17721\ 77221\ 36512\ 27521\ 22057\ 4343$
	36	15	$g_{14}(x) \cdot (253) = 31460\ 74666\ 52207\ 50447\ 64574\ 72173$ 5
	29	21	$g_{15}(x) \cdot (361) = 40311\ 44613\ 67670\ 60366\ 75301\ 41176$ 155
	22	23	$g_{21}(x) \cdot (247) = 12337\ 60704\ 04722\ 52243\ 54456\ 26637$ $64704\ 3$
	15	27	$g_{23}(x) \cdot (271) = 22057\ 04244\ 56045\ 54770\ 52301\ 37622$ $17604\ 353$
	8	31	$g_{27}(x) \cdot (375) = 70472\ 64052\ 75103\ 06514\ 76224\ 27156$ $77331\ 30217$
	1	63	$g_{31}(x) \cdot (221) = 17777\ 77777\ 77777\ 77777\ 77777\ 77777$ $77777\ 77777\ 777$
255	247	1	$g_1(x) = p(x) = 435$
	239	2	$g_1(x) \cdot (567) = 26754\ 3$
	231	3	$g_2(x) \cdot (763) = 15672\ 0665$
	223	4	$g_3(x) \cdot (551) = 75626\ 64137\ 5$
	215	5	$g_4(x) \cdot (675) = 23157\ 56472\ 6421$
	207	6	$g_5(x) \cdot (747) = 16176\ 56056\ 76362\ 27$
	199	7	$g_6(x) \cdot (453) = 76330\ 31270\ 42072\ 2341$
	191	8	$g_7(x) \cdot (727) = 26634\ 70176\ 11533\ 37145\ 67$
	187	9	$g_8(x) \cdot (23)\ = 52755\ 31354\ 00013\ 22236\ 351$
	179	10	$g_9(x) \cdot (545) = 22624\ 71071\ 73404\ 32416\ 30045\ 5$
	171	11	$g_{10}(x) \cdot (613) = 15416\ 21421\ 23423\ 56077\ 06163\ 0637$
	163	12	$g_{11}(x) \cdot (543) = 75004\ 15510\ 07560\ 25515\ 74724\ 51460$ 1
	155	13	$g_{12}(x) \cdot (433) = 37575\ 13005\ 40766\ 50157\ 22506\ 46467$ 7633
	147	14	$g_{13}(x) \cdot (477) = 16421\ 30173\ 53716\ 55253\ 04165\ 30544$ $10117\ 11$
	139	15	$g_{14}(x) \cdot (615) = 46140\ 17320\ 60175\ 56157\ 07227\ 30247$ $45356\ 7445$
	131	18	$g_{15}(x) \cdot (455) = 21571\ 33314\ 71510\ 15126\ 12502\ 77442$ $14202\ 41654\ 71$

Table A-1 (continued)

n	k	t	Generator polynomial, $g_t(x)$ (octal)
255	123	19	$g_{18}(x) \cdot (537) = 12061\ 40522\ 42066\ 00371\ 72103\ 26516$ $14122\ 62725\ 06267$
	115	21	$g_{19}(x) \cdot (771) = 60526\ 66557\ 21002\ 47263\ 63640\ 46002$ $76352\ 55631\ 34727\ 37$
	107	22	$g_{21}(x) \cdot (703) = 22205\ 77232\ 20662\ 56312\ 41730\ 02353$ $47420\ 17657\ 47501\ 54441$
	99	23	$g_{22}(x) \cdot (471) = 10656\ 66725\ 34731\ 74222\ 74141\ 62015$ $74332\ 25241\ 10764\ 32303\ 431$
	91	25	$g_{23}(x) \cdot (651) = 67502\ 65030\ 32744\ 41727\ 23631\ 72473$ $25110\ 75550\ 76272\ 07243\ 44561$
	87	26	$g_{25}(x) \cdot (37)\ \ = 11013\ 67634\ 14743\ 23643\ 52316\ 34307$ $17204\ 62067\ 22545\ 27331\ 17213\ 17$
	79	27	$g_{26}(x) \cdot (607) = 66700\ 03563\ 76575\ 00020\ 27034\ 42073$ $66174\ 62101\ 53267\ 11766\ 54134\ 2355$
	71	29	$g_{27}(x) \cdot (661) = 24024\ 71052\ 06443\ 21515\ 55417\ 21123$ $31163\ 20544\ 42503\ 62557\ 64322\ 17060$ 35
	63	30	$g_{29}(x) \cdot (515) = 10754\ 47505\ 51635\ 44325\ 31521\ 73577$ $07003\ 66611\ 17264\ 55267\ 61365\ 67025$ 43301
	55	31	$g_{30}(x) \cdot (717) = 73154\ 25203\ 50110\ 01330\ 15275\ 30603$ $20543\ 25414\ 32675\ 50105\ 57044\ 42603$ $54736\ 17$
	47	42	$g_{31}(x) \cdot (735) = 25335\ 42017\ 06264\ 65630\ 33041\ 37740$ $62331\ 75123\ 33414\ 54460\ 45005\ 06602$ $45525\ 43173$
	45	43	$g_{42}(x) \cdot (7)\ \ \ = 15202\ 05605\ 52341\ 61131\ 10134\ 63764$ $23701\ 56367\ 00244\ 70762\ 37303\ 32021$ $57025\ 05154\ 1$
	37	45	$g_{43}(x) \cdot (643) = 51363\ 30255\ 06700\ 74141\ 77447\ 24543$ $75304\ 20735\ 70617\ 43234\ 32347\ 64435$ $47374\ 03044\ 003$
	29	47	$g_{45}(x) \cdot (765) = 30257\ 15536\ 67307\ 14655\ 27064\ 01236$ $13771\ 15342\ 24232\ 42011\ 74114\ 06025$ $46574\ 10403\ 56503\ 7$
	21	55	$g_{47}(x) \cdot (637) = 12562\ 15257\ 06033\ 26560\ 01773\ 15360$ $76121\ 03227\ 34140\ 56530\ 74542\ 52115$ $31216\ 14466\ 51347\ 3725$
	13	59	$g_{55}(x) \cdot (573) = 46417\ 32005\ 05256\ 45444\ 26573\ 71425$ $00660\ 04330\ 67744\ 54765\ 61403\ 17467$ $72135\ 70261\ 34460\ 50054\ 7$

Table A-1 (continued)

n	k	t	Generator polynomial, $g_t(x)$ (octal)
	9	63	$g_{59}(x) \cdot (31) = 15726\ 02521\ 74724\ 63201\ 03104\ 32553$
			$55134\ 61416\ 23672\ 12044\ 07454\ 51127$
			$66115\ 54770\ 55616\ 77516\ 057$
255	1	127	$g_{63}(x) \cdot (561) = 77777\ 77777\ 77777\ 77777\ 77777\ 77777$
			$77777\ 77777\ 77777\ 77777\ 77777\ 77777$
			$77777\ 77777\ 77777\ 77777\ 77777$

Table A-2. Selected Generator Polynomials of Nonprimitive BCH Codes

n	k	t	m	Generator polynomial (octal)
17	9	2	8	727
21	12	2	6	1663
23	12	3	11	5343
33	22	2	10	5145
33	12	4	10	3777
41	21	4	20	66471 33
47	24	5	23	43073 357
65	53	2	12	10761
65	40	4	12	35430 3067
73	46	4	9	17177 73537

Appendix B. Code Generators for Convolutional Codes

This appendix presents a series of tables of good code generators for convolutional codes. Separate tables provide generators specifically optimized for use with Viterbi decoders, table look-up decoders, threshold decoders, and sequential decoders.

B.1. Viterbi Decoding

Tables B-1 and B-2 list the optimum short-constraint-length convolutional codes for $R = 1/3$ and $R = 1/2$.[91] The code generators are given in octal notation. For each code the free distance, d, is given and the total information weight of all paths with weight d through $d + 4$. The constraint length is designated by v to facilitate comparisons among the various code rates (remember that the number of states is 2^v so that all codes with the same v have roughly the same decoder complexity even if the rates differ).

For the higher-rate codes, lists of the best punctured codes[55] will be provided rather than the optimum codes. This is done because the punctured code approach is usually preferred because of implementation considerations. The best punctured codes with two unique generators (which can be decoded by a modified $R = 1/2$ decoder) are given in Tables B-3 and B-4. The convention used in these tables is that the output symbols corresponding to the first two generators are transmitted on the same branch while the outputs corresponding to each of the other generators are successively transmitted on separate branches. Table B-5 presents lists of generators which can be used to produce either $R = 1/2$, $2/3$, or $3/4$ codes depending on the way in which they are sampled. These sets of generators could be used to produce a selectable rate decoder (1/2, 2/3, and 3/4) with only slightly more complexity than a $R = 1/2$ decoder. Other lists of codes may be found in the literature. Larsen[113] provides a list of $R = 1/3$ and 1/4 code generators with maximum free distance with $v \leq 14$. For higher rates Paaske[114] provides code generators with maximum free distance for $R = 2/3$ ($v \leq 10$) and $R = 3/4$ ($v \leq 9$). The latter codes are typically about 0.1 dB better than punctured codes, but they do not have the corresponding implementation advantages.

Optimum codes for M-ary orthogonal signaling are different from those of the preceding tables because Hamming distance is no longer the proper criterion of optimality. Table B-6 lists the generators and weight structure for some optimum codes while Table B-7 lists generators for some near-optimum codes ($M = 4$ and 8).[115] Note that one of the codes in Table B-6 is a dual-3 code.[106,116] For this code, bits are shifted into the encoder in groups of three bits (or one 8-ary symbol). The value of v for this code is one 8-ary symbol. For each new symbol shifted in, two coded symbols are computed (according to the two sets of generators in parentheses).

B.2. Table Look-up Decoding

For table look-up decoders the one criterion of optimality is $d(L)$, the minimum distance between the all-zero path and the L-branch truncated lower half code tree. However, if for a given L there are several codes with the same $d(L)$, then the code with the minimum number of paths with weight $d(L)$ should be selected. This criterion was used by Bussgang[69] in finding the codes in Table B-8. Only one generator is given because the codes are systematic.

B.3. Threshold Decoding

Code generators for self-orthogonal convolutional codes are easily constructed using difference triangles. The manner in which this is done is described in Section 7.1.3.1. Extensive tables of such difference triangles were compiled by Robinson and Bernstein[71] and Klieber.[72] The difference triangles of most interest are given in Tables B-9 through B-15. Using these tables one may construct codes of $R = 1/n$ varying in rate from 1/8 through 1/2 and codes of $R = (n-1)/n$ varying in rate from 2/3 through 7/8. The value of J given is that for the $R = (n-1)/n$ code. The corresponding value for the $R = 1/n$ code is $J^* = (n-1)J$. The parameter v_{max} corresponds to the maximum element and is also the degree of the maximum degree generator polynomial.

Another interesting class of codes are the so-called *trial-and-error codes* found by Massey,[13] and shown in Table B-16. These codes can achieve a given value of J with a shorter constraint length than the self-orthogonal codes. The manner in which they should be used is described in Section 7.1.3.1.

B.4. Sequential Decoding

With sequential decoding, decoder complexity is not strongly dependent on v. Thus, it is not necessary to optimize the code for a given value of v. Instead, one can increase v as much as necessary to achieve the desired level of performance. Since finding the optimum code (in the sense of the best weight structure) is computationally infeasible for large values of v, most researchers have been content to find codes with large values of d_f. The generators for a number of long codes suitable for sequential decoding are shown in Table B-17.

The first code shown is a $R = 1/2$, $v = 48$ systematic code found by Forney.[78] Of course, since the code is systematic, only the parity generator, G^2, need be shown. The code was constructed by starting with a known good short code, and sequentially adding terms to the generator in such a fashion to make the code minimum distance grow rapidly. At each step the minimum distance of the new code was calculated by using the Fano algorithm in a novel manner. With this technique a Hamming distance metric is used, and the decoder is started in the lower half tree and never allowed to enter the upper half tree. Then using metric increments of -1 for each disagreement and 0 for each agreement with a threshold spacing of 1, the first path found at any depth L into the tree will be a minimum-weight path at that depth. The code minimum distance was found by checking the minimum weight at $L = v + 1$. The resulting code generator is nested. That is, the generator may be terminated at any point (keeping the left-hand segment), and the resulting code will be good.

Bahl and Jelinek[117] found a good $R = 1/2$, $v = 23$ nonsystematic code with complementary generators. That is, all bits (except for the first and last) in generator G^1 are complements of the corresponding bits of G^2. The first and last bits in both generators are ones. This code, being nonsystematic, has a better distance structure than the previous code and may be preferred in situations where it is very desirable to minimize v.

Another useful nonsystematic code is the so-called quick-look code found by Massey and Costello.[79] These codes eliminate one of the disadvantages of nonsystematic codes (that the transmitted information sequence is not easily recovered from the received sequence). The resulting error rate of the recovered estimate of $I(x)$ is only twice the channel error rate (an "error amplication factor" of 2). Note that the penalty paid is that the distance structure is not as good as the Bahl–Jelinek code.

The code found by Layland and Lushbaugh[118] has a large minimum free distance ($d_f = 28$ at $v = 31$). In hardware tests this code provided an undetected bit error rate of approximately 10^{-5} at $E_b/N_0 = 2\,dB$ with three-bit soft decisions. (They also found almost identical performance with the Massey and Costello code.[79])

Johannesson[119] constructed a set of codes that have *optimal distance profiles* (ODP). With each of these codes the growth of the minimum-weight path at each depth L into the tree was optimized. This property is useful with sequential decoding because it facilitates early rejection of bad paths by the decoder. This will reduce the computation load on the decoder.

The nested generators found by Cain[73] can provide either $R = 1/2$, $2/3$, or $3/4$ codes. Likewise, $\mathbf{G}^1$ and $\mathbf{G}^2$ are used to generate a parity sequence from $I^1(x)$ and $I^2(x)$ for $R = 2/3$, and $\mathbf{G}^1$, $\mathbf{G}^2$, and $\mathbf{G}^3$ are used to produce a parity sequence for $R = 3/4$. Since these codes are systematic, considerably larger values of d_f can be obtained at high code rates for a given value of v with nonsystematic codes.

Table B-1. Generators (in Octal) and Weight Structure of the Best $R = 1/3$ Codes

v	Generators	d	w_d	w_{d+1}	w_{d+2}	w_{d+3}	w_{d+4}
2	7, 7, 5	8	3	0	15	0	58
3	17, 15, 13	10	6	0	6	0	58
4	37, 33, 25	12	12	0	12	0	56
5	75, 53, 47	13	1	8	26	20	19
6	171, 165, 133	15	7	8	22	44	22
7	367, 331, 225	16	1	0	24	0	113

Table B-2. Generators (in Octal) and Weight Structure of the Best $R = 1/2$ Codes

v	Generators	d	w_d	w_{d+1}	w_{d+2}	w_{d+3}	w_{d+4}
2	7, 5	5	1	4	12	32	80
3	17, 15	6	?	7	18	49	130
4	35, 23	7	4	12	20	72	225
5	75, 53	8	2	36	32	62	332
6	171, 133	10	36	0	211	0	1404
7	371, 247	10	2	22	60	148	340
8	753, 561	12	33	0	281	0	2179

**Table B-3. Generators (in Octal) and Weight Structure of the Best $R = 2/3$
Punctured Codes with Only Two Different Generators**

v	Generators	d	w_d	w_{d+1}	w_{d+2}	w_{d+3}	w_{d+4}
2	7, 5, 7	3	1	10	54	226	853
3	15, 13, 15	4	8	34	180	738	2989
4	31, 33, 31	5	25	112	357	1858	8406
5	73, 41, 73	6	75	0	1571	0	31474
6	163, 135, 163	6	1	81	402	1487	6793
7	337, 251, 337	8	395	0	6695	0	235288
8	661, 473, 661	8	97	0	2863	0	56633

**Table B-4. Generators (in Octal) and Weight Structure of the Best $R = 3/4$
Punctured Codes with Only Two Different Generators**

v	Generators	d	w_d	w_{d+1}	w_{d+2}	w_{d+3}	w_{d+4}
2	5, 7, 5, 7	3	15	104	540	2520	11048
3	15, 17, 15, 17	4	124	0	4504	0	124337
4	25, 37, 37, 37	4	22	0	1687	0	66964
5	61, 53, 53, 53	5	78	572	3831	24790	152108
6	135, 163, 163, 163	6	919	0	31137	0	1142571
6	121, 165, 121, 165	5	21	252	1903	11995	72115
7	205, 307, 307, 307	6	117	0	8365	0	319782
8	515, 737, 737, 737	6	12	342	1996	12296	78145

**Table B-5. Sets of Two Different Generators Which Produce
Good $R = 1/2, 2/3,$ and $3/4$ Codes [a]**

v	$R = 1/2$ Generators	$R = 2/3$ Generators	$R = 3/4$ Generators
2	7, 5(5)	7, 5, 7(3)	7, 5, 5, 7(3)
3	15, 17(6)	15, 17, 15(4)	15, 17, 15, 17(4)
4	31, 33(7)	37, 33, 31(5)	31, 33, 31, 31(3)
4	37, 25(6)	37, 25, 37(4)	37, 25, 37, 37(4)
5	57, 65(8)	57, 65, 57(6)	65, 57, 57, 65(4)
6	133, 171(10)	133, 171, 133(6)	133, 171, 133, 171(5)
6	135, 147(10)	135, 147, 147(6)	135, 147, 147, 147(6)
7	237, 345(10)	237, 345, 237(7)	237, 345, 237, 345(6)

[a] d_f shown in parentheses.

Table B-6. Generators (in Octal) and Weight Structure of Some Optimum Codes for M-ary Orthogonal Signaling

M	v	Generators	d_f	w_d	w_{d+1}	w_{d+2}	w_{d+3}	w_{d+4}
4	6	176, 133	7	7	39	104	352	> 1348
8	3	11, 15, 13	4	1	2	7	13	--
8	6	176, 155, 127	7	1	4	8	49	> 92
8	Dual-3	(44, 22, 11) (46, 21, 14)	4	7	28	154	644	2765

Table B-7. Near Optimal Generators (in Octal) for M-ary Orthogonal Signaling

v	Generators ($M = 4$)	Generators ($M = 8$)
2	7, 5	1, 3, 5
3	13, 5	11, 13, 15
4	13, 25	31, 33, 15
5	57, 75	31, 73, 55

Table B-8. Best $R = 1/2$ Codes to Achieve Odd Values of $d(L)$

L	v	$d(L)$	Generator (octal)
2	1	3	3
6	5	5	65
11	8	7	671
16	12	9	16651

Table B-9. Difference Triangles for Generating $R = 1/2$ CSOCs

J	v	First row of triangle
2	1	(1)
4	6	(2, 3, 1,)
6	17	(2, 5, 6, 3, 1)
8	35	(7, 3, 6, 2, 12, 1, 4)
10	55	(2, 12, 7, 8, 3, 13, 4, 5, 1)

Table B-10. Difference Triangles for $n = 3 (R = 2/3$ and $R = 1/3)$

J	v_{max}	First row of each triangle
2	2	(1) (2)
4	13	(8, 1, 3) (6, 5, 2)
6	39	(14, 7, 2, 13, 3) (8, 19, 1, 4, 6)
8	78	(19, 2, 24, 18, 10, 1, 4) (27, 12, 8, 14, 3, 6, 7)
10	130	(1, 5, 19, 7, 40, 28, 8, 12, 10) (23, 16, 18, 3, 14, 27, 2, 9, 4)

Table B-11. Difference Triangles for $n = 4$ $(R = 3/4$ and $R = 1/4)$

J	v_{max}	First row of each triangle
2	3	(1) (2) (3)
4	19	(3, 12, 4) (8, 9, 1) (6, 5, 2)
6	61	(12, 4, 26, 6, 13) (3, 2, 9, 20, 24) (15, 7, 1, 17, 10)
8	127	(16, 15, 3, 14, 27, 10, 39) (28, 7, 12, 11, 13, 9, 20) (21, 4, 1, 55, 6, 2, 38)
10	202	(7, 20, 49, 37, 24, 18, 1, 14, 32) (8, 30, 10, 11, 23, 29, 35, 4, 2) (12, 13, 3, 50, 5, 17, 9, 36, 54)

Table B-12. Difference Triangles for $n = 5 (R = 4/5$ and $R = 1/5)$

J	v_{max}	First row of each triangle
2	4	(1) (2) (3) (4)
4	26	(16, 4, 1) (2, 8, 15) (14, 3, 9) (11, 7, 6)
6	82	(5, 21, 25, 4, 14) (6, 1, 34, 19, 12) (8, 3, 13, 20, 38) (10, 22, 15, 2, 28)
8	178	(19, 40, 9, 17, 3, 15, 38) (39, 48, 30, 21, 10, 6, 8) (2, 11, 12, 71, 22, 50, 4) (7, 58, 5, 27, 1, 46, 34)

Table B-13. Difference Triangles for $n = 6 (R = 5/6$ and $R = 1/6)$

J	v_{max}	First row of each triangle
2	5	(1) (2) (3) (4) (5)
4	32	(1, 14, 16) (4, 8, 13) (7, 2, 18) (5, 6, 17) (10, 19, 3)
6	104	(38, 2, 22, 9, 30) (10, 4, 7, 67, 15) (6, 41, 17, 1, 35) (5, 23, 20, 12, 44) (45, 8, 26, 3, 13)

Table B-14. Difference Triangles for $n = 7$ ($R = 6/7$ and $R = 1/7$)

J	v_{max}	First row of each triangle
2	6	(1) (2) (3) (4) (5) (6)
4	40	(3, 16, 12) (20, 10, 8) (22, 4, 13) (1, 14, 21) (29, 5, 6) (23, 2, 7)
6	125	(42, 45, 3, 16, 12) (49, 43, 23, 2, 7) (1, 14, 21, 37, 47) (20, 10, 8, 33, 52) (55, 27, 29, 5, 6) (62, 24, 22, 4, 13)

Table B-15. Difference Triangles for $n = 8$ ($R = 7/8$ and $R = 1/8$)

J	v_{max}	First row of each triangle
2	7	(1) (2) (3) (4) (5) (6) (7)
4	47	(3, 16, 23) (21, 13, 9) (29, 4, 14) (25, 11, 1) (15, 5, 26) (2, 6, 24) (7, 10, 28)
6	146	(2, 6, 24, 56, 54) (3, 16, 33, 26, 68) (11, 1, 50, 23, 46) (21, 4, 14, 43, 44) (5, 15, 27, 37, 60) (58, 38, 10, 7, 28) (41, 36, 31, 9, 13)

Table B-16. $R = 1/2$ Trial-and-Error Codes

J	v	Code generator	Orthogonal estimates
2	1	$1 + x$	s_0, s_1
4	5	$1 + x^3 + x^4 + x^5$	$s_0, s_3, s_4, s_1 + s_5$
6	11	$1 + x^6 + x^7 + x^9 + x^{10} + x^{11}$	$s_0, s_6, s_7, s_9, s_1 + s_3 + s_{10}, s_4 + s_8 + s_{11}$
8	21	$1 + x^{11} + x^{13} + x^{16} + x^{17} + x^{19} + x^{20} + x^{21}$	$s_0, s_{11}, s_{13}, s_{16}, s_{17}, s_2 + s_3 + s_6 + s_{19}, s_4 + s_{14} + s_{20}, s_1 + s_5 + s_8 + s_{15} + s_{21}$
10	35	$1 + x^{18} + x^{19} + x^{27} + x^{28} + x^{29} + x^{30} + x^{32} + x^{33} + x^{35}$	$s_0, s_{18}, s_{19}, s_{27}, s_1 + s_9 + s_{28}, s_{10} + s_{20} + s_{29}, s_{11} + s_{30} + s_{31}, s_{13} + s_{21} + s_{23} + s_{32}, s_{14} + s_{33} + s_{34}, s_2 + s_3 + s_{16} + s_{24} + s_{26} + s_{35}$

Table B-17. Good Long-Constraint-Length Convolutional Codes

Code	d_f	Generators (octal)
$R = 1/2$, $v = 48$, nested, systematic (Forney[78])	18 at $v = 32$	$\mathbf{G}^2 = 71547370131746504$
$R = 1/2$, $v = 23$, complementary, nonsystematic (Bahl and Jelinek[117])	24	$\mathbf{G}^1 = 51202215$ $\mathbf{G}^2 = 66575563$
$R = 1/2$, $v = 46$, nested, "quick-look," nonsystematic (Massey and Costello[79])	23 at $v = 31$	$\mathbf{G}^1 = 5335336767373553$ $\mathbf{G}^2 = 7335336767373553$
$R = 1/2$, $v = 31$, nonsystematic (Layland and Lushbaugh[118])	28	$\mathbf{G}^1 = 42545013236$ $\mathbf{G}^2 = 70436206116$
$R = 1/2$, $v = 23$, ODP, "quick-look", nonsystematic (Johannesson[119])	≥ 19	$\mathbf{G}^1 = 74041567$ $\mathbf{G}^2 = 54041567$
$R = 1/2$, $v = 35$, ODP, systematic (Johannesson[119])	≥ 19	$\mathbf{G}^2 = 714461626555$
$R = 1/2, 2/3, 3/4$, nested, systematic (Cain[73])	19 at $v = 35$ ($R = 1/2$) > 13 at $v = 60$ ($R = 2/3$) 9 at $v = 63$ ($R = 3/4$)	$\mathbf{G}^1 = 756563114066$ $\mathbf{G}^2 = 63772271634$ $\mathbf{G}^3 = 45272374$

References

1. Peterson, W. W., and Weldon, E. J., *Error-Correcting Codes*, Cambridge, Massachusetts: MIT Press, 1972.
2. Berlekamp, E. R., *Algebraic Coding Theory*, New York: McGraw-Hill, 1968.
3. Gallager, R. G., *Information Theory and Reliable Communication*, New York: Wiley, 1968.
4. Shannon, C. E., A mathematical theory of communication, *Bell Syst. Tech. J.*, **27**, 379–423 (pt. I) ; 623–656 (pt. II), 1948.
5. Justessen, J., A class of constructive asymptotically good algebraic codes, *IEEE Trans. Inf. Theory*, **IT-18**, 652–656, September 1972.
6. Wozencraft, J. M., and Jacobs, I. M., *Principles of Communication Engineering*, New York: Wiley, 1965.
7. Marcum, J. I., Statistical theory of target detection by pulsed radar, *IEEE Trans. Inf. Theory*, **IT-6**, 59–267, April 1960.
8. MacWilliams, F. J., and Sloane, N. J. A., *The Theory of Error-Correcting Codes*, New York: North-Holland, 1977.
9. Hamming, R. W., Error detecting and error correcting codes, *Bell Syst. Tech. J.*, **29**, 147–160, April 1960.
10. MacWilliams, F. J., A theorem of the distribution of weights in a systematic code, *Bell Syst. Tech. J.*, **42**, 79–94, 1963.
11. Golay, M. J. E., Notes on digital coding, *Proc. IRE*, **37**, 657, 1949.
12. Meggitt, J. E., Error correcting codes and their implementation for data transmission systems, *IRE Trans. Inf. Theory*, **IT-7**, 234–244, October 1961.
13. Massey, J. L., *Threshold Decoding*, Cambridge, Massachusetts: MIT Press, 1963.
14. Prange, E., The use of information sets in decoding cyclic codes, *IRE Trans. Inf. Theory*, **IT-8**, S5–S9, 1962.
15. Schönhein, J., On coverings, *Pac. J. Math.*, **14**, 1405–1411, 1964.
16. Baumert, L. D., McEliece, R. J., and Solomon, G., Decoding with multipliers, JPL Deep Space Network Progress Report, 42–34, pp. 42–46.
17. MacWilliams, F. J., Permutation decoding of systematic codes, *Bell Syst. Tech. J.*, **43**, 485–505, 1963.
18. Omura, J. K., *A Probabilistic Decoding Algorithm for Binary Group Codes*, Stanford Research Institute, Menlo Park, California, March 1969.
19. Weldon, E. J., Difference-set cyclic codes, *Bell System Technical Journal*, **45**, 1045–1055, 1966.
20. Hartman, C. R. P., and Rudolph, L. D., An optimum symbol-by-symbol decoding rule

for linear codes, *IEEE Trans. Inf. Theory*, **IT-22**, 514–517, September 1976.

21. Weldon, E. J., Jr., Decoding binary block codes on Q-ary output channels, *IEEE Trans. Inf. Theory*, **IT-17**, 713–718, November 1971.

22. Forney, G. D., Study of correlation coding, Technical Report No. RADC-TR-67-410, Technical Report to Rome Air Development Center, Griffis Air Force Base, New York, 1967.

23. Davis, R. C., Approximate APP threshold decoding, A Radiation, Inc. Technical Memorandum. March 1971 (Harris Corporation, Government Systems Group, Melbourne, Florida).

24. Greenberger, H., An iterative algorithm for decoding block codes transmitted over a memoryless channel, *JPL DSN Progress Report 42-47*, Jet Propulsion Laboratory, California Institute of Technology, Pasadena, California, July and August 1978.

25. Forney, G. D., Jr., Generalized minimum distance decoding, *IEEE Trans. Inf. Theory*, **IT-12**, 125–131, April 1966.

26. Chase, D., A class of algorithms for decoding block codes with channel measurement information, *IEEE Trans. Inf. Theory*, **IT-18**, 170–182, Janurary 1972.

27. Baumert, L. D., and McEliece, R. J., Soft decision decoding of block codes, *DSN Progress Report 42-47*, Jet Propulsion Laboratory, California Institute of Technology, Pasadena, California, July and August 1978.

28. Rudolph, L. D., Hartmann, C. R. P., Hwang, T. Y., and Duc, N. Q., Algebraic analog decoding of linear binary codes, *IEEE Trans. Inf. Theory*, **IT-25**, 430–440, July 1979.

29. Chase, D., Weng, L. J., Bello, P. A., Joffe, A., Pinto, R., Goldfine, D., and Boardman, C., Demod/decoder integration, RADC-TR-78-70, Technical Report to Rome Air Development Center, Griffis Air Force Base, New York, pp. 3–71, 1967.

30. Boyd, R. W., Aided hard decision decoding, Harris Corporation, Government Systems Group, Melbourne, Florida, May 1979.

31. Greenberger, H. J., Approximate maximum-likelihood decoding of block codes, *JPL Publication 78-107*, Jet Propulsion Laboratory, California Institute of Technology, Pasadena, California, February 1979.

32. Baumert, L. D., and Welch, L. R., Minimum weight codewords in the (128, 64) BCH code, *The Deep Space Network Progress Report 42-42*, Jet Propulsion Laboratory, Pasadena, California, pp. 92–94, December 1977.

33. Gore, W. C., Transmitting binary symbols with Reed–Solomon codes, *Proceedings of the Princeton Conference on Information Sciences and Systems*, Princeton, New Jersey, pp. 495–597, 1973.

34. Michelson, A., A fast transform in some Galois fields and an application to decoding Reed–Solomon codes, *IEEE International Symposium on Information Theory*, Ronneby, Sweden, 1976.

35. Lempel, A. and Winograd, S., A new approach to error correcting codes, *IEEE Trans. Inf. Theory*, **IT-23**, 503–508, July 1977.

36. Chien, R. T., and Choy, D. M., Algebraic generalization of BCH–Goppa–Helgert codes, *IEEE Trans. Inf. Theory*, **IT-21**, 70–79, January 1975.

37. Blahut, R. E., Transform techniques for error control codes, *IBM J. of Res. Dev.*, **23**, 299–315, May 1979.

38. Pollard, J. M., The fast Fourier transform in a finite field, *Math. Comp.*, **25**, 365–374, 1971.

39. Hocquenghem, A., "Codes correcteurs d'erreurs," *Chiffres*, **2**, 147–156, 1959.

40. Bose, R. C., and Ray-Chaudhuri, D. K., On a class of error correcting binary group codes, *Inf. Control*, **3**, 68–79, 1960.

41. Bose, R. C., and Ray-Chaudhuri, D. K., Further results on error correcting binary group codes, *Inf. Control*, **3**, 279–290, 1960.

42. Reed, I. S., and Solomon, G., Polynomial codes over certain finite fields, *J. Soc. Ind. Appl. Math.*, **8**, 300–304, 1960.

43. Reed, I. S., Scholtz, R. A., Truong, T. K., and Welch, L. R., The fast decoding of Reed–Solomon codes using Fermat theoretic transforms and continued fractions, *IEEE Trans. Inf. Theory*, **IT-24**, 100–106, January 1978.

44. Chien R. T., Cyclic decoding procedures for the Bose–Chaudhuri–Hocquenghem codes, *IEEE Trans. Inf. Theory*, **IT-10**, 357–363, October 1964.

45. Sugiyama, Y., Kasahara, M., Hirasawa, S., and Namekawa, T., A method for solving key equation for decoding Goppa codes, *Inf. Control*, **27**, 87–99, 1975.

46. Massey, J. L., Shift-register synthesis and BCH decoding, *IEEE Trans. Inf. Theory*, **IT-15**, 122–127, January 1969.

47. Boyd, R. W., Cain, J. B., Holt, B., and LaBanca, D. L., High speed concatenated decoding technique for satellite communications, *Proc. Nat. Electron. Conf.*, pp. 332–337, 1978.

48. Forney, G. D., *Concatenated Codes*, Cambridge, Massachusetts: MIT Press, 1966.

49. Goppa, V. D., A new class of linear error-correcting codes, *Probl. Peredach. Inform.*, **6**, 24–30, September 1970.

50. Berlekamp, E. R., Goppa codes, *IEEE Trans. Inf. Theory*, **IT-19**, 590–592, September 1973.

51. Forney, G. D., Review of random tree codes, Appendix A, "Study of Coding Systems Design for Advanced Solar Missions," NASA Contract NAS2-3637, Codex Corporation, December 1967.

52. Viterbi, A. J., Error bounds for convolutional codes and an asymptotically optimum decoding algorithm, *IEEE Trans. Inf. Theory*, **IT-13**, 260–269, April 1967.

53. Pollack, M., and Wiebenson, W., Solutions of the shortest-route problem—A review, *Oper. Res.*, **8**, 224–230, March 1960.

54. Omura, J. K., On the Viterbi decoding algorithm, *IEEE Trans. Inf. Theory*, **IT-15**, 177–179, January 1969.

55. Cain, J. B., Clark, G. C., and Geist, J. M., Punctured convolutional codes of rate $(n-1)/n$ and simplified maximum likelihood decoding, *IEEE Trans. Inf. Theory*, **IT-25**, 97–100, January 1979.

56. Mason, S. J., and Zimmerman, H. J., *Electronic Circuits, Signals, and Systems*, New York: Wiley, 1960.

57. Clark, G. C., Davis, R. C., Herndon, J. C., and McRae, D. D., Interim report on convolution coding research, Memorandum Report No. 38, Advanced Systems Operations, Radiation Systems Division, September 1969.

58. Viterbi, A. J., Convolutional codes and their performance in communication systems, *IEEE Trans. Commun. Technol.*, **COM-19**, 751–772, October 1971.

59. Gilhousen, K. S., Heller, J. A., Jacobs, I. M., and Viterbi, A. J., Coding Systems Study for High Data Rate Telemetry Links, Prepared for Ames Research Center under Contract No. NAS2-6024 by Linkabit Corporation, San Diego, California, January 1971.

60. Massey, J. L., and Sain, M. K., Inverses of linear sequential circuits, *IEEE Trans. Comput.*, **C-17**, 330–337, April 1968.

61. Heller, J. A., Sequential decoding: Short constraint length convolutional codes, *JPL Space Program Summary 37-54*, **3**, 171–174, December 1968.

62. Heller, J. A., and Jacobs, I. M., Viterbi decoding for satellite and space communications, *IEEE Trans. Commun. Technol.*, **COM-19**, 835–848, October 1971.

63. Elias, P., Coding for noisy channels, *IRE Convention Record*, 1955.

64. Wozencraft, J. M., Sequential decoding for reliable communications, Technical Report No. 325, Research Laboratory of Electronics, MIT, August 1957.

65. Morrissey, T. W., Analysis of decoders for convolutional codes by stochastic sequential machine methods, *IEEE Trans. Inf. Theory*, **IT-16**, 460–469, July 1970.

66. Cain, J. B., and Clark, G. C., Some results on the error propagation of convolutional feedback decoders, *IEEE Trans. Inf. Theory*, **IT-18**, 681–683, September 1972.

67. Massey, J. L., and Liu, R. W., Application of Lyapunov's direct method to the error propagation effect in convolutional codes, *IEEE Trans. Inf. Theory*, **IT-10**, 248–250, July 1964.

68. Heller, J. A., Feedback decoding of convolutional codes, in *Advances in Communication Systems*, Vol. 4, A. J. Viterbi (ed.), New York: Academic Press, pp. 261–278, 1975.

69. Bussgang, J. J., Some properties of binary convolutional code generators, *IEEE Trans. Inf. Theory*, **IT-11**, 90–100, January 1965.

70. Odenwalder, J. P., Error control coding handbook, Final Report under Contract No. F44620-76-C0056, Linkabit Corp., July 1976.

71. Robinson, J. P., and Bernstein, A. J., A class of binary recurrent codes with limited error propagation, *IEEE Trans. Inf. Theory*, **IT-13**, 106–113, January 1967.

72. Klieber, E. J., Some difference triangles for constructing self orthogonal codes, *IEEE Trans. Inf. Theory*, **IT-16**, 237–238, March 1970.

73. Cain, J. B., Utilization of low-redundancy convolutional codes, *National Telecommunications Conference Record*, pp. 21B-1 through 21B-8, November 1973.

74. Fano, R. M., A heuristic discussion of probabilistic decoding, *IEEE Trans. Inf. Theory*, **IT-9**, 64–74, April 1963.

75. Zigangirov, K. Sh., Some sequential decoding procedures, *Probl. Peredach. Inform.*, **2**, 13–25, 1966.

76. Jelinek, F., A fast sequential decoding algorithm using a stack, *IBM J. Res. Dev.*, **13**, 675–685, 1969.

77. Massey, J. L., Variable-length codes and the Fano metric, *IEEE Trans. Inf. Theory*, **IT-18**, 196–198, January 1972.

78. Forney, G. D., Use of sequential decoder to analyze convolutional code structure, *IEEE Trans. Inf. Theory*, **IT-16**, 793–795, November 1970.

79. Massey, J. L., and Costello, D. J., Nonsystematic convolutional codes for sequential decoding in space applications, *IEEE Trans. Commun. Technol.*, **COM-19**, 806–813, October 1971.

80. Jacobs, I. M., and Berlekamp, E. R., A lower bound to the distribution computation for sequential decoding, *IEEE Trans. Inf. Theory*, **IT-13**, 167–174, April 1967.

81. Savage, J. E., The distribution of the sequential decoding computation time, *IEEE Trans. Inf. Theory*, **IT-12**, 143–147, April 1966.

82. Jordan, K. L., The performance of sequential decoding in conjunction with efficient modulation, *IEEE Trans. Commun. Technol.*, **COM-14**, 283–297, June 1966.

83. Yudkin, H. L., "Channel State Testing in Information Decoding," Sc. D. thesis, MIT Department of Electrical Engineering, February 1965.

84. Bucher, E.A., "Error Mechanisms for Convolutional Codes," Ph. D. thesis, MIT, Department of Electrical Engineering, September 1968.

85. Forney, G. D., and Langelier, R. M., A high speed sequential decoder for satellite communications, *ICC-'68 Conference Record*, pp. 39-9 through 39-17, 1968.

86. Forney, G. D., and Bower, E. K., A high-speed sequential decoder: Prototype design and test, *IEEE Trans. Commun. Technol.*, **COM-19**, 821–835, October 1971.

87. Viterbi, A. J., On coded phase-coherent communications, *IRE Trans. Space Electron. Telem.*, **SET-7**, 3–14, March 1961.

88. Golomb, S. W., ed., *Digital Communication with Space Applications*, Englewood Cliffs, New Jersey: Prentice-Hall, 1964.

89. Lindsey, W. C., and Simon, M. K., *Telecommunication Systems Engineering*, Englewood Cliffs, New Jersey: Prentice-Hall, 1973.

90. Boyd, R. W., Cain, J. B., and Clark, G. C., A concatenated coding approach for high data rate applications, *NTC '77 Conference Record*, pp. 36: 2-1 through 36: 2-7, December 1977.

91. Odenwalder, J. P., "Optimal Decoding of Convolutional Codes," Ph. D. dissertation, School of Engineering and Applied Science, UCLA, 1970.

92. Odenwalder, J. P., Concatenated Reed–Solomon/Viterbi channel coding for advanced planetary missions: Analysis, simulations, and tests, Linkabit Corporation. Final Report on Contract No. 953866, December 1974.

93. Ramsey, J. L., Realization of optimum interleavers, *IEEE Trans. Inf. Theory*, **IT-16**, 338–345, May 1970.

94. Forney, G. D., Burst-correcting codes for the classic bursty channel, *IEEE Trans. Commun. Technol.*, **COM-19**, 772–781, October 1971.

95. Viterbi, A. J., Odenwalder, J. P., Bar-David, I., and Kumm, K. M., RFI/coding sensitivity analysis for tracking and data relay satellite system (TDRSS), Phase II Final Report, Linkabit Corporation, January 1979.

96. Richer, I., A simple interleaver for use with Viterbi decoding, *IEEE Trans. Commun.*, **COM-26**, 406–408, March 1978.

97. Coding for Navy UHF satellite communications, Final Report on Contract No. N00039-77-C-0259, Harris Corporation, August 1978.

98. Fire, P., A class of multiple-error-correcting binary codes for non-independent errors, Sylvania Report RSL-E-2, 1959.

99. Tong, S. Y., Burst trapping techniques for a compound channel, Bell Telephone Laboratories Technical Memorandum, 1968.

100. Kohlenberg, A., and Forney, G. D., Convolutional coding for channels with memory, *IEEE Trans. Inf. Theory*, **IT-14**, 618–626, September 1968.

101. Cain, J. B., and Boyd, R. W., Convolutional code performance with PSK signaling in nonstationary Gaussian noise, *NTC '78 Conference Record*, pp. 2.5.1–2.5.6, December 1978.

102. Trumpis, B. D., and McAdam, P. L., Performance of convolutional codes on burst noise channels, *NTC '77 Conference Record*, pp. 36: 3-1 through 36: 3-14, December 1977.

103. Geist, J. M., and Cain, J. B., Viterbi decoder performance in Gaussian noise and periodic erasure bursts, *IEEE Trans. Commun.*, **COM-28**, August 1980.

104. Berlekamp, E. R., "Final Report on a New Coding System Designed to Combat Both Gaussian Noise and Radio Frequency Interference," Cyclotomics, Inc., February 1979.

105. Van Trees, H. L., *Detection, Estimation, and Modulation Theory Part I*, New York: Wiley, 1968.

106. Viterbi, A. J., and Jacobs, I. M., Advances in coding and modulation for noncoherent channels affected by fading, partial band, and multiple access interference, *Advances in Communications Systems*, Vol. 4, New York: Academic Press, pp. 279–308, 1975.

107. Huth, G. K., Optimization of coded spread spectrum system performance, *IEEE Trans. Commun.*, **COM-25**, 763–770, August 1977.

108. Viterbi, A. J., Spread spectrum communications—Myths and realities, *IEEE Commun. Magazine*, **17**, 11–18, May 1979.

109. Ungerboeck, G., Trellis coding with expanded channel-signal sets, *1977 International Symposium on Information Theory*, October 1977.

110. Ungerboeck, G., Channel coding with multilevel/phase signals, *IEEE Trans. Inf. Theory*, (to appear).

111. Anderson, J. B., and Taylor, D. P., A bandwidth efficient class of signal-space codes, *IEEE Trans. Inf. Theory*, **IT-24**, 703–712, November 1978.

112. Pelchat, M. G., and Geist, J. M., Surprising properties of two-level bandwidth compaction codes, *IEEE Trans. Commun.*, **COM-23**, 878–883, September 1975.

113. Larsen, K. J., Short convolutional codes with maximal free distance for rates 1/2, 1/3, and 1/4, *IEEE Trans. Inf. Theory*, **IT-19**, 371–372, May 1973.

114. Paaske, E., Short binary convolutional codes with maximal free distance for rates 2/3 and 3/4, *IEEE Trans. Inf. Theory*, **IT-20**, 683–689, September 1974.

115. Trumpis, B. D., "Convolutional Coding for M-ary Channels," Ph. D. thesis, UCLA, 1975.

116. Odenwalder, J. P., Dual-K convolutional codes for noncoherently demodulated channels, *ITC Conference Record*, 1976.
117. Bahl, L. R., and Jelinek, F., Rate 1/2 convolutional codes with complementary generators, *IEEE Trans. Inf. Theory*, **IT-17**, 718–727, November 1971.
118. Layland, J. W., and Lushbaugh, W. A., A flexible high-speed sequential decoder for deep space channels, *IEEE Trans. Commun. Technol.*, **COM-19**, 813–820, October 1971.
119. Johannesson, R., Robustly optimal rate one-half binary convolutional codes, *IEEE Trans. Inf. Theory*, **IT-21**, 464–468, July 1975.

Index

A posteriori probability threshold
  decoding, 142–153, 157–159, 180,
  290–297, 329
Add–compare–select, 258–261
Adder, modulo-2, 69
AGC effects, 264–265
Algorithm
  Berlekamp, 195, 201–214, 218–219,
    223–225
  Euclid's, 69, 195–201, 216–218, 223–225
  Fano, 297–326
  Hartmann–Rudolph, 154–162, 179–180
  modified Omura, 128–131
  Omura, 120–124, 127, 175–176
  stack, 326–329
  threshold decoding, 131–139, 142–153,
    158–161, 180, 283–297, 329
  Viterbi, 227–266, 298, 337–344,
    354–364, 373–375, 378–391
Anderson, J. B., 381, 390
Approximate APP threshold decoding,
  148–150, 295–297
Associative law, 52, 63
Asymptotic coding gain, 35–37, 383
Augmented code, 85
Automatic resynchronization, 318
Autonomous behavior, 276

Backsearch depth, 320
Bahl, L. R., 401, 407
Bandwidth-constrained channels, coding
  for, 381–392
Bar-David, I., 348, 363
Basis vector, 10, 56

Baumert, L. D., 111, 143, 172–173, 179
Baye's rule, 26, 145, 155
BCH bound, 186, 187, 223, 393–394
BCH code, 86–87, 94–96, 98–102,
  139–140, 181–225
  decoding of, 98–102
  designed distance, 187
  erasures and errors correction, 214–225
  error correction, 181–225
  minimum distance of, 86
  narrow-sense, 86
  nonprimitive, 86
  performance, 219–222
  primitive, 86, 186
  tables of codes, 393–397
Bello, P. A., 158
Berlekamp algorithm: *see* Algorithm,
  Berlekamp
Berlekamp, E. R., vi, 38, 46, 80, 85, 93,
  113, 195, 205, 213, 218, 223, 305
  306, 363
Bernstein, A. J., 283, 285, 286, 400
Binary code, 10
Binary symmetric channel, 15, 47
Binomial coefficient, 5
Blahut, R. E., 182, 187, 218
Block code, 9–10
  bounds on distance for, 37–40
  decoding table, 4–6
  dual, 57–58
  linear, 9–10
  performance computations, 18–26,
    89–90, 136, 158–160, 168, 171, 179,
    219–222, 333–344
  tables of codes, 134, 393–397

Block interleaving, 346–347
Block length, 6–7
Boardman, C., 158
Bose, R. C. 185
Bose–Chaudhuri code: *see* BCH code
Bose–Chaudhuri–Hocquenghem code:
  see BCH code
Bound
  BCH, 186, 187, 223, 393–394
  Gilbert, 37, 39–40, 53–54
  Hamming, 6, 37–40
  Plotkin, 37, 39–40
  random-coding, 41–46
  union, 23–24
  Varsharmov–Gilbert, 37. 39–40, 53–54
Bounded distance decoding, 19
Bounds on distance for block code, 37–40
Bounds on probability of error
  block code, 18–26
  convolutional code, 242–246, 247–253,
    274–276, 292–297, 313–318
Bower, E. K., 319, 322
Boyd, R. W., 169–172, 213, 337, 339, 353
Branch, 229
Branch metric, 232–235, 256–258, 298
Branch metric computer, 256–258
BSC, 15, 47
Bucher, E. A., 315
Buffer overflow problem, 304–313,
  316–318
Burst-correcting code, 10, 352–353
Burst erasure process, 353–366
  burst length, 353
  duty factor, 353
  interleaving implications, 364–366
  periodic, 353, 358–363
  random, 353, 363
Burst error: *see* Burst noise channels
Burst noise channels, 352–366
Bussgang, J. J., 281, 400

Cain, J. B., 213, 237, 238, 276, 277, 278,
  294, 337, 339, 353, 359, 399, 402,
  407
Canonical form, echelon, 52
Canonical self-orthogonal codes, 285
Capacity, channel, 1, 43
Catastrophic error propagation, 246
Channel, 1, 11

Channel (*cont.*)
  binary symmetric, 15, 47
  coherent Gaussian, 12–13, 30–31
  discrete memoryless, 15–16
  noncoherent Gaussian, 13–14, 34
Channel capacity, 1, 43
Chase algorithm, 167–172
  variations, 169–172
Chase, D., 143, 158, 167, 338
Check symbol, 50
Chien, R. T., 182, 193
Chien search, 192–194
Chinese remainder theorem, 80
Choy, D. M., 182
Clark, G. C., 237, 238, 239, 276–278,
  337, 399
Closure, 51–52, 63
Code invariance, 112–114
Code length, 6–7
Code preserving permutation, 112–114
Code rate, 6
Code sequence, 4–7
Code tree, 229–231
Code word, 4
Codes
  block, 9–10
  burst-error-correcting, 10, 352–353
  convolutional, 9–10
  group, 10, 52
  linear, 9–10
  modifications of, 84–85
  nonlinear, 9–10
  nonsystematic, 6
  random-error-correcting, 10
  systematic, 6
  tree, 9–10
Coding gain, 35
  asymptotic, 35–37
Coding problem, 1–17
Communication system, digital, 8–17
Commutative, 63
Computational cutoff rate, 307
Computational problem, sequential
  decoder, 304–313
Concatenated code, 331–339
Congruent, 65
Constraint length
  definite decoding, 273
  encoding, 9, 228–231
  feedback decoding, 273

Convolutional code, 9-10
  bounds on probability of error, 243-246,
    247-253, 274-276, 292-297, 313-318
  encoding, 228-229
  error correction, 231-235, 268-274,
    278-292, 299-304, 318-330
  nonsystematic, 229
  self-orthogonal, 285-289
  systematic, 229
  tables of codes, 399-407
  trial-and-error, 289
Correct decoding
  probability of, 20
Correction: *see* Decoding
Coset, 58-59
Coset leader, 58-59
Costello, D. J., 304, 315, 401, 402, 407
Covering, 108, 109
  bounds on size, 110
  construction of, 110-119
CSOC, 285
Cyclic code, 72-75
  decoding, 97-140
  encoding, 72-75
  extended, 84
  shortened, 85

Davis, R. C., 142, 148, 239, 260, 261, 291
Decoder: *see* Decoding
Decoding
  BCH code, 181-225
  block code, 4-5
  Chase, 167-172
  cyclic code, 97-140
  definite, 269-276
  feedback, 269-276
  Hartmann-Rudolph, 154-162, 179-180
  information set, 102-131
  majority-logic, 131-137
  maximum-likelihood, 5
  Meggitt, 97-102
  permutation, 112-119
  sequential, 297-330
  stack, 326-330
  table look-up, 4-5, 278-283
  threshold, 131-138, 142-153, 158-161,
    180, 283-297, 329
  Viterbi, 227-266, 298, 337-344, 354-
    364, 373-375, 378-391

Decoding (*cont.*)
  Weldon, 162-166
Decoding constraint length, 273
Decoding table, 4-5, 18, 19
Definite decoding, 269-276, 286-292
Definite decoding minimum distance, 273
Deinterleaving: *see* Interleaving
Demodulation
  coherent, 12, 13
  noncoherent, 13-14, 34
Demodulator quantization, 15-16, 30-34
Diagram
  encoder state, 238-240
Difference-set code, 133-136
Difference triangle, 285
Differentially coherent PSK, 14
Diffuse code, 353
Digital communication system, 8-17
Discrepancy, 205
Distance
  designed, 187
  definite decoding minimum, 273
  Euclidean, 27, 382
  feedback decoding minimum, 273
  Hamming, 4
  minimum, 4
Distribution of computation, 304-307,
  312-313
Distributive law, 63
DPSK, 14
Dual code, 57-58
Duc, N. Q., 157

Echelon canonical form, 52
Elias, P., 267
Encoder, 9-10, 68-75, 228-231
Encoder state diagram, 238-240
Encoding, 6-10
  convolutional code, 228-231
  cyclic code, 68-75
Encoding constraint length, 9, 228-231
Envelope detector, 13-14
Erasure, 47
Erasure and error, 214-225, 353-366
Erasure channel, 47
Erasure-locator polynomial, 215
Error, 1, 2
  bit, 18
  probability of, 18-26

Error (*cont.*)
  sequence, 18
Error-detecting code, 24–26
Error-evaluator polynomial, 191–194
Error-locator polynomial, 191–194
Error correcting capability, 4, 37–40
Error correction: *see* Decoding
Error detection, 24–26
Error exponent, 43–44
Error pattern, 58
Error propagation
  catastrophic, 246
  limited, 275–278
  type-I unlimited, 276–278
  type-II unlimited, 276–278
Estimate size, 293
Euclidean distance, 27, 382
Euclid's algorithm, 69, 195–201, 216–218,
  223–225
Existence of good codes: *see* Gilbert bound
Exponent error, 43–44
Exponential bound parameter, 42–45
Expurgated code, 85
Extended code, 84
Extension field, 63

Fano, R. M., 297
Fano algorithm, 297–330
Feedback decoding, 269–283
  minimum distance, 273
Field, 63–67, 75–78
  extension, 63
  Galois, 63–67, 75–78
  prime, 63
Finite field: *see* Field
Finite field Fourier transforms, 183–185
Finite-state machine, 238–242
Fire, P., 352
First decoding error, 275
Forney, G. D., 142, 143, 216, 231, 291,
  304, 315, 318, 319, 322, 329, 331,
  334, 347, 353, 391, 401, 407
Free distance, 9, 240

Gallager, R. G., vi, 38, 46, 205, 299, 353
Galois field
  computation, 63–67
  definition, 64–67
  logarithmic representation, 65–67

Galois field (*cont.*)
  polynomial representation, 63–67
  properties, 63–67, 75–78
GCD, 195–201
Geist, J. M., 237, 238, 359, 390, 399
Generalized parity check, 50–51
Generating function, 239–240
Generator matrix, 56–57
Generator polynomial
  block code, 62–63
  convolutional code, 228
Genie-aided feedback decoder, 274, 275
GF(q): *see* Galois field
Gilbert bound, 37, 39–40, 53, 54
Gilhousen, K. S., 241, 322, 323
Golay code, 88, 93, 168
Golay, M. J. E., 88
Goldfine, D. 158
Golomb, S. W. 334
Goppa, V. D., 223
Gore, W. C., 182
Graph, signal flow, 238–242
Gray code mapping, 383–384
Greatest common divisor, 195–201
Greenberger, H., 142, 161, 172, 174
Group code, 10, 50–51
Guard space, 362
Guess-and-restart, 318

Hamming bound, 6, 37–40
Hamming code, 54–56, 71
Hamming distance, 4
Hamming, R. W., 54
Hamming weight, 10
Hartmann, C. R. P., 142, 154–157
Hartmann–Rudolph algorithm, 154–162,
  179–180
Heller, J. A., 241, 247, 265, 279, 281,
  282, 322, 323
Herndon, J. C., 239
Hirasawa, S., 195, 223
Hocquenghem, A., 185
Holt, B., 213, 337, 339
Huth, G. K., 376
Hwang, T. Y., 157

Identity element, 52, 63
Identity matrix, 52
Inpulse response, 10, 229

Information sequence, 50
Information set, 102
  decoding, 102–131, 172–178
Information symbol, 2, 50
Inner code, 332
Integers modulo q, 63
Interlacing: *see* Interleaving
Interleaving, 345–352
  application of, 364–366
  block, 346–347
  convolutional, 347–349
  external, 345–352
  internal, 283
  periodic, 346–349
  pseudorandom, 349–352
Invariance
  code, 112–114
Inverse, 52, 63
Irreducible polynomial 64–67
Iterated code, 61–62

Jacobs, I. M., 13, 219, 241, 265, 305, 306,
  322, 323, 376, 400
Jelinek, F., 298, 326, 401, 407
Joffe, A., 158
Johannesson, R., 402, 407
Jordan, K. L., 309
Justesen, J., 7

Kasahara, M., 195, 223
Key equation, 189–193
  solution of, 195–208
Klieber, E. J., 283, 400
Kohlenberg, A., 353, 391
Kumm, K. M., 348, 363

LaBanca, D. L., 213, 337, 339
Langelier, R. M. 318
Larsen, K. J., 400
Layland, J. W., 402, 407
Least common multiple (LCM), 79
Lempel, A., 182
Length
  block, 6–7
  constraint: *see* Constraint length
Lengthened code, 85
Lindsey, W. C., 335
Linear block code, 9–10, 50–51

Linear code, 9–10, 50–51
Linear combination, 52–54
Linear congruential sequence, 350
Linear tree code, 9–10
Linearly independent, 53
Liu, R. W., 276
Lushbaugh, W. A., 402, 407

MacWilliams, F. J., 38, 46, 58, 85, 93, 114
MacWilliams identity, 58
McAdam, P. L., 354
McEliece, R. J., 111, 143, 172, 173
McRae, D. D. 239
Majority-logic decodable code, 131–140,
  283–289, 400–401, 404–406
Majority-logic decoding, 133–140
Mapping by set partitioning, 388–390
Marcum, J. I., 34
Mason, S. J., 239, 240
Massey, J. L., 97, 142, 205, 213, 246, 276,
  283, 289, 290, 292, 302, 304, 315,
  401, 402, 407
Matrix
  generator, 56–57
  generator polynomial, 236
  identity, 52
  parity-check 52–53
  semi-infinite, 229
Maximal-length codes, 89–90
Maximal-length feedback shift register, 90
Maximum-distance-separable code, 188,
  333
Maximum-length sequence, 90
Maximum-length-sequence code, 89–90
Maximum-likelihood decoding, 5
Meggitt decoder, 97–102
Meggitt, J. E., 97
Memoryless channel, 15
Metric, 26–34
Metric bias, 298, 303
Michelson, A., 182
Minimum distance, 4
  feedback decoding, 273
  definite decoding, 273
Minimum function, 76
Minimum polynomial, 76
Minimum weight, 53
Modified Omura algorithm, 128–131
Modified syndrome, 216, 217
Modulation, 10–14

Modulo q, 10, 63–64
Morrissey, T. W., 275
Multiplication
    finite field, 63–67
    polynomial, 68–72

(n,k) code, 6
n-tuple, 5
Namekawa, T., 195–223
Narrow-sense BCH code, 86
Nested code generators, 304
Node, 229–231
Noise, 11–14
Noise averaging, 2
Nonbinary code, 52, 400, 404
    convolutional, 250–253, 309–310,
        378–380, 400, 404
    Reed–Solomon, 87, 187, 188, 200–201,
        208, 217–219, 331–341
Noncoherent combining, 34
Nonlinear code, 9–10
Nonresidues, 92–93
Nonsystematic convolutional code, 229
Nontransparent code, 256

Odenwalder, J. P., 281, 339, 340, 348,
    363, 400
Omura algorithm, 119–124, 175–176
Omura, J. K., 119–120, 232
Optimum code, 246, 247, 399–400,
    402–404
Optimum decoding: see Maximum-likeli-
    hood decoding and Hartman–Rudolph
    algorithm
Optimum symbol-by-symbol decoding,
    154–162
Order of element, 77
Orthogonal estimate, 131–137
Orthogonal paritycheck, 131–137
Orthogonal signaling, 11, 13–14, 34
Orthogonalizable L-step, 135–137
Orthogonalizable one-step, 131–137
Outer code, 332
Overall parity check, 84–85

Paaske, E., 400
PAM, 386–390
Pareto distribution, 304–307

Parity check
    generalized, 50–51
    orthogonal, 131–137
    overall, 84–85
Parity-check code: see Linear code
Parity-check matrix, 52–53
Parity-check polynomial, 73
Parity-check symbol, 50–51
Parity set, 102
Partial syndrome decoding, 177–178
Path metric, 231–235, 258–261, 298,
    302–303
Path metric range, 259–260
Path metric updating, 258–261
Pelchat, M. G., 390
Penny-weighing problem, 59–61
Peterson, W. W., vi, 38, 46, 64, 85, 93,
    137, 351
Perfect code, 6, 54–56, 88
Permutation
    code preserving, 112–114
Permutation decoding, 112–119
Pinto, R., 158
Plotkin bound, 37, 39–40
Pollack, M., 232
Pollard, J. M., 183
Polynomial
    degree, 63
    generator: see Generator polynomial
    irreducible, 64–67
    minimum: see Minimum function
    parity-check, 73
    primitive, 76
Polynomial code, 62–63
Polynomial division, 69–72
Polynomial multiplication, 68–72
Polynomial representation, Galois field,
    63–67
Prange, E., 97
Premultiplication by x^{n-k}, 69–71
Prime, 63
Prime field, 63
Primitive BCH code, 86–87, 186
Primitive field element, 65
Primitive polynomial, 76
Primitive root, 76
Probability of correct decoding, 20
Probability of detected error, 19–22,
    24–26
Probability of error bounds
    block code, 18–26

Probability of error bounds (*cont.*)
   convolutional code, 242–246, 247–253,
     274–276, 281–283, 313–318
Probability of first error, 274–276,
   281–283
Propagation, error: *see* Error propagation
Pseudo-cyclic code, 73
Pseudo-random sequence, 90
PSK modulation, 11–14
Punctured code, 84–85

Quadratic residue, 92–93
Quadratic residue code, 92–93
Quantization, 15–17, 30–34
Quasi-cyclic code, 73
Quick-look codes, 304, 401, 407

R_0 parameter, 41–46, 307–310, 363
Ramsey, J. L., 347
Random code, 41
Random-coding bound, 41–46
Random erasure process, 354–357
Random-error-correcting ability, 4
Ray–Chaudhuri, D. K., 185
Recurrent code: *see* Convolutional code
Recursive extension, 189
Redundant symbol, 2
Reed, I. S., 187, 191
Reed–Muller code, 91–92
Reed–Solomon code, 87, 200–201, 208,
   217–219, 332–341
Register exchange, 261
Remainder theorem, Chinese, 80
Richer, I., 350, 351
Robinson, J. P., 283, 285, 286, 400
Roots of polynomial, 75–79
Rudolph, L. D., 142, 154, 157
Running threshold, 299

Sain, M. K., 246, 276
Savage, J. E., 307
Scholtz, R. A., 191
Self-orthogonal convolutional code,
   285–289
Semi-infinite matrix, 229
Sequence
   code, 3–7
   maximum-length, 90

Sequence (*cont.*)
   pseudo-random, 90
Sequential decoding, 297–330
Sequential-state machine, 238–242
Shannon, C. E., 1
Shannon's channel capacity, 1, 43
Shift register, 69
Shonheim, J., 110
Shortened code, 85
Simon, M. K., 335
Simplex, 89–90
Single-error-correcting block code, 54–56,
   71–72
Size, 293
Sloane, N. J. A., 38, 46, 93
Soft decision decoding, 26–34
Solomon, G., 111, 187
Spectrum, 10, 54
Sphere-packing bound, 6, 37–40
Spread spectrum systems, 366–380
   coded, 370–375, 378–380
   direct sequence, 367–375
   frequency hop, 375–380
Stack decoding, 326–330
Standard array, 58–59
State diagram encoder, 238–240
Sugiyama, Y., 195, 223
Superchannel, 332
Supercoder, 332
Superdecoder, 332
Survivor, 232
Symbol, 2
   check, 50–51
   information, 6
   parity-check, 50–51
   redundant, 2
Synchronization
   branch, 255–256, 279
   interleaver, 347–352
Syndrome
   block codes, 58–59, 79–84
   convolutional codes, 270–272
Systematic code, 6, 229

Taylor, D. P., 381, 390
Threshold, 299
Threshold decoding
   APP, 142–153, 158–161, 290–297
   for block codes, 131–138, 142–153,
     158–161

Threshold decoding (*cont.*)
  for convolutional codes, 283–297
  majority-logic, 131–138
Threshold spacing, 299
Trace-back method, 261–262
Tong, S. Y., 353
Transfer function, distance, 239–240
Transparent code, 255–256
Tree code, 9–10
  linear, 10
Trellis, 231–232
Trial-and-error codes, 289, 401
Triangle, difference, 285
Trumpis, B. D., 354, 400, 404
Truong, T. K., 191

Undetected error, probability of, 19–22
Ungerboeck, G., 381, 388, 389
Union bound, 23–24, 243–346
Upper bound
  Hamming, 6, 37–40
  Plotkin, 37, 39–40
  random-coding, 41–46

Van Trees, H. L., 372
Varsharmov–Gilbert bound, 37, 39–40,
  53–54
Vector, 10, 56
  basis, 10, 56

Vector (*cont.*)
  spectrum of, 183–184
Vector space, 9–10, 56
Viterbi, A. J., 232, 239, 241, 244, 266,
  322, 323, 334, 348, 363, 376, 380,
  400
Viterbi decoding: *see* Decoding, Viterbi

Weight
  Hamming, 10
  minimum, 53–54
Weight distribution, 54
Weight enumerator, 54
Weight spectrum, 10, 54
Weight structure, 10, 54
Welch, L. R., 179, 191
Weldon algorithm, 162–166
Weldon, E. J., vi, 38, 46, 64, 85, 93, 133,
  137, 142, 162, 351
Weng, L. J., 158
Wiebenson, W., 232
Winograd, S., 182
Wozencraft, J. M., 12, 219, 267, 297

Yudkin, H. L., 315

Zech logarithm, 67
Zigangirov, K. Sh., 298, 326
Zimmerman, H. J., 239, 240